W9-CUS-601

Rules of Exponents

1. $x^m \cdot x^n = x^{m+n}$ **product rule**

2. $\dfrac{x^m}{x^n} = x^{m-n}$, $x \neq 0$ **quotient rule**

3. $(x^m)^n = x^{m \cdot n}$ **power rule**

4. $x^0 = 1$, $x \neq 0$ **zero exponent rule**

5. $x^{-m} = \dfrac{1}{x^m}$, $x \neq 0$ **negative exponent rule**

6. $\left(\dfrac{ax}{y}\right)^m = \dfrac{a^m x^m}{y^m}$, $y \neq 0$ **expanded power rule**

FOIL (*First*, *Outer*, *Inner*, *Last*) method of multiplying binomials: $(a+b)(c+d) = ac + ad + bc + bd$

Product of sum and difference of two quantities:
$(a+b)(a-b) = a^2 - b^2$

Squares of binomials: $(a+b)^2 = a^2 + 2ab + b^2$
$(a-b)^2 = a^2 - 2ab + b^2$

If $a \cdot b = c$, then a and b are **factors** of c.
Difference of two squares: $a^2 - b^2 = (a+b)(a-b)$
Sum of two cubes: $a^3 + b^3 = (a+b)(a^2 - ab + b^2)$
Difference of two cubes: $a^3 - b^3 = (a-b)(a^2 + ab + b^2)$

To Factor a Polynomial

1. Determine if the polynomial has a greatest common factor other than 1. If so factor out the GCF from every term in the polynomial.

2. If the polynomial has two terms, determine if it is a difference of two squares or a sum or difference of two cubes. If so, factor using the appropriate formula.

3. If the polynomial has 3 terms, factor the trinomial using one of the procedures discussed.

4. If the polynomial has more than 3 terms, then try factoring by grouping.

5. As a final step, examine your factored polynomial to see if any factors listed have a common factor and can be factored further. If you find a common factor, factor it out at this point.

Quadratic equation: $ax^2 + bx + c = 0$, $a \neq 0$.
Zero-factor Property: If $ab = 0$, then $a = 0$ or $b = 0$.

To Solve A Quadratic Equation Using Factoring

1. Write the equation in standard form with the squared term positive. This will result in the one side of the equation being equal to 0.

2. Set each factor containing a variable equal to zero and find the solution.

To Reduce Rational Expressions

1. Factor both numerator and denominator as completely as possible.

2. Divide both the numerator and denominator by any common factors.

To Multiply Rational Expressions

1. Factor all numerators and denominators as completely as possible.

2. Divide out common factors.

3. Multiply numerators together and multiply denominators together.

To Add or Subtract Two Rational Expressions

1. Determine the least common denominator (LCD).

2. Rewrite each fraction as an equivalent fraction with the LCD.

3. Add or subtract the numerators while maintaining the LCD.

4. When possible, factor the remaining numerator and reduce the fraction.

To Solve Equations Containing Fractions

1. Determine the LCD of all fractions in the equation.

2. Multiply **both** sides of the equation by the LCD. This will result in every term in the equation being multiplied by the LCD.

3. Remove any parentheses and combine like terms on each side of the equation.

4. Solve the equation.

5. Check your solution in the original equation.

Elementary Algebra
for College Students

Elementary Algebra for College Students

SECOND EDITION

Allen R. Angel

Monroe Community College

Prentice Hall
Englewood Cliffs, New Jersey 07632

Library of Congress Cataloging-in-Publication Data

ANGLE, ALLEN R.,   date
    Elementary algebra for college students/Allen R. Angel-2nd ed.

p.——cn.
    Rev. ed. of: Elementary algebra. c1985.
Includes index.
ISBN 0-13-252644-1   ISBN 0-13-252735-9
                            (annotated instructor's ed.)
    1. Algebra.   I Angel, Allen R., date   Elementary algebra.
II. Title.
QA152.2.A54 1988
512.9—dc 19

Editorial/production supervision: Virginia Huebner
Interior design: Judith A. Matz-Coniglio
Cover design: Judith A. Matz-Coniglio
Manufacturing buyer: Paula Massenaro
Page layout: Meryl Poweski
Photo Research: Teri Stratford
Photo Editor: Lorinda Morris-Nantz
Cover Art: Detail of "Untitled" by Ronald Ghiz

1, Steve Gardiner/*Taurus Photos* ● *21*, Paul Shambroom/*Photo Researchers* ●
25, Alex Von Koschembahr/*Photo Researchers* ● *51*, Allen R. Angel ● *85*, Hewlett-Packard ●
94, Allen R. Angel ● *123*, American Stock Exchange ● *128*, Allen R. Angel ●
164, Ted Eckhart/*Photo Researchers* ● *199*, Allen R. Angel ● *243*, Sylvie Chappaz/*Photo Researchers* ●
276, Philippe Blondel, *All Sport*/Vandystadt ● *301*, E. A. Heiniger/*Photo Researchers* ● *302, 309*, Allen R. Angel ●
338, Georg Gerster/*Photo Researchers* ● *347*, Page Poore

Printed in the United States of America
10 9 8 7 6 5 4 3 2

ISBN   0-13-252644-1   01

Prentice-Hall International (UK) Limited, *London*
Prentice-Hall of Australia Pty. Limited, *Sydney*
Prentice-Hall Canada Inc., *Toronto*
Prentice-Hall Hispanoamericana, S.A., *Mexico*
Prentice-Hall of India Private Limited, *New Delhi*
Prentice-Hall of Japan, Inc., *Tokyo*
Prentice-Hall of Southeast Asia Pte. Ltd., *Singapore*
Editora Prentice-Hall do Brasil, Ltda., *Rio de Janeiro*

To my wife, Kathy
and my sons, Robert and Steven

Contents

3   Formulas and Applications of Algebra   94

4   Exponents and Polynomials   128

5   Factoring   164

6 Rational Expressions and Equations 199

7 Graphing Linear Equations 243

8 Systems of Linear Equations 276

9 Roots and Radicals 309

Preface

This book was written for college students and other adults who have never been exposed to algebra or those who have been exposed but need a refresher course. My primary goal was to write a book that students can read, understand, and enjoy. To achieve this goal I have used short sentences, clear explanations, and many detailed worked-out examples. I have tried to make the book relevant to college students by using practical applications of algebra throughout the text.

Features of the Text

Four-color Format: Color is used pedagogically in the following ways:

Important definitions and procedures are color screened;

Color screening or color type is used to make other important items stand out;

Errors that students commonly make are given in colored boxes as warnings for students;

Artwork is enhanced and clarified with use of multiple colors;

Other important items such as the Helpful Hints, Just For Fun Problems, and Calculator Corners are enhanced with color;

The four-color format allows for all these, and other features, to be presented in different forms and colors for easy identification by students;

The four-color format helps make the text more appealing and interesting to students.

Practical Applications: Practical applications of algebra are stressed throughout the text. Students need to learn how to translate word problems into algebraic symbols. The problem-solving approach used throughout this text gives students ample practice in setting up and solving word problems.

Detailed Worked Out Examples: A wealth of examples have been worked out in a step-by-step detailed manner. Important steps are highlighted in color, and no steps are omitted until after the student has seen a sufficient number of similar examples.

Exercise Sets: Each exercise set is graded in difficulty. The early problems help develop the students' confidence, and then they are eased gradually into the more difficult problems. A sufficient number and variety of examples are given in the section for the student to successfully complete even the more difficult exercises. The number of exercises in each section is more than ample for student assignments and practice.

Keyed Section Objectives: Each section opens with a list of skills that the student should learn in that section. The objectives are then keyed to the appropriate portions of the sections with symbols such as ■.

Common Student Errors: Errors that students often make are illustrated. The reasons these procedures are wrong are explained, and the correct procedure for working the problem is illustrated. These common student error boxes will help prevent your students from making those errors we see so often.

Helpful Hints: The helpful hint boxes offer useful suggestions for problem solving and other varied topics. They are set off in a special manner so that students will be sure to read them.

Just for Fun Problems: At the end of many exercise sets are Just For Fun problems. These offer more challenging problems for the bright students in your class who want something extra. These problems present additional applications of algebra, material to be presented later in the text, or material to be covered in a later mathematics course.

Calculator Corners: The Calculator Corners, placed at appropriate locations in the text, are written to reinforce the algebraic topics presented in the section.

Chapter Summaries: At the end of each chapter is a chapter summary which includes a glossary and important chapter facts.

Review Exercises: At the end of each chapter are review exercises that cover all types of exercises presented in the chapter. The review exercises are keyed to the sections where the material was first introduced.

Practice Tests: The comprehensive end-of-chapter practice test will enable the students to see how well they are prepared for the actual class test. The Instructor's Resource Manual includes several forms of each chapter test that are similar to the student's practice test.

Readability: One of the most important features of the text is its readability. The book is very readable even for those with weak reading skills. Short clear sentences are used and more easily recognized and understood words are used whenever possible. Because so many of our students now taking algebra are from different countries, this feature has become increasingly important.

Accuracy: Accuracy in a mathematics text is essential. To insure accuracy in this book, no fewer than five mathematicians from around the country have read the galleys carefully for typographical errors, and have checked all the answers.

Prerequisite

This text assumes no prior knowledge of algebra. However, a working knowledge of arithmetic skills is important. Fractions are reviewed early in the text and decimals and percent are reviewed in the appendix.

Mode of Instruction

The format of this book—pedagogical use of four colors; short but complete sections; clear explanations; important points stressed in color; many detailed step-by-step worked-out examples; ample and graded exercise sets; Common Student Errors; Helpful Hints; chapter summaries, review exercises and practice test; answers to odd exercises and all Just For Fun problems, review exercises and practice tests—makes this text suitable for many types of instructional modes including lecture, modified lecture, learning laboratory, and self-paced instruction. Many student supplements are available to assist the student in the learning process. Please see "Available Supplements For Students", which follows.

Changes in the Second Edition

Many users of the first edition, and others, have indicated that they prefer a hardbound book for non-programmed courses. Therefore, the second edition of this algebra series has been written as a hardcover text. (The first edition in paperback will still remain in print for those who prefer the paperback version.)

Another major change is the new four-color format. Many reviewers and users of the text felt that additional color would enhance the book in many ways, particularly in increased clarity of the book's many distinguishing features.

A third major change is the deletion of Chapter 0. The book now begins with Chapter 1 "Real Numbers". A review of fractions is now included in Chapter 1, and decimals and percent are covered in the appendix.

Other changes include: increased number of worked-out examples and exercises; reordering of certain topics to allow for a smoother flow of material; reorganization of certain sections for greater clarity; addition of sections covering scientific notation, sum and difference of two cubes, a general factoring procedure, reducing rational expressions, and fractional exponents and higher roots (optional).

Available Supplements for the Instructor

Annotated Instructor's Edition: Contains the answers to all exercises clearly illustrated next to each exercise in the student's edition. This saves time and insures accuracy in classroom lectures.

Instructor's Solution Manual: Contains complete and detailed solutions to all even-numbered exercises in the text. Solutions to all odd-numbered problems are found in the student's Solution Manual.

Instructor's Resource Manual: Contains 8 forms of each chapter test. Five are of the open-ended question type, similar to the practice test in the student's edition, and three are in multiple choice format.

Software Testing Package (for IBM and Apple): This flexible package allows you to construct your own exams, or will generate any number of individualized exams to your specifications. Multiple-choice items are also available.

Available Supplements for Students

Student's Study Guide: Includes additional worked-out examples, additional drill problems and practice test and their answers. Important points are emphasized.

Student's Solution Manual: Includes detailed step-by-step solutions to all odd-numbered problems in the exercise sets.

"How to Study Math":   Designed to help your students overcome math anxiety and to offer helpful hints regarding study habits. This useful booklet is available free with each copy sold. To request copies for your

students in quantity, contact your local Prentice Hall representative.

Video Tapes and Audio Tapes: Professionally done by qualified individuals, these tapes will enhance student's learning by providing further reinforcement of key concepts. Free with a qualified adoption; contact your Prentice Hall representative for details.

Tutorial Software: More for your students. The program, available for the Apple and IBM PC will provide your students with a unique form of drill and practice. Free with a qualified adoption; contact your local Prentice-Hall representative for details.

Acknowledgements

Writing a textbook is a long and time-consuming project. Many people deserve thanks for encouraging and assisting me with this project. Most importantly I would like to thank my wife, Kathy; and sons, Robert and Steven. Without their constant encouragement and understanding, this project would not have become a reality.

I would like to thank my colleagues at Monroe Community College for helping with this project, especially: Larry Clar, Gary Egan, Huebert Haefner, and Annette Leopard. Judith Conturo Karas did an excellent job of typing the manuscript.

I would like to thank my students, and students and faculty from around the country, for using the first edition and offering valuable suggestions for this edition.

I would like to thank the following individuals for working with me on the various supplements for the book:

Instructor's Solution Manual
Test Item File
Students Solution Manual
Instructor's Resource Manual
   Julie Monte   *Daytona Beach Junior College*
   Lea Pruet   *Daytona Beach Junior College*
   Joan Dykes   *Edison Community College*

Student's Study Guide
   Francis Mandery   *Community College of the Finger Lakes*

Video Tapes

 Roger Breen   *Florida Community College at Jacksonville*

 Margaret Greene   *Florida Community College at Jacksonville*

I would like to thank my editor at Prentice-Hall, Priscilla McGeehon; executive editor, Robert Sickles; production editor, Virginia Huebner, and Judith A. Matz-Coniglio, designer.

I would like to thank the following reviewers and proofreaders for their valuable suggestions and their conscientiousness.

 Helen Burrier   *Kirkwood Community College*
 Frank Cerrato   *City College of San Francisco*
 Peter Freedhand   *New York University*
 Margaret Greene   *Florida Community College at Jacksonville*
 Judy Kasabian   *El Camino College*
 Lois Miller   *Golden West College*
 Julie Monte   *Daytona Beach Junior College*
 Cathy Pace   *Louisiana Tech University*
 Ken Seydel   *Skyline College*
 Tommy Thompson   *Brookhaven College*
 John Wenger   *Loop College*
 Brenda Wood   *Florida Community College at Jacksonville*

Finally, I would like to thank the following people who were helpful in advising me on issues of content and organization:

 Ronald Bohuslov   *Merritt College*
 Francine Bortzel   *Seton Hall University*
 Dale Ewen   *Parkland College*
 Mark Gidney   *Lees McRae College*
 Robert Gesell   *Cleary College*
 Larry Hoehn   *Austin Peay State University*
 Herbert Kasube   *Bradley University*
 Adele Legere   *Oakton Community College*
 Melvin Kirkpatrick   *Roane State Community College*
 Glenn Lipely   *Malone College*
 Charles Luttrell   *Frederick Community College*
 Merwin Lyng   *Mayville State College*
 P. William Magliaro   *Bucks County Community College*
 Jack McCown   *Central Oregon Community College*
 John Michaels   *SUNY at Brockport*
 Matthew Pickard   *University of Puget Sound*
 C. V. Peele   *Marshall University*
 James Perkins   *Piedmont Community College*
 Raymond Pluta   *Castleton State College*
 Jon Plachy   *Metropolitan State College*
 Dolores Schaffner   *University of South Dakota*
 Edith Silver   *Mercer County Community College*
 Fay Thames   *Lamar University*
 Karl Zilm   *Lewis and Clark Community College*

To The Student

Algebra is a course that cannot be learned by observation. To learn algebra you must become an active participant. You must read the text, pay attention in class, and, most importantly, you must work the exercises. The more exercises you work the better.

If you purchased this text new then you should have received a complementary copy of "How to Study Mathematics." I suggest you read, and follow, the instructions in the booklet very carefully. You will find them very helpful.

This text was written with you in mind. Short, clear sentences were used and many examples were given to illustrate specific points. The text stresses useful applications of algebra. Hopefully, as you progress through the course you will come to realize that algebra is not just another math course that you are required to take, but a course that offers a wealth of useful information and applications.

This text makes full use of color. The different colors are used to highlight important information. Important procedures, definitions and formulas are placed within colored boxes.

The boxes marked Common Student Errors should be studied carefully. These boxes point out errors that students commonly make, and the correct procedures for doing these problems. The boxes marked Helpful Hints should also be studied carefully for they also stress important information.

At the end of many exercise sets are Just for Fun Problems. These exercises are not for everyone. They are for those students who are doing well in the course and are looking for more of a challenge. These exercises often present additional applications of algebra, material that will be presented in a later section, or material that will be presented in a later course.

At the end of each chapter are a chapter summary, a set of review exercises, and a chapter practice test. Before each examination you should review these sections carefully and take the practice test. If you do well on the practice test you should do well on the class test. The questions in the review exercises are marked to indicate the section in which that material was first introduced. If you have a problem with a review exercise question, reread the section indicated.

In the back of the text there is an answer section which contains the answers to the odd-numbered exercises, Just For Fun problems, all review exercises, and all practice tests. The answers should be used only to check your work.

Various supplements are available to help you achieve success in this course. They include: student's study guide, student's solution manual, video tapes, audio tapes, and tutorial software. Ask your instructor which of these are available for your use.

I have tried to make this text as clear and error free as possible. No text is perfect, however. If you find an error in the text, or an example or section that you believe can be improved, I would greatly appreciate hearing from you. If you enjoy the text, I would also appreciate hearing from you.

ALLEN R. ANGEL

1 Real Numbers

See Section 1.5, Exercise 105

Fractions

1. ▪ *Learn multiplication symbols.*
2. ▪ *Identify factors.*
3. ▪ *Reduce fractions to lowest terms.*
4. ▪ *Multiply fractions.*
5. ▪ *Divide fractions.*
6. ▪ *Add and subtract fractions.*

Students taking algebra for the first time often ask: "What is the difference between arithmetic and algebra?" When doing arithmetic, all the quantities used in the calculations are known. In algebra, however, one or more of the quantities are often unknown and must be found.

EXAMPLE 1   A recipe calls for 3 cups of flour. Mrs. Clark has 2 cups of flour. How many additional cups does she need?

Solution: The answer is 1 cup.   ▪

Although very elementary, this is an example of an algebraic problem. The unknown quantity is the number of additional cups of flour needed.

Since an understanding of fractions is essential to an understanding of algebra we will begin the course with a review of fractions. You will need to know how to reduce a fraction to its lowest terms and how to add, subtract, multiply and divide fractions. We will review these topics in this section. We will also explain the meaning of factors.

▪ In algebra we often use letters called **variables** to represent numbers. A letter commonly used for a variable is the letter x. So that we do not confuse the variable x with the times sign we use different notation to indicate multiplication.

Multiplication Symbols

If a and b stand for (or represent) any two mathematical quantities then each of the following may be used to indicate the product of a and b ("a times b").

$$ab \qquad a \cdot b \qquad a(b) \qquad (a)b \qquad (a)(b)$$

Examples

3 times 4	3 times x	x times y
may be written	may be written	may be written
3(4)	3x	xy
(3)4	(3)x	$(x)y$
(3)(4)	(3)(x)	$(x)(y)$
3 · 4	3 · x	$x \cdot y$

2   The numbers or variables multiplied in a multiplication problem are called **factors.**

> If $a \cdot b = c$ then a and b are **factors** of c.

For example, in $3 \cdot 5 = 15$, the numbers 3 and 5 are factors of the product 15. As a second example consider $2 \cdot 15 = 30$. The numbers 2 and 15 are factors of the product 30. Note that 30 has many other factors. Since $5 \cdot 6 = 30$, the numbers 5 and 6 are also factors of 30. Since $3x$ means 3 times x, both the 3 and the x are factors of $3x$.

3   Now we have the necessary information to discuss fractions. The top number of a fraction is called the **numerator** and the bottom number is called the **denominator.** In the fraction $\frac{3}{5}$, the 3 is the numerator and the 5 is the denominator.

A fraction is **reduced to its lowest terms** when the numerator and denominator have no common factors other than 1. To reduce a fraction to its lowest terms follow these steps.

To Reduce a Fraction to Its Lowest Terms

1. Find the largest number that will divide (without remainder) into both the numerator and the denominator. This number is called the **greatest common factor.**
2. Divide both the numerator and the denominator by the greatest common factor.

EXAMPLE 2   Reduce $\dfrac{10}{25}$ to its lowest terms.

Solution: The largest number that divides both 10 and 25 is 5. Therefore 5 is the greatest common factor. Divide both the numerator and the denominator by 5 to reduce the fraction to its lowest terms.

$$\frac{10}{25} = \frac{10 \div 5}{25 \div 5} = \frac{2}{5} \quad \blacksquare$$

EXAMPLE 3   Reduce $\dfrac{6}{18}$ to its lowest terms.

Solution: Both 6 and 18 can be divided by 1, 2, 3, and 6. The largest of these numbers, 6, is the greatest common factor. Divide both the numerator and the denominator by 6.

$$\frac{6}{18} = \frac{6 \div 6}{18 \div 6} = \frac{1}{3} \quad \blacksquare$$

Note that in Example 3 the numerator and the denominator could have both been written with a factor of 6. Then the common factor 6 divided out.

$$\frac{6}{18} = \frac{1 \cdot \cancel{6}}{3 \cdot \cancel{6}} = \frac{1}{3}$$

When you work with fractions you should give your answers in lowest terms.

Multiplication of
Fractions

◢ To multiply two or more fractions, multiply their numerators together, then multiply their denominators together, as illustrated below.

Multiplication of Fractions

$$\frac{a}{b} \cdot \frac{c}{d} = \frac{ac}{bd}$$

EXAMPLE 4   Multiply $\frac{6}{13}$ by $\frac{5}{12}$.

Solution: $\dfrac{6}{13} \cdot \dfrac{5}{12} = \dfrac{6 \cdot 5}{13 \cdot 12} = \dfrac{30}{156} = \dfrac{5}{26}$ ■

In Example 4, reducing $\frac{30}{156}$ to its lowest terms, $\frac{5}{26}$, is for many students more difficult than the multiplication itself. When multiplying fractions, to help avoid having to reduce an answer to its lowest terms, we often divide both a numerator and denominator by a common factor. **This process can be used only when multiplying fractions; it cannot be used when adding or subtracting fractions.**

EXAMPLE 5   Divide by a common factor, and then multiply.

$$\frac{6}{13} \cdot \frac{5}{12}$$

Solution: Since the numerator 6 and the denominator 12 can both be divided by the common factor 6, we divide out, as illustrated below.

$$\frac{6}{13} \cdot \frac{5}{12} = \frac{\overset{1}{\cancel{6}}}{13} \cdot \frac{5}{\underset{2}{\cancel{12}}} = \frac{1 \cdot 5}{13 \cdot 2} = \frac{5}{26} \qquad ■$$

Note that the answer obtained in Example 5 is identical to the answer obtained in Example 4.

EXAMPLE 6   Multiply $\dfrac{27}{40} \cdot \dfrac{16}{9}$.

Solution: $\dfrac{27}{40} \cdot \dfrac{16}{9} = \dfrac{\overset{3}{\cancel{27}}}{40} \cdot \dfrac{16}{\underset{1}{\cancel{9}}}$ divide both 27 and 9 by 9

$\qquad\quad = \dfrac{\overset{3}{\cancel{27}}}{\underset{5}{\cancel{40}}} \cdot \dfrac{\overset{2}{\cancel{16}}}{\underset{1}{\cancel{9}}}$ divide both 40 and 16 by 8

$\qquad\quad = \dfrac{3 \cdot 2}{5 \cdot 1} = \dfrac{6}{5}$ ■

To multiply an integer by a fraction, write the integer with a denominator of 1, then multiply.

EXAMPLE 7   Multiply $5 \cdot \dfrac{2}{15}$.

Solution: $\dfrac{5}{1} \cdot \dfrac{2}{15} = \dfrac{\overset{1}{\cancel{5}}}{1} \cdot \dfrac{2}{\underset{3}{\cancel{15}}} = \dfrac{2}{3}$ ■

Division of Fractions

5 To divide one fraction by another, invert the divisor (second fraction if written with ÷) and proceed as in multiplication.

> **Division of Fractions**
>
> $$\frac{a}{b} \div \frac{c}{d} = \frac{a}{b} \cdot \frac{d}{c} = \frac{ad}{bc}$$

EXAMPLE 8   Evaluate $\dfrac{3}{5} \div \dfrac{5}{6}$.

Solution: $\dfrac{3}{5} \div \dfrac{5}{6} = \dfrac{3}{5} \cdot \dfrac{6}{5} = \dfrac{3 \cdot 6}{5 \cdot 5} = \dfrac{18}{25}$ ■

EXAMPLE 9   Evaluate $\dfrac{4}{7} \div \dfrac{5}{12}$.

Solution: $\dfrac{4}{7} \div \dfrac{5}{12} = \dfrac{4}{7} \cdot \dfrac{12}{5} = \dfrac{48}{35}$ ■

EXAMPLE 10   Evaluate $\dfrac{3}{8} \div 9$.

Solution: Write the integer 9 as $\frac{9}{1}$.

$\dfrac{3}{8} \div 9 = \dfrac{3}{8} \div \dfrac{9}{1} = \dfrac{3}{8} \cdot \dfrac{1}{\underset{3}{\cancel{9}}} = \dfrac{1}{24}$ ■

Addition and Subtraction of Fractions

6 *Only fractions that have the same* (or common) *denominator can be added or subtracted.* To add (or subtract) fractions with the same denominator, add (or subtract) the numerators and keep the common denominator.

> **Addition and Subtraction of Fractions**
>
> $$\frac{a}{c} + \frac{b}{c} = \frac{a+b}{c} \quad \text{or} \quad \frac{a}{c} - \frac{b}{c} = \frac{a-b}{c}$$

EXAMPLE 11   Evaluate $\dfrac{9}{15} + \dfrac{2}{15}$.

Solution: $\dfrac{9}{15} + \dfrac{2}{15} = \dfrac{9+2}{15} = \dfrac{11}{15}$ ▪

EXAMPLE 12   Evaluate $\dfrac{8}{13} - \dfrac{5}{13}$.

Solution: $\dfrac{8}{13} - \dfrac{5}{13} = \dfrac{8-5}{13} = \dfrac{3}{13}$ ▪

To add (or subtract) fractions with unlike denominators, we must first rewrite each fraction with the same, or a common, denominator. The smallest number that is divisible by two or more denominators is called the **lowest common denominator.**

EXAMPLE 13   Add $\dfrac{1}{2} + \dfrac{1}{5}$.

Solution: We cannot add these fractions until we rewrite them with a common denominator. Since the lowest number that both 2 and 5 divide (without remainder) is 10, we will rewrite both fractions with the lowest common denominator of 10.

$$\frac{1}{2} = \frac{1}{2} \cdot \frac{5}{5} = \frac{5}{10} \quad \text{and} \quad \frac{1}{5} = \frac{1}{5} \cdot \frac{2}{2} = \frac{2}{10}$$

Now add

$$\frac{1}{2} + \frac{1}{5} = \frac{5}{10} + \frac{2}{10} = \frac{7}{10} \quad ▪$$

Note that multiplying both the numerator and denominator by the same number is the same as multiplying by 1. Thus the value of the fraction does not change.

EXAMPLE 14   Subtract $\dfrac{3}{4} - \dfrac{2}{3}$.

Solution: The lowest common denominator is 12. Therefore we rewrite both fractions with a denominator of 12.

$$\frac{3}{4} = \frac{3}{4} \cdot \frac{3}{3} = \frac{9}{12} \quad \text{and} \quad \frac{2}{3} = \frac{2}{3} \cdot \frac{4}{4} = \frac{8}{12}$$

Now subtract

$$\frac{3}{4} - \frac{2}{3} = \frac{9}{12} - \frac{8}{12} = \frac{1}{12} \quad ▪$$

Consider the number $5\frac{2}{3}$. This number means $5 + \frac{2}{3}$.

$$5\frac{2}{3} = 5 + \frac{2}{3} = \frac{15}{3} + \frac{2}{3} = \frac{17}{3}$$

We will use this concept to work Example 15.

EXAMPLE 15   Add $2\frac{1}{4} + \frac{1}{2}$.

Solution: $2\frac{1}{4} = 2 + \frac{1}{4} = \frac{8}{4} + \frac{1}{4} = \frac{9}{4}$

Change $2\frac{1}{4}$ to $\frac{9}{4}$, then add.

$$2\frac{1}{4} + \frac{1}{2} = \frac{9}{4} + \frac{1}{2}$$
$$= \frac{9}{4} + \frac{2}{4}$$
$$= \frac{11}{4} \text{ or } 2\frac{3}{4} \quad \blacksquare$$

COMMON STUDENT ERROR

It is important that the student realize that dividing out a common factor in the numerator of one fraction and the denominator of a different fraction can be performed only when multiplying fractions. **This process cannot be performed when adding or subtracting fractions.**

Correct	*Wrong*
Multiplication problems	Addition problems

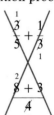

Exercise Set 1.1

Reduce each fraction to its lowest terms. If a fraction is presently in its lowest terms, so state.

1. $\dfrac{10}{25}$ **2.** $\dfrac{12}{20}$ **3.** $\dfrac{9}{12}$ **4.** $\dfrac{4}{5}$

5. $\dfrac{18}{36}$ **6.** $\dfrac{9}{30}$ **7.** $\dfrac{15}{35}$ **8.** $\dfrac{36}{72}$

9. $\dfrac{40}{64}$ **10.** $\dfrac{15}{120}$ **11.** $\dfrac{8}{15}$ **12.** $\dfrac{5}{35}$

13. $\dfrac{72}{96}$ **14.** $\dfrac{14}{28}$

Find the product or quotient. Write answers in lowest terms.

15. $\dfrac{1}{2} \cdot \dfrac{3}{5}$

16. $\dfrac{3}{5} \cdot \dfrac{4}{7}$

17. $\dfrac{5}{8} \cdot \dfrac{2}{7}$

18. $\dfrac{5}{12} \cdot \dfrac{6}{5}$

19. $\dfrac{3}{8} \cdot \dfrac{2}{9}$

20. $\dfrac{15}{16} \cdot \dfrac{4}{3}$

21. $\dfrac{1}{3} \div \dfrac{1}{5}$

22. $\dfrac{2}{3} \cdot \dfrac{3}{5}$

23. $\dfrac{5}{12} \div \dfrac{4}{3}$

24. $\dfrac{4}{9} \div \dfrac{16}{5}$

25. $\dfrac{10}{3} \div \dfrac{5}{9}$

26. $\dfrac{12}{5} \div \dfrac{3}{7}$

27. $\dfrac{4}{9} \cdot \dfrac{15}{16}$

28. $\dfrac{3}{10} \cdot \dfrac{5}{12}$

29. $\dfrac{4}{15} \div \dfrac{12}{13}$

30. $\dfrac{15}{16} \div \dfrac{1}{2}$

31. $\dfrac{12}{7} \cdot \dfrac{19}{24}$

32. $\dfrac{28}{13} \cdot \dfrac{2}{7}$

33. $\dfrac{9}{5} \cdot \dfrac{20}{3}$

34. $\dfrac{24}{5} \div \dfrac{8}{15}$

Add or subtract. Write answers in lowest terms.

35. $\dfrac{2}{5} + \dfrac{1}{5}$

36. $\dfrac{3}{10} + \dfrac{5}{10}$

37. $\dfrac{5}{12} - \dfrac{2}{12}$

38. $\dfrac{18}{36} - \dfrac{1}{36}$

39. $\dfrac{9}{13} + \dfrac{4}{13}$

40. $\dfrac{9}{10} - \dfrac{3}{10}$

41. $\dfrac{21}{29} - \dfrac{18}{29}$

42. $\dfrac{1}{3} + \dfrac{1}{5}$

43. $\dfrac{2}{5} + \dfrac{5}{6}$

44. $\dfrac{1}{9} - \dfrac{1}{18}$

45. $\dfrac{4}{12} - \dfrac{2}{15}$

46. $\dfrac{5}{6} - \dfrac{3}{7}$

47. $\dfrac{2}{10} + \dfrac{1}{15}$

48. $\dfrac{5}{8} - \dfrac{1}{6}$

49. $\dfrac{8}{9} - \dfrac{4}{6}$

50. $\dfrac{3}{8} + \dfrac{5}{12}$

51. $\dfrac{5}{6} + \dfrac{9}{24}$

52. $\dfrac{7}{15} - \dfrac{12}{30}$

53. $\dfrac{11}{12} - \dfrac{3}{4}$

54. $2\dfrac{1}{2} + \dfrac{1}{4}$

55. $3\dfrac{1}{4} + \dfrac{2}{3}$

56. $\dfrac{3}{10} + 2\dfrac{1}{3}$

57. $2\dfrac{1}{2} + 1\dfrac{1}{3}$

58. $\dfrac{4}{5} - \dfrac{2}{7}$

59. $4\dfrac{2}{3} - 1\dfrac{1}{5}$

60. $3\dfrac{1}{8} - \dfrac{3}{4}$

61. $1\dfrac{4}{5} - \dfrac{3}{4}$

62. $2\dfrac{2}{3} + 1\dfrac{3}{5}$

Solve each problem.

63. John, a dressmaker, wishes to make 5 identical dresses. If each dress needs $2\dfrac{3}{4}$ yards of material, how much material will John need?

64. A board is $22\dfrac{1}{2}$ feet long. What is the length of each piece when cut in five equal lengths? (Ignore the thickness of the cuts.)

65. A length of $3\dfrac{1}{16}$ inches is cut from a piece of wood $16\dfrac{3}{4}$ inches long. What is the length of the remaining piece of wood?

66. At the beginning of the day a stock was selling for $11\dfrac{7}{8}$ dollars. At the close of the session the same stock was selling for $13\dfrac{3}{4}$. How much did the stock gain that day?

67. A plumber connected two pieces of pipe measuring $3\dfrac{3}{8}$ feet and $5\dfrac{1}{16}$ feet, respectively. What is the total length of these two pieces of pipe?

68. A recipe calls for $2\dfrac{1}{2}$ cups of flour and another $1\dfrac{1}{3}$ cups of flour to be added later. How much flour does the recipe require?

69. At high tide the water level at a measuring stick is $20\dfrac{3}{4}$ feet. At low tide the water level dropped to $8\dfrac{7}{8}$ feet. How much did the water level fall?

JUST FOR FUN

1. Use the directions shown to find the amount of each ingredient needed to make three servings of Minute Rice.

Amounts of RICE & WATER Use equal amounts rice and water. Minute Rice doubles in volume.			
TO MAKE	RICE & WATER (equal measures)	SALT	BUTTER OR MARGARINE (if desired)
2 servings	$\frac{2}{3}$ cup	$\frac{1}{4}$ tsp.	1 tsp.
4 servings	$1\frac{1}{3}$ cups	$\frac{1}{2}$ tsp.	2 tsp.

MINUTE® is a registered trademark of General Foods Corporation, White Plains, N.Y.

1.2
The Real Number System

☐1 *Identify some important sets of numbers.*
☐2 *Know the structure of the real numbers.*

We will be talking about and using various types of numbers throughout the text. This section introduces you to some of those numbers, and to the structure of the real number system. This section is a quick overview. Some of the sets of numbers we mention in this section, such as rational and irrational numbers, are discussed in greater depth later in the text.

☐1 A **set** is a collection of **elements** listed within braces. The set {a, b, c, d, e} consists of five elements, namely a, b, c, d, and e. There are many different sets of numbers. Two important sets are the natural numbers and the whole numbers.

Natural numbers: {1, 2, 3, 4, 5, . . .}
Whole numbers: {0, 1, 2, 3, 4, 5, . . .}

The three dots indicate that the numbers continue in the same manner and there is no last element in the set. An aid in understanding sets of numbers is the real number line (Fig. 1.1).

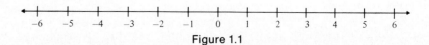

Figure 1.1

The real number line continues indefinitely in both directions. The numbers to the right of 0 are positive and those to the left of 0 are negative. Zero is neither positive nor negative (Fig. 1.2).

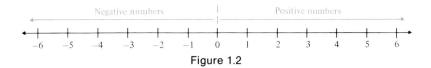

Figure 1.2

Figure 1.3 illustrates the natural numbers marked on the number line. The natural numbers are also called the **positive integers** or the **counting numbers.**

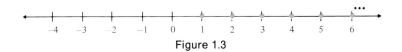

Figure 1.3

Another important set of numbers is the integers.

Integers: $\underbrace{\{\ldots, -5, -4, -3, -2, -1,}_{\text{negative integers}} 0, \underbrace{1, 2, 3, 4, 5, \ldots\}}_{\text{positive integers}}$

The integers consist of the negative integers, 0, and the positive integers. The integers are marked on the number line in Fig. 1.4.

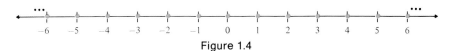

Figure 1.4

Can you think of any numbers that are not integers? You probably said "fractions" or "decimal numbers." Fractions and decimal numbers belong to the set of rational numbers. The set of *rational numbers* consists of all the numbers that can be expressed as a quotient of two integers, with the denominator not 0.

Rational numbers: {quotient of two integers, denominator not 0}

The fraction $\frac{1}{2}$ is a quotient of two integers with the denominator not 0. Thus $\frac{1}{2}$ is a rational number. The decimal number 0.4 can be written $\frac{4}{10}$ and is therefore a rational number. All integers are also rational numbers since they can be written with a denominator of 1. For example $3 = \frac{3}{1}$, $-12 = \frac{-12}{1}$, and $0 = \frac{0}{1}$. Some rational numbers are illustrated on the number line in Fig. 1.5.

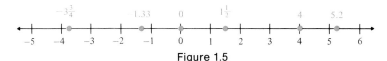

Figure 1.5

Most of the numbers that we use are rational numbers; however, there are some numbers that are not rational. Numbers such as the square root of 2, written $\sqrt{2}$, are not rational numbers. Any number that can be represented on the number line that is not a rational number is called an **irrational number.** The $\sqrt{2}$ is *approximately* 1.41. Some irrational numbers are illustrated on the number line in Fig. 1.6. Rational and irrational numbers will be discussed further in later chapters.

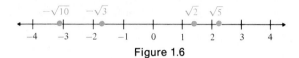

Figure 1.6

2 Notice that many different types of numbers can be illustrated on the number line. Any number that can be represented on the number line is a *real number*.

Real numbers: {all numbers that can be represented on the real number line}

All the numbers mentioned thus far—the natural numbers, the whole numbers, the integers, the rational numbers, and the irrational numbers—are real numbers. There are some types of numbers that are not real numbers, but these numbers are beyond the scope of this book. Figure 1.7 illustrates the relationship between the various sets of numbers and the set of real numbers.

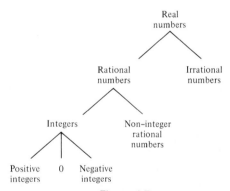

Figure 1.7

In Fig. 1.7 we can see that when we combine the rational numbers and the irrational numbers we get the real numbers. When we combine the integers with the non-integer rational numbers (such as $\frac{1}{2}$ and 0.42) we get the rational numbers. When we combine the positive integers (or natural numbers), 0, and the negative integers we get the integers.

Consider the positive integer 5. If we follow the positive integer branch in Fig. 1.7 upward, we see that the number 5 is also an integer, a rational number, and a real number. Now consider the number $\frac{1}{2}$. It belongs to the non-integer rational numbers. If we follow this branch upward, we can see that $\frac{1}{2}$ is also a rational number and a real number.

EXAMPLE 1   Consider the following set of numbers:

$$\left\{-6, -0.5, 4\frac{1}{2}, -96, \sqrt{3}, 0, 9, -\frac{4}{7}, -2.9, \sqrt{7}, -\sqrt{5}\right\}$$

List the elements of the set that are

(a) Natural numbers.   (b) Whole numbers.
(c) Integers.   (d) Rational numbers.
(e) Irrational numbers.   (f) Real numbers.

Solution: (a) 9

(b) 0, 9

(c) $-96, -6, 0, 9$

(d) $-96, -6, -2.9, -\frac{4}{7}, -0.5, 0, 4\frac{1}{2}, 9$

(e) $-\sqrt{5}, \sqrt{3}, \sqrt{7}$

(f) $-96, -6, -2.9, -\sqrt{5}, -\frac{4}{7}, -0.5, 0, \sqrt{3}, \sqrt{7}, 4\frac{1}{2}, 9$ ■

Exercise Set 1.2

List the elements of each set.

1. Integers
2. Counting numbers
3. Natural numbers
4. Positive integers
5. Negative integers
6. Whole numbers

State whether each statement is true or false.

7. -4 is a negative integer.
8. 0 is a whole number.
9. 0 is an integer.
10. -1 is an integer.
11. $\frac{1}{2}$ is an integer.
12. 0.5 is an integer.
13. $\sqrt{7}$ is a rational number.
14. $\sqrt{7}$ is a real number.
15. $-\frac{3}{5}$ is a rational number.
16. 0 is a rational number.
17. $-19\frac{1}{5}$ is an irrational number.
18. -7 is a real number.
19. -0.06 is a real number.
20. $2\frac{5}{8}$ is an irrational number.
21. 0 is a positive integer.
22. The natural numbers, counting numbers, and positive integers are different names for the same set of numbers.
23. When zero is added to the set of counting numbers, the whole numbers are formed.

24. When the negative integers, the positive integers, and 0 are combined, the integers are formed.
25. Any number to the left of zero on the number line is a negative number.
26. Every integer is a rational number.
27. Every integer is an irrational number.
28. Every rational number is a real number.
29. Every irrational number is a real number.
30. The number 0 is an irrational number.
31. Some real numbers are not rational numbers.
32. Some rational numbers are not real numbers.
33. Every natural number is positive.
34. Every integer is positive.
35. No rational numbers are integers.
36. All real numbers can be represented on the number line.
37. Irrational numbers cannot be represented on the number line.
38. Some rational numbers are negative integers.

39. Consider the set of numbers

$$\left\{-6, 7, 12.4, -\frac{9}{5}, -2\frac{1}{4}, \sqrt{3}, 0, 9, \sqrt{7}, 0.35\right\}$$

List those numbers that are
(a) Positive integers.
(b) Whole numbers.
(c) Integers.
(d) Rational numbers.
(e) Irrational numbers.
(f) Real numbers.

40. Consider the set of numbers

$$\left\{-\frac{5}{3}, 0, -2, 5, 5\frac{1}{2}, \sqrt{2}, -\sqrt{3}, 1.63, 207\right\}$$

List those numbers that are
(a) Positive integers.
(b) Whole numbers.
(c) Integers.
(d) Rational numbers.
(e) Irrational numbers.
(f) Real numbers.

41. Consider the set of numbers

$$\left\{\frac{1}{2}, \sqrt{2}, -\sqrt{2}, 4\frac{1}{2}, \frac{5}{12}, -1.67, 5, -300, -9\frac{1}{2}\right\}$$

List those numbers that are
(a) Positive integers.
(b) Whole numbers.
(c) Negative integers.
(d) Integers.
(e) Rational numbers.
(f) Irrational numbers.
(g) Real numbers.

In each of the following exercises give three examples of numbers that satsify the conditions.

42. A real number but not an integer.
43. A rational number but not an integer.
44. An integer but not a negative integer.
45. A real number but not a rational number.
46. An irrational number and a positive number.
47. A integer and a rational number.

48. A negative integer and a real number.
49. A negative integer and a rational number.
50. A real number but not a rational number.
51. A rational number but not a negative number.
52. An integer but not a positive integer.
53. A real number but not an irrational number.

1.3

Inequalities

1 *Determine which is the greater of two numbers.*
2 *Find the absolute value of a number.*

1 The number line (Fig. 1.8) can be used to explain inequalities. When comparing two numbers, **the number to the right on the number line is the greater number, and the number to the left is the lesser number.** The symbol $>$ is used to represent the words "greater than." The symbol $<$ is used to represent the words "less than."

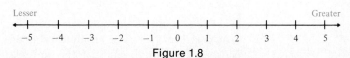

Figure 1.8

The number 3 is greater than the number 2, written $3 > 2$, because 3 is to the right of 2 on the number line. Similarly, we can see that $0 > -1$ because 0 is to the right of -1 on the number line.

Instead of stating that 3 is greater than 2, we could state that 2 is less than 3, $2 < 3$, because 2 is to the left of 3 on the number line. Similarly, $-1 < 0$ because -1 is to the left of 0.

The student should remember that the point of the inequality symbol will always point to the lesser number.

EXAMPLE 1   Insert either $>$ or $<$ in the space between the paired numbers to make the statement correct.

(a) $-4 \quad -2$ (b) $-\frac{3}{2} \quad 2.5$ (c) $\frac{1}{2} \quad \frac{1}{4}$ (d) $-2 \quad 4$

Solution: The points given above are shown on the number line (Fig. 1.9).

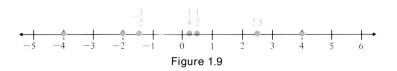

Figure 1.9

(a) $-4 < -2$, since -4 is to the left of -2.
(b) $-\frac{3}{2} < 2.5$, since $-\frac{3}{2}$ is to the left of 2.5.
(c) $\frac{1}{2} > \frac{1}{4}$, since $\frac{1}{2}$ is to the right of $\frac{1}{4}$.
(d) $-2 < 4$, since -2 is to the left of 4.   ■

EXAMPLE 2   Insert either $>$ or $<$ in the space between the paired numbers to make the statement correct.

(a) $-1 \quad -2$ (b) $-1 \quad 0$ (c) $-2 \quad 2$ (d) $-4.09 \quad -4.9$

Solution: The points given above are shown on the number line (Fig. 1.10).

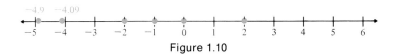

Figure 1.10

(a) $-1 > -2$, since -1 is to the right of -2.
(b) $-1 < 0$, since -1 is to the left of 0.
(c) $-2 < 2$, since -2 is to the left of 2.
(d) $-4.09 > -4.9$, since -4.09 is to the right of -4.9 ■

2   The concept of absolute value can be explained with the help of the number line shown in Fig. 1.11. The **absolute value** of a number can be considered the distance between the number and 0 on the number line. Thus the absolute value of 3, symbolized by $|3|$, is 3 since it is 3 units from 0 on the number line. Similarly, the absolute value of the number -3, symbolized by $|-3|$, is also 3 since -3 is 3 units from 0.

$$|3| = 3 \quad \text{and} \quad |-3| = 3$$

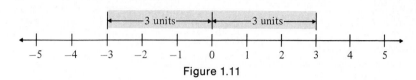

Figure 1.11

Since the absolute value of a number measures the distance (without regard to direction) of a number from 0 on the number line, **the absolute value of every number will be either positive or zero.**

Number	Absolute value of number
6	$|6| = 6$
-6	$|-6| = 6$
0	$|0| = 0$
$-\dfrac{1}{2}$	$\left|-\dfrac{1}{2}\right| = \dfrac{1}{2}$

EXAMPLE 3   Insert either $>$, $<$, or $=$ to make the statement true.

(a) $|3| \quad 3$ (b) $|-2| \quad |2|$ (c) $-2 \quad |-4|$

(d) $|-5| \quad 0$ (e) $|12| \quad |-18|$

Solution: (a) $|3| = 3$.

(b) $|-2| = |2|$, since both $|-2|$ and $|2|$ equal 2.

(c) $-2 < |-4|$, since $|-4| = 4$.

(d) $|-5| > 0$, since $|-5| = 5$.

(e) $|12| < |-18|$, since $|12| = 12$ and $|-18| = 18$.   ■

Exercise Set 1.3

Insert either $<$ or $>$ to make each expression true.

1. 2   3

2. 4 -2

3. -3 0

4. -6 -4

5. $\dfrac{1}{2}$ $-\dfrac{2}{3}$

6. $\dfrac{3}{5}$ $\dfrac{4}{5}$

7. 0.2   0.4

8. -0.2 -0.4

9. $\dfrac{2}{5}$ -1

10. 0 -0.9

11. 4 -4

12. $-\dfrac{3}{4}$ -1

13. -2.1 -2

14. -1.83 -1.82

15. $\dfrac{5}{9}$ $-\dfrac{5}{9}$

16. -9 -12

17. $-\dfrac{3}{2}$ $\dfrac{3}{2}$

18. -4.09 -5.3

19. 0.49 0.43

20. -1.0 -0.7

21. 5 -7

22. 0.001 0.002

23. -0.006 -0.007

24. $\dfrac{1}{2}$ $-\dfrac{1}{2}$

25. $\dfrac{3}{5}$ 1

26. $\dfrac{5}{3}$ $\dfrac{3}{5}$

27. $-\dfrac{2}{3}$ -3

28. -5 -2

Insert either $<$, $>$, or $=$ to make each expression true.

29. 8 $|-7|$

30. $|-8|$ $|-7|$

31. $|0|$ $\dfrac{2}{3}$

32. $|-4|$ -3

33. $|-3|$ $|-4|$

34. $|-1.9|$ -1.8

35. 4 $\left|-\dfrac{9}{2}\right|$

36. -5 $|5|$

37. $\left|-\dfrac{6}{2}\right|$ $\left|-\dfrac{2}{6}\right|$

38. $\left|\dfrac{2}{5}\right|$ $|-0.40|$

39. What numbers are 4 units from 0 on the number line?

41. What numbers are 2 units from 0 on the number line?

40. What numbers are 5 units from 0 on the number line?

1.4

Addition of Real Numbers

1️⃣ *Add real numbers using the number line.*

2️⃣ *Identify opposites or additive inverses.*

3️⃣ *Add using absolute values.*

1️⃣ There are many practical uses for negative numbers. A submarine going below sea level, a bank account that has been overdrawn, a business spending more than it makes, and a temperature below zero are some examples.

To add numbers, we make use of the number line. Represent the first number to be added (first *addend*) by an arrow starting at 0. The arrow will move to the right if the number is positive. If the number is negative, the arrow will move to the left. From the tip of the first arrow, draw a second arrow to represent the second addend. The second arrow moves to the right or left, as explained above. The sum of the two numbers is found at the tip of the second arrow.

EXAMPLE 1   Evaluate $3 + (-4)$ using the number line.

Solution: The first arrow starts at 0 and moves 3 units to the right (Fig. 1.12).

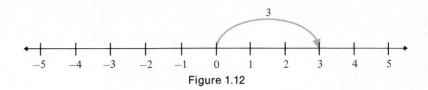

Figure 1.12

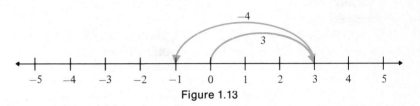

Figure 1.13

The second arrow starts at 3 and moves 4 units to the left, since the second addend is negative (Fig. 1.13). The tip of the second arrow is at -1. Thus

$$3 + (-4) = -1 \quad \blacksquare$$

EXAMPLE 2   Evaluate $-4 + 2$ using the number line.

Solution: See Fig. 1.14.

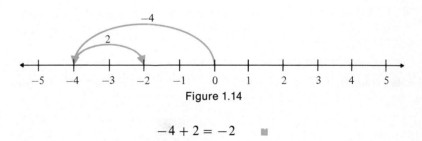

Figure 1.14

$$-4 + 2 = -2 \quad \blacksquare$$

EXAMPLE 3   Evaluate $-3 + (-2)$ using the number line.

Solution: See Fig. 1.15.

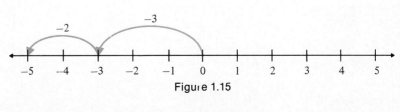

Figure 1.15

$$-3 + (-2) = -5 \quad \blacksquare$$

EXAMPLE 4   Add $5 + (-5)$.

Solution: See Fig. 1.16.

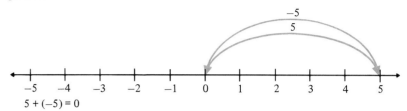

$$5 + (-5) = 0$$

Figure 1.16

$$5 + (-5) = 0 \quad \blacksquare$$

2

> Any two numbers whose sum is zero are said to be **opposites** (or **additive inverses**) of each other. In general, if we let a represent any real number, then its opposite is $-a$ and $a + (-a) = 0$.

In Example 4 the sum of 5 and -5 is zero. Thus -5 is the opposite of 5 and 5 is the opposite of -5.

EXAMPLE 5   Find the opposite of each number.

(a) 3   (b) -4

Solution: (a) The opposite of 3 is -3, since $3 + (-3) = 0$
(b) The opposite of -4 is 4, since $-4 + 4 = 0$. $\blacksquare$

EXAMPLE 6   A submarine dives 250 feet. A short while later it dives an additional 190 feet. Find the depth of the submarine with respect to sea level (assume that depths below sea level are indicated by negative numbers).

Solution: It may be helpful to use a vertical number line (Fig. 1.17) to explain this problem.

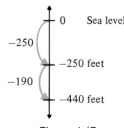

Figure 1.17

$$-250 + (-190) = -440 \text{ feet.} \quad \blacksquare$$

EXAMPLE 7   The ABC Company had a loss of $4000 for the first 6 months of the year, and a profit of $15,500 for the second 6 months of the year. Find the net profit or loss for the year.

Solution: As shown in Fig. 1.18, the net profit for the year is $11,500.

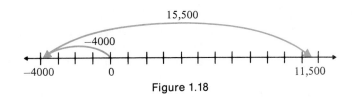

Figure 1.18

3   Now that we have had some practice adding signed numbers on the number line we will give a rule (in two parts), using absolute value, for the addition of signed numbers. Remember that the absolute value of a nonzero number will always be positive.

> **To add real numbers with the same sign** (either both positive or both negative), add their absolute values. The sum has the same sign as the numbers being added.

EXAMPLE 8   Add $4 + 8$.

Solution: Since both numbers have the same sign, both positive, we add their absolute values: $|4| + |8| = 4 + 8 = 12$. Since both numbers being added are positive, the sum is positive. Thus $4 + 8 = 12$. ▪

EXAMPLE 9   Add $-6 + (-9)$.

Solution: Since both numbers have the same sign, both negative, we add their absolute values: $|-6| + |-9| = 6 + 9 = 15$. Since both numbers being added are negative, their sum is negative. Thus $-6 + (-9) = -15$. ▪

The sum of two positive numbers will always be positive and the sum of two negative numbers will always be negative.

> **To add two signed numbers with different signs** (one positive and the other negative), find the difference between the larger absolute value and the smaller absolute value. The answer has the sign of the number with the larger absolute value.

EXAMPLE 10   Add $10 + (-6)$.

Solution: Since the two numbers being added have different signs, find the difference between the larger absolute value and the smaller: $|10| - |-6| = 10 - 6 = 4$. Since $|10|$ is greater than $|-6|$, the sum is positive. Thus $10 + (-6) = 4$. ▪

EXAMPLE 11   Add $12 + (-18)$.

Solution: Since the numbers being added have different signs, find the difference between the larger absolute value and the smaller: $|-18| - |12| = 18 - 12 = 6$. Since $|-18|$ is greater than $|12|$, the sum is negative. Thus $12 + (-18) = -6$. ∎

EXAMPLE 12   Add $-24 + 19$.

Solution: Since the two numbers being added have different signs, find the difference between the larger absolute value and the smaller: $|-24| - |19| = 24 - 19 = 5$. Since $|-24|$ is greater than $|19|$, the sum is negative. Thus $-24 + 19 = -5$. ∎

The sum of two signed numbers with different signs may be either positive or negative. The sign of the sum will be the same as the sign of the number with the larger absolute value.

Exercise Set 1.4 _____

State the opposite of each number.

1. 18	**2.** -7	**3.** -32	**4.** 3
5. 0	**6.** 6	**7.** $\dfrac{5}{3}$	**8.** $-\dfrac{1}{2}$
9. $\dfrac{3}{5}$	**10.** -1	**11.** 0.63	**12.** -0.721
13. $2\dfrac{1}{2}$	**14.** $-3\dfrac{1}{4}$	**15.** -3.1	**16.** 5.26

Add as indicated.

17. $4 + 3$	**18.** $-4 + 3$	**19.** $4 + (-3)$	**20.** $4 + (-2)$
21. $-4 + (-2)$	**22.** $-3 + (-5)$	**23.** $6 + (-6)$	**24.** $-6 + 6$
25. $-4 + 4$	**26.** $-3 + 5$	**27.** $-8 + (-2)$	**28.** $6 + (-5)$
29. $-3 + 3$	**30.** $-8 + 2$	**31.** $-3 + (-7)$	**32.** $0 + (-3)$
33. $0 + 0$	**34.** $0 + (-0)$	**35.** $-6 + 0$	**36.** $-9 + 13$
37. $22 + (-19)$	**38.** $-13 + (-18)$	**39.** $-45 + 36$	**40.** $40 + (-25)$
41. $18 + (-9)$	**42.** $-7 + 7$	**43.** $-14 + (-13)$	**44.** $-27 + (-9)$
45. $-35 + (-9)$	**46.** $34 + (-12)$	**47.** $4 + (-30)$	**48.** $-16 + 9$
49. $-35 + 40$	**50.** $-12 + 17$	**51.** $180 + (-200)$	**52.** $-33 + (-92)$
53. $-105 + 74$	**54.** $183 + (-183)$	**55.** $184 + (-93)$	**56.** $-42 + 129$
57. $-452 + 312$	**58.** $-94 + (-98)$	**59.** $-496 + 496$	

Set up an expression that can be used to solve each problem, and then solve.

60. Mr. Thorp owed $38 on his bank credit card. He charged another item costing $121. Find the amount that Mr. Thorp owed the bank.

61. Mr. Weber charged $193 worth of goods on his charge card. Find his balance after he made a payment of $112.

62. Mrs. Petrie paid $1424 in federal income tax. When she was audited, Mrs. Petrie had to pay an additional $503. What was her total tax?

63. Mr. Terrell hiked down to the base of the Grand Canyon, a distance of 1620 meters. The next day he climbed up 213 meters. Find his distance from the rim of the canyon.

64. A car accelerates to a speed of 60 miles per hour. It then decelerates by 20 miles per hour. Find the speed.

65. An airplane at an altitude of 2400 feet above sea level drops a package into the ocean. The package settles at a point 200 feet below sea level. How far did the object fall?

66. A football team loses 18 yards on one play, then loses 3 yards on the following play. What is the total loss in yardage?

See Exercise 63.

1.5
Subtraction of Real Numbers

❶ *Subtract real numbers.*

❶ Any subtraction problem can be rewritten as an addition problem using the additive inverse.

> **Subtraction of Real Numbers**
>
> In general, if a and b represent any two real numbers, then
> $$a - b = a + (-b)$$

The rule above says that to subtract b from a, add the opposite or additive inverse of b to a.

EXAMPLE 1   Evaluate $9 - (+4)$.

Solution: In this example we are subtracting a positive 4 from 9. To accomplish this using the rule just given we must add the opposite of $+4$, which is -4, to 9.

$$9 - (+4) = 9 + (-4) = 5$$

subtract   positive 4   add   negative 4

Note that $9 + (-4)$ was evaluated using the procedures for *adding* real numbers presented in the previous section. ■

21

Often in a subtraction problem, when the number being subtracted is a positive number, the $+$ sign preceding the number being subtracted is not illustrated. For example, in the subtraction $9 - 4$,

$$9 - 4 \text{ means } 9 - (+4)$$

Thus to evaluate $9 - 4$ we must add the opposite of 4 (or $+4$), which is -4, to 9.

$$9 - 4 = 9 + (-4) = 5$$

subtract   positive 4   add   negative 4

This procedure is illustrated in Example 2.

EXAMPLE 2   Evaluate $5 - 3$.

Solution: We must subtract a positive 3 from 5. To change this problem to an addition problem, add the opposite of 3, which is -3, to 5.

subtraction   addition
problem   problem
$$5 - 3 = 5 + (-3) = 2$$

subtract   positive 3   add   negative 3   ■

EXAMPLE 3   Evaluate $4 - 9$.

Solution: Add the opposite of 9, -9, to 4.

$$4 - 9 = 4 + (-9) = -5 \quad ■$$

EXAMPLE 4   Evaluate $-4 - 2$.

Solution: Add the opposite of 2, -2, to -4.

$$-4 - 2 = -4 + (-2) = -6 \quad ■$$

EXAMPLE 5   Evaluate $4 - (-2)$.

Solution: We are asked to subtract a negative 2 from 4. To do this, add the opposite of -2, 2, to 4.

$$4 - (-2) = 4 + 2 = 6$$

subtract   negative 2   add   positive 2   ■

EXAMPLE 6   Evaluate $-6 - (-3)$.

Solution: Add the opposite of -3, 3, to -6.

$$-6 - (-3) = -6 + 3 = -3 \quad ■$$

EXAMPLE 7   Subtract 12 from 3.

Solution: $3 - 12 = 3 + (-12) = -9 \quad ■$

EXAMPLE 8   Subtract 5 from 5.

Solution: $5 - 5 = 5 + (-5) = 0$ ■

EXAMPLE 9   Subtract -6 from 4.

Solution: $4 - (-6) = 4 + 6 = 10$ ■

HELPFUL HINT

By examining Example 9 we see that

$$4 - (-6) = 4 + 6$$

two negative   + sign
signs in a row

Whenever we are *subtracting a negative number*, we can replace the two negative signs in a row with a plus sign.

EXAMPLE 10   Evaluate: (a) $8 - (-5)$ (b) $-3 - (-9)$.

Solution: (a) $8 - (-5) = 8 + 5 = 13$
(b) $-3 - (-9) = -3 + 9 = 6$ ■

EXAMPLE 11   Mary Jo Morin's checkbook indicated a balance of $125 before she wrote a check for $183. Find the balance in her checkbook.

Solution: $125 - 183 = 125 + (-183) = -58$. The negative indicates a deficit. Therefore Mary Jo has a deficit of $58.   ■

EXAMPLE 12   Janet has made $4200 in the stock market, whereas Peter has lost $3000. How much farther ahead is Janet than Peter in their financial positions?

Solution: Janet's gain is represented as a positive number. Peter's loss is represented by a negative number.

$$4200 - (-3000) = 4200 + 3000 = 7200$$

Janet is therefore $7200 ahead of Peter.   ■

EXAMPLE 13   Evaluate each of the following.

(a) $12 + (-4)$ (b) $-16 - 3$ (c) $5 + (-4)$
(d) $6 - (-5)$ (e) $-12 - (-3)$ (f) $8 - 13$

Solution: Parts (a) and (c) are addition problems while the rest are subtraction problems.

(a) $12 + (-4) = 8$ (b) $-16 - 3 = -16 + (-3) = -19$
(c) $5 + (-4) = 1$ (d) $6 - (-5) = 6 + 5 = 11$
(e) $-12 - (-3) = -12 + 3 = -9$ (f) $8 - 13 = 8 + (-13) = -5$ ■

Now let's look at examples that combine addition and subtraction.

EXAMPLE 14   Evaluate $-13 + 4 - 2$.

Solution: We work from left to right.

$$-13 + 4 - 2 = -9 - 2$$
$$= -9 + (-2)$$
$$= -11 \quad \blacksquare$$

EXAMPLE 15   Evaluate $-3 - (-4) + (-10) + (-5)$.

Solution: Again we work from left to right.

$$-3 - (-4) + (-10) + (-5) = -3 + 4 + (-10) + (-5)$$
$$= 1 + (-10) + (-5)$$
$$= -9 + (-5)$$
$$= -14 \quad \blacksquare$$

Exercise Set 1.5 _____

Evaluate each expression.

1. $6 - 3$
2. $-6 - 4$
3. $4 - 5$
4. $5 - 3$
5. $3 - 3$
6. $-4 - 2$
7. $(-7) - (-4)$
8. $-4 - (-3)$
9. $-3 - 3$
10. $-4 - 4$
11. $3 - (-3)$
12. $4 - 4$
13. $0 - 6$
14. $6 - 6$
15. $0 - (-6)$
16. $9 - (-3)$
17. $-3 - 5$
18. $-5 - (-3)$
19. $-5 + 7$
20. $-7 - 9$
21. $5 - 3$
22. $5 - 12$
23. $6 - (-3)$
24. $6 - 10$
25. $8 - 8$
26. $-8 - 8$
27. $-8 - 10$
28. $4 - 12$
29. $-5 - (-3)$
30. $7 - 9$
31. $(-4) - (-4)$
32. $15 - 8$
33. $6 - 6$
34. $(-8) - (-12)$
35. $8 - 8$
36. $-6 - (-2)$
37. $4 - 5$
38. $-9 - 2$
39. $-2 - 3$
40. $9 - (-12)$
41. $-25 - 16$
42. $-20 - (-15)$
43. $37 - 40$
44. $40 - 37$
45. $-100 - 80$
46. $80 - 100$
47. $-20 - 90$
48. $-50 - (-40)$
49. $70 - (-70)$
50. $130 - (-90)$
51. $87 - 87$
52. $93 - (-93)$
53. $-45 - 37$
54. $-53 - (-7)$
55. Subtract 4 from 9.
56. Subtract 9 from 4.
57. Subtract 3 from -15.
58. Subtract -4 from -5.
59. Subtract 8 from -8.
60. Subtract 10 from -20.
61. Subtract 8 from 18.
62. Subtract 5 from -5.
63. Subtract -3 from -5.
64. Subtract 10 from -3.
65. Subtract -4 from 9.
66. Subtract 18 from -18.
67. Subtract 18 from 18.
68. Subtract 5 from 5.
69. Subtract 12 from 8.
70. Subtract -9 from 12.
71. Subtract -15 from -4.
72. Subtract -12 from 3.
73. Subtract -36 from 45.
74. Subtract 17 from -12.

Evaluate each expression.

75. $6 + 5 - (+4)$
76. $9 - (+6) - (+5)$
77. $-3 + (-4) + 5$
78. $9 - 7 + (-2)$
79. $-13 - (+5) + 3$
80. $7 - (+4) - (-3)$
81. $-9 - (-3) + 4$
82. $15 + (-7) - (-3)$
83. $5 - (+3) + (-2)$
84. $12 + (-5) - (-4)$
85. $25 + (+12) - (-6)$
86. $-7 + 6 - 3$

87. $-4 - 7 + 5$

90. $-36 - 5 + 9$

93. $-9 - 4 - 8$

96. $(-4) + (-3) + 5 - 7$

99. $32 + 5 - 7 - 12$

101. $-7 - 4 - 3 + 5$

88. $20 - 4 - 25$

91. $45 - 3 - 7$

94. $25 - 19 + 27$

97. $-9 - 3 - (-4) + 5$

100. $-19 + (-3) - (-5) - (-2)$

102. $19 + 4 - 20 - 25$

89. $-4 + 7 - 12$

92. $-2 + 7 - 9$

95. $-4 - 13 + 5$

98. $17 + (-3) - 9 - (-7)$

Solve the following problems.

103. An airplane is 2000 feet above sea level. A submarine is 1500 feet below sea level. How far above the submarine is the airplane?

104. A Jeep travels 162 miles due east. It then turns around and travels 83 miles due west. What is the Jeep's distance from its starting point?

105. The highest point on Earth, Mt. Everest, is 29,028 feet above sea level. The lowest point on Earth, the Marianas Trench, is 36,198 feet below sea level. How far above the Marianas Trench is Mt. Everest?

106. The greatest change in temperature within a 24-hour period occurred at Browning, Montana, on January 23, 1916. The temperature fell from 44°F to −56°F. How much did the temperature drop?

See Exercise 104.

JUST FOR FUN

Find the sum.

1. $1 - 2 + 3 - 4 + 5 - 6 + 7 - 8 + 9 - 10$

2. $1 - 2 + 3 - 4 + 5 - 6 + \cdots + 99 - 100$

3. $-1 + 2 - 3 + 4 - 5 + 6 - \cdots - 99 + 100$

1.6

Multiplication and Division of Real Numbers

◻ *Multiply real numbers.*

◻ *Divide real numbers.*

◻ *Remove negative signs from denominators.*

◻ *Learn the differences between $\frac{1}{0}$, $\frac{0}{1}$, and $\frac{0}{0}$.*

Multiplication of Real Numbers

◻ The following rules are used in determining the sign of the product when two numbers are multiplied.

Multiplication of Real Numbers

1. The product of two numbers with **like signs** is a **positive number**.

2. The product of two numbers with **unlike** signs is a **negative number**.

By the above rule the product of two positive numbers or two negative numbers will be a positive number. The product of a positive number and a negative number will be a negative number.

EXAMPLE 1   Evaluate $3(-5)$.

Solution: Since the numbers have unlike signs, the product is negative.

$$3(-5) = -15 \qquad \blacksquare$$

EXAMPLE 2   Evaluate $(-6)(7)$.

Solution: Since the numbers have unlike signs, the product is negative.

$$(-6)(7) = -42 \qquad \blacksquare$$

EXAMPLE 3   Evaluate $(-7)(-5)$.

Solution: Since the numbers have like signs, both negative, the product is positive.

$$(-7)(-5) = 35 \qquad \blacksquare$$

EXAMPLE 4   Evaluate each expression.

(a) $-6 \cdot 3$ (b) $(-4)(-8)$ (c) $4(-9)$
(d) $0 \cdot 4$ (e) $0(-2)$ (f) $-3(-6)$

Solution: (a) $-6 \cdot 3 = -18$ (b) $(-4)(-8) = 32$ (c) $4(-9) = -36$
(d) $0 \cdot 4 = 0$ (e) $0(-2) = 0$ (f) $-3(-6) = 18$

Note that zero times any real number equals zero.   ▪

EXAMPLE 5   Multiply $\left(\dfrac{-1}{8}\right)\left(\dfrac{-3}{5}\right)$.

Solution: $\left(\dfrac{-1}{8}\right)\left(\dfrac{-3}{5}\right) = \dfrac{(-1)\cdot(-3)}{8\cdot 5} = \dfrac{3}{40}$ ▪

EXAMPLE 6   Evaluate $\left(\dfrac{3}{20}\right)\left(\dfrac{-3}{10}\right)$.

Solution: $\left(\dfrac{3}{20}\right)\left(\dfrac{-3}{10}\right) = \dfrac{3(-3)}{20\cdot 10} = \dfrac{-9}{200}$ ▪

Sometimes you may be asked to perform more than one multiplication in a given problem. When this happens the sign of the final product can be determined by counting the number of *negative* numbers being multiplied. **The product of an even number of negative numbers will always be positive. The product of an odd number of negative numbers will always be negative.** Can you explain why?

EXAMPLE 7   Evaluate $(-2)(3)(-2)(-1)$.

Solution: Since there are three negative numbers (an odd number of negatives), the product will be negative as illustrated below.

$$(-2)(3)(-2)(-1) = (-6)(-2)(-1)$$
$$= (12)(-1)$$
$$= -12 \quad \blacksquare$$

EXAMPLE 8   Evaluate $(-3)(2)(-1)(-2)(-4)$.

Solution: Since there are four negative numbers (an even number), the product will be positive.

$$(-3)(2)(-1)(-2)(-4) = (-6)(-1)(-2)(-4)$$
$$= (6)(-2)(-4)$$
$$= (-12)(-4)$$
$$= 48 \quad \blacksquare$$

Division of Real Numbers

2   The rules for dividing numbers are very similar to those used in multiplying numbers.

Division of Real Numbers

1. The quotient of two numbers with **like** signs is a **positive** number.

2. The quotient of two numbers with **unlike** signs is a **negative** number.

Therefore the quotient of two positive numbers or two negative numbers will be a positive number. The quotient of a positive and negative number will be a negative number.

EXAMPLE 9   Evaluate $\dfrac{20}{-5}$.

Solution: Since the numbers have unlike signs, the quotient is negative.

$$\frac{20}{-5} = -4 \quad \blacksquare$$

EXAMPLE 10   Evaluate $\dfrac{-36}{4}$.

Solution: Since the numbers have unlike signs, the quotient is negative.

$$\frac{-36}{4} = -9 \quad \blacksquare$$

EXAMPLE 11   Evaluate $\dfrac{-30}{-5}$.

Solution: Since the numbers have like signs, both negative, the quotient is positive.

$$\frac{-30}{-5} = 6 \quad \blacksquare$$

EXAMPLE 12   Evaluate $-16 \div (-2)$.

Solution: $\dfrac{-16}{-2} = 8$ ∎

EXAMPLE 13   Evaluate $\dfrac{-2}{3} \div \dfrac{-5}{7}$.

Solution: Invert the *divisor*, $\dfrac{-5}{7}$, and then multiply.

$$\frac{-2}{3} \div \frac{-5}{7} = \left(\frac{-2}{3}\right)\left(\frac{7}{-5}\right)$$

$$= \frac{-14}{-15}$$

$$= \frac{14}{15} \quad ∎$$

3   We now know that the quotient of a positive and a negative number is a negative number. The fractions $-\frac{3}{4}, \frac{-3}{4}$, and $\frac{3}{-4}$ all represent the same negative number, negative three-fourths.

> If a and b represent any real numbers, $b \neq 0$, then
>
> $$\frac{a}{-b} = \frac{-a}{b} = -\frac{a}{b}$$

In mathematics we generally do not write a fraction with a negative sign in the denominator. When a negative sign appears in a denominator we can move it to the numerator, or place it in front of the fraction. For example the fraction $\frac{5}{-7}$ should be written as either $-\frac{5}{7}$ or $\frac{-5}{7}$.

EXAMPLE 14   Evaluate $\dfrac{2}{5} \div \dfrac{-8}{15}$.

Solution: $\dfrac{2}{5} \div \dfrac{-8}{15} = \dfrac{\overset{1}{2}}{\underset{1}{5}} \cdot \dfrac{\overset{3}{15}}{\underset{4}{-8}}$

$$= \frac{1(3)}{1(-4)}$$

$$= \frac{3}{-4} = -\frac{3}{4} \quad ∎$$

HELPFUL HINT

> **For multiplication and division**
>
> $\left.\begin{array}{l} (+)(+) = + \\ (-)(-) = + \end{array}\right\}$ like signs = positive
>
> $\left.\begin{array}{l} (+)(-) = - \\ (-)(+) = - \end{array}\right\}$ unlike signs = negative

COMMON STUDENT ERROR

At this point some students begin confusing problems like $-2-3$ with $(-2)(-3)$, and problems like $2-3$ with problems like $2(-3)$. If you do not understand the difference between problems like $-2-3$ and $(-2)(-3)$, make an appointment to see your instructor as soon as possible.

Addition problems	*Multiplication problems*
$-2-3 = -2+(-3) = -5$	$(-2)(-3) = 6$
$2-3 = 2+(-3) = -1$	$(2)(-3) = -6$

The operations on real numbers are summarized in Table 1.1.

Table 1.1   SUMMARY OF OPERATIONS ON REAL NUMBERS

Signs of Numbers	Addition	Subtraction	Multiplication	Division
Both Numbers Are Positive	**Sum Is Always Positive**	**Difference May Be Either Positive or Negative**	**Product Is Always Positive**	**Quotient Is Always Positive**
Examples				
6 and 2	$6+2=8$	$6-2=4$	$6 \cdot 2 = 12$	$6 \div 2 = 3$
2 and 6	$2+6=8$	$2-6=-4$	$2 \cdot 6 = 12$	$2 \div 6 = \frac{1}{3}$
One Number Is Positive and the Other Number Is Negative	**Sum May Be Either Positive or Negative**	**Difference May Be Either Positive or Negative**	**Product Is Always Negative**	**Quotient Is Always Negative**
Examples				
6 and -2	$6+(-2)=4$	$6-(-2)=8$	$6(-2)=-12$	$6 \div (-2) = -3$
-6 and 2	$-6+2=-4$	$-6-(2)=-8$	$-6(2)=-12$	$-6 \div 2 = -3$
Both Numbers Are Negative	**Sum Is Always Negative**	**Difference May Be Either Positive or Negative**	**Product Is Always Positive**	**Quotient Is Always Positive**
Examples				
-6 and -2	$-6+(-2)=-8$	$-6-(-2)=-4$	$-6(-2)=12$	$-6 \div (-2) = 3$
-2 and -6	$-2+(-6)=-8$	$-2-(-6)=4$	$-2(-6)=12$	$-2 \div (-6) = \frac{1}{3}$

4 Now let us look at division involving the number 0.

What is $\frac{0}{1}$ equal to? Note that $\frac{6}{3} = 2$ because $3 \cdot 2 = 6$. We can follow the same procedure to determine the value of $\frac{0}{1}$. Suppose that $\frac{0}{1}$ is equal to some number which we will designate by a question mark.

$$\text{If} \quad \frac{0}{1} = ? \quad \text{then} \quad 1 \cdot ? = 0$$

Since only $1 \cdot 0 = 0$, ? must be 0. Thus $\frac{0}{1} = 0$. Using the same technique, we can show that zero divided by any nonzero number is zero.

$$\frac{0}{a} = 0, \qquad a \neq 0$$

What is $\frac{1}{0}$ equal to?

$$\text{If} \quad \frac{1}{0} = ? \quad \text{then} \quad 0 \cdot ? = 1$$

But since 0 multiplied by any number will be 0, there is no value that can replace ? We say that $\frac{1}{0}$ is *undefined*. Using the same technique we can show that any real number, except 0, divided by 0 is undefined.

$$\frac{a}{0} \text{ is \textbf{undefined},} \qquad a \neq 0$$

What is $\frac{0}{0}$ equal to?

$$\text{If} \quad \frac{0}{0} = ? \quad \text{then} \quad 0 \cdot ? = 0$$

But since the product of any number and 0 is 0, the ? can be replaced by any real number. For this reason we say that $\frac{0}{0}$ is *indeterminate*.

$$\frac{0}{0} \text{ is \textbf{indeterminate}}$$

Summary of Division Involving Zero

$$\frac{0}{a} = 0, a \neq 0 \qquad \frac{a}{0} \text{ is undefined, } a \neq 0 \qquad \frac{0}{0} \text{ is indeterminate}$$

Exercise Set 1.6

Find the product.

1. $(-4)(-3)$
2. $-4 \cdot 2$
3. $3(-3)$
4. $6(-2)$
5. $(-4)(8)$
6. $(-3)(2)$
7. $9(-1)$
8. $-1(8)$
9. $-4(-3)$
10. $0(4)$
11. $-9(-4)$
12. $(-12)(-3)$
13. $-6 \cdot 5$
14. $-9(-3)$
15. $5(-12)$
16. $(-9)(-8)$
17. $-4(0)$
18. $0(8)$
19. $(-4)(-4)$
20. $(-6)(-6)$
21. $8 \cdot 3$
22. $-4(-6)$
23. $5(-3)$
24. $-4 \cdot 7$
25. $8(12)$
26. $(-5)(-6)$
27. $-9(-9)$
28. $(15)(-4)$
29. $-2(5)$
30. $6(-12)$
31. $(-6)(2)(-3)$
32. $5(-2)(-8)$
33. $0(3)(8)$
34. $2(-3)(7)$
35. $(-1)(-1)(-1)$
36. $2(4)(-2)(-5)$
37. $-5(-3)(8)(-1)$
38. $(-3)(-4)(-5)(-1)$
39. $4(3)(1)(-1)$
40. $(-3)(2)(5)(3)$
41. $(-4)(3)(-7)(1)$
42. $(-1)(3)(0)(-7)$

Find the product.

43. $\left(\dfrac{-1}{2}\right)\left(\dfrac{3}{5}\right)$

44. $\left(\dfrac{2}{3}\right)\left(\dfrac{-3}{5}\right)$

45. $\left(\dfrac{-8}{9}\right)\left(\dfrac{-7}{12}\right)$

46. $\left(\dfrac{-5}{12}\right)\left(\dfrac{-6}{11}\right)$

47. $\left(\dfrac{6}{-3}\right)\left(\dfrac{4}{-2}\right)$

48. $\left(\dfrac{8}{-11}\right)\left(\dfrac{6}{-5}\right)$

49. $\left(\dfrac{5}{-7}\right)\left(\dfrac{6}{8}\right)$

50. $\left(\dfrac{9}{10}\right)\left(\dfrac{7}{-8}\right)$

Find the quotient.

51. $\dfrac{6}{2}$

52. $9 \div (-3)$

53. $-16 \div (-4)$

54. $\dfrac{-24}{8}$

55. $\dfrac{-36}{-9}$

56. $-45 \div 5$

57. $\dfrac{-16}{4}$

58. $\dfrac{36}{-2}$

59. $\dfrac{18}{-1}$

60. $\dfrac{-12}{-1}$

61. $-15 \div (-3)$

62. $12 \div (-6)$

63. $\dfrac{-6}{-1}$

64. $\dfrac{60}{-12}$

65. $\dfrac{-25}{-5}$

66. $\dfrac{36}{-4}$

67. $\dfrac{1}{-1}$

68. $\dfrac{-1}{1}$

69. $\dfrac{-48}{12}$

70. $\dfrac{50}{-5}$

71. $\dfrac{-18}{-2}$

72. $\dfrac{100}{-5}$

73. $\dfrac{0}{-1}$

74. $-200 \div (-20)$

75. $(-30) \div (-30)$
78. Divide -16 by -2.
81. Divide -30 by -10.
84. Divide -25 by -5.
87. Divide -90 by -2.

76. $(-180) \div 20$
79. Divide 20 by -5.
82. Divide -180 by 30.
85. Divide 80 by -20.
88. Divide 125 by -25.

77. Divide 0 by 3.
80. Divide 30 by -10.
83. Divide -60 by 5.
86. Divide -60 by 12.

Find the quotient.

89. $\dfrac{5}{12} \div \left(\dfrac{-5}{9}\right)$

90. $(-3) \div \dfrac{5}{19}$

91. $\dfrac{3}{-10} \div (-8)$

92. $\dfrac{-4}{9} \div \left(\dfrac{-6}{7}\right)$

93. $\dfrac{-15}{21} \div \left(\dfrac{-15}{21}\right)$

94. $\dfrac{8}{-15} \div \left(\dfrac{-9}{10}\right)$

95. $(-12) \div \dfrac{5}{12}$

96. $\dfrac{-16}{3} \div \left(\dfrac{5}{-9}\right)$

97. $6 \div \left(\dfrac{-5}{6}\right)$

98. $-12 \div \left(\dfrac{-2}{3}\right)$

Indicate whether each of the following is 0, undefined, or indeterminate.

99. $0 \div 6$

100. $-4 \div 0$

101. $\dfrac{0}{0}$

102. $\dfrac{-2}{0}$

103. $\dfrac{0}{1}$

104. $0 \div (-2)$

105. $8 \div 0$

106. $\dfrac{0}{4}$

107. $\dfrac{0}{-6}$

108. $\dfrac{0}{-1}$

Answer true or false.

109. The product of an even number of negative numbers is a positive number.

110. The product of an odd number of negative numbers is a negative number.

111. Zero divided by 0 is 1.

112. Six divided by 0 is 0.

113. Zero divided by 0 is 0.

114. Zero divided by 1 is undefined.

115. One divided by 0 is undefined.

116. Zero divided by 0 is indeterminate.

JUST FOR FUN

Find the quotient.

1. $\dfrac{1 - 2 + 3 - 4 + 5 - \cdots + 99 - 100}{1 - 2 + 3 - 4 + 5 - \cdots + 99 - 100}$

2. $\dfrac{-1 + 2 - 3 + 4 - 5 + \cdots - 99 + 100}{1 - 2 + 3 - 4 + 5 - \cdots + 99 - 100}$

1.7

An Introduction to Exponents

1 *Identify exponents.*

2 *Evaluate expressions containing exponents.*

3 *Learn the difference between $-x^2$ and $(-x)^2$.*

---

1 To understand certain topics in algebra, you must understand exponents. Exponents will be discussed in more detail in Chapter 4.

In the expression 4^2, the 4 is called the **base**, and the 2 is called the **exponent**. 4^2 is read "4 squared" or "4 to the second power," and means

$$\underbrace{4 \cdot 4}_{\text{2 factors of 4}} = 4^2$$

The number 4^3 is read "4 cubed," or "4 to the third power," and means

$$\underbrace{4 \cdot 4 \cdot 4}_{\text{3 factors of 4}} = 4^3$$

In general, the number b to the nth power, written b^n, means

$$\underbrace{b \cdot b \cdot b \cdot \cdots \cdot b}_{n \text{ factors of } b} = b^n$$

Thus $b^4 = b \cdot b \cdot b \cdot b$ or $bbbb$ and $x^3 = x \cdot x \cdot x$ or xxx.

EXAMPLE 1   **2** Evaluate each expression.

(a) 3^2 (b) 2^5 (c) 1^5 (d) 4^3 (e) $(-3)^2$ (f) $(-2)^3$ (g) $(\frac{2}{3})^2$

Solution: (a) $3^2 = 3 \cdot 3 = 9$
(b) $2^5 = 2 \cdot 2 \cdot 2 \cdot 2 \cdot 2 = 32$
(c) $1^5 = 1 \cdot 1 \cdot 1 \cdot 1 \cdot 1 = 1$ (1 raised to any real power equals 1; why?)
(d) $4^3 = 4 \cdot 4 \cdot 4 = 64$
(e) $(-3)^2 = (-3)(-3) = 9$
(f) $(-2)^3 = (-2)(-2)(-2) = -8$
(g) $(\frac{2}{3})^2 = (\frac{2}{3})(\frac{2}{3}) = \frac{4}{9}$ ■

COMMON STUDENT ERROR

Students should realize that $a^b \neq ba$.
2^5 means $2 \cdot 2 \cdot 2 \cdot 2 \cdot 2$, not $5 \cdot 2$. Thus $2^5 = 32$, and not 10.

Other examples of exponential notation are:

(a) $x \cdot x \cdot x \cdot x = x^4$ **(b)** $aabbb = a^2b^3$
(c) $x \cdot x \cdot y = x^2y$ **(d)** $aaabb = a^3b^2$
(e) $xyxx = x^3y$ **(f)** $xyzzy = xy^2z^2$
(g) $3 \cdot x \cdot x \cdot y = 3x^2y$ **(h)** $5xyyyy = 5xy^4$
(i) $3 \cdot 3 \cdot x \cdot x = 3^2x^2$ **(j)** $5 \cdot 5 \cdot 5 \cdot xxy = 5^3x^2y$

Notice that in parts (e) and (f), the order of the factors does not matter.

It is not necessary to write exponents of 1. Thus when writing xxy, we write x^2y and not x^2y^1. **Whenever we see a letter or number without an exponent, we always assume that letter or number has an exponent of 1.**

EXAMPLE 2   Write each expression as a product of factors.

(a) x^2y (b) xy^3z (c) $3x^2yz^3$ (d) 2^3xy (e) $3^2x^3y^2$

Solution: (a) $x^2y = xxy$ (b) $xy^3z = xyyyz$
(c) $3x^2yz^3 = 3xxyzzz$ (d) $2^3xy = 2 \cdot 2 \cdot 2xy$
(e) $3^2x^3y^2 = 3 \cdot 3xxxyy$ ■

3 Note that **an exponent refers to only the number or letter that directly precedes it unless parentheses are used to indicate otherwise.** For example, in the expression $3x^2$, only the x is squared. In the expression $-x^2$ only the x, not the minus sign, is squared. To help explain this concept we will write $-x^2$ as $-1x^2$. This can be done since any real number may be multiplied by 1 without affecting its value.

$$-x^2 = -1x^2$$

By looking at $-1x^2$ it should be clear that only the x is squared, not the -1. If the entire expression $-x$ was to be squared, we would need to use parentheses and write $(-x)^2$. Note the following difference

$$-x^2 = -(x)(x)$$
$$(-x)^2 = (-x)(-x)$$

Consider the expressions -3^2 and $(-3)^2$. How do they differ?

$$-3^2 = -(3)(3) = -9$$
$$(-3)^2 = (-3)(-3) = 9$$

EXAMPLE 3   Evaluate each expression.

(a) -5^2 (b) $(-5)^2$ (c) -2^3 (d) $(-2)^3$

Solution: (a) $-5^2 = -(5)(5) = -25$ (b) $(-5)^2 = (-5)(-5) = 25$
(c) $-2^3 = -(2)(2)(2) = -8$ (d) $(-2)^3 = (-2)(-2)(-2) = -8$ ■

EXAMPLE 4   Evaluate (a) -2^4 and (b) $(-2)^4$.

Solution: (a) $-2^4 = -(2)(2)(2)(2) = -16$ (b) $(-2)^4 = (-2)(-2)(-2)(-2) = 16$ ■

EXAMPLE 5   Evaluate (a) x^2 and (b) $-x^2$ for $x = 3$.

Solution: Substitute 3 for x.

(a) $x^2 = 3^2 = 3 \cdot 3 = 9$ (b) $-x^2 = -3^2 = -(3)(3) = -9$ ■

EXAMPLE 6   Evaluate (a) y^2 and (b) $-y^2$ for $y = -4$.

Solution: Substitute -4 for y.

(a) $y^2 = (-4)^2 = (-4)(-4) = 16$
(b) $-y^2 = -(-4)^2 = -(-4)(-4) = -16$ ■

Note that $-x^2$ will always be a negative number for any nonzero value of x, and $(-x)^2$ will always be a positive number for any nonzero value of x. Can you explain why? See exercises 84 and 85.

COMMON STUDENT ERROR

When asked to evaluate $-x^2$ for any real number x, many students will incorrectly treat $-x^2$ as $(-x)^2$.

Evaluate $-x^2$ when $x = 5$

Correct   *Wrong*

$-5^2 = -(5)(5)$ $-5^2 = (-5)(-5)$

$= -25$ $= 25$

To evaluate 2^3 on the calculator, press

$$\boxed{c}\,2\,\boxed{\times}\,2\,\boxed{\times}\,2\,\boxed{=}\,8$$

If your calculator contains a $\boxed{y^x}$ key, use this key to evaluate 2^3 in the following manner:

$$\boxed{c}\,2\,\boxed{y^x}\,3\,\boxed{=}\,8$$

To evaluate an exponential expression, key in the base, press the y^x key, then key in the exponent. After the $\boxed{=}$ is pressed, the answer is displayed. To evaluate 4^8, use $\boxed{c}\,4\,\boxed{y^x}\,8 = 65536$.

Exercise Set 1.7

Evaluate each expression.

1. 3^2	**2.** 4^2	**3.** 2^3	**4.** 1^5
5. 3^3	**6.** -5^2	**7.** 6^3	**8.** $(-2)^2$
9. $(-2)^3$	**10.** -3^4	**11.** $(-1)^3$	**12.** 6^2
13. 3^3	**14.** 2^5	**15.** -6^2	**16.** 5^3
17. $(-6)^2$	**18.** $(-3)^3$	**19.** 2^4	**20.** $(-3)^4$
21. 5^1	**22.** -3^2	**23.** $(-2)^4$	**24.** -1^4
25. -2^4	**26.** $(-1)^4$	**27.** $(-4)^3$	**28.** $3^2(4)^2$
29. $5^2 \cdot 3^2$	**30.** $(-1)^4(3)^3$	**31.** $5(4^2)$	**32.** $2^3 \cdot 5^1$
33. $2^1 \cdot 4^2$	**34.** $(-2)^4(-1)^3$	**35.** $3(-5^2)$	**36.** $9(-2)^2$
37. $3^2 \cdot 2^4$			

Express in exponential form.

38. $x \cdot x \cdot y \cdot y$	**39.** $x \cdot y \cdot z \cdot z$	**40.** $xyyyz$
41. $xxxxz$	**42.** $yyzzz$	**43.** $aabbab$
44. $xyxyz$	**45.** $x \cdot x \cdot y \cdot z \cdot z$	**46.** $a \cdot x \cdot a \cdot x \cdot y$
47. $x \cdot x \cdot x \cdot y \cdot y$	**48.** $x \cdot y \cdot y \cdot z \cdot z \cdot z$	**49.** $xyyyy$
50. $3xyy$	**51.** $5 \cdot 5yyz$	**52.** $2 \cdot 2 \cdot 2 \cdot xxyyy$

Express as a product of factors.

53. $x^2 y$	**54.** $y^2 z$	**55.** xy^3
56. $x^2 yz$	**57.** $xy^2 z^3$	**58.** $2x^2 y^2$
59. $3^2 yz$	**60.** $2^3 y^3$	**61.** $2^3 x^3 y$
62. $3^3 xy^3$	**63.** $(-2)^2 y^3 z$	**64.** $(-1)^2 x^3 y^2$

Evaluate (a) x^2 and (b) $-x^2$, for each of the following values of x.

65. 3	**66.** 2	**67.** 4
68. 1	**69.** -2	**70.** 5
71. 7	**72.** 8	**73.** -1
74. -5	**75.** $-\dfrac{1}{2}$	**76.** $\dfrac{3}{4}$

Answer true or false.

77. $(-4)^{20}$ is a negative number.

78. $(-4)^{19}$ is a negative number.

79. $-(-3)^{15}$ is a negative number.

80. $-(-2)^{14}$ is a negative number.

84. Explain why $-x^2$ will always be a negative number for any nonzero value of x.

81. x^2y means x^2y^1.

82. $3xy^4$ means $3^1x^1y^4$.

83. $2x^5y$ means $2^1x^5y^1$.

85. Explain why $(-x)^2$ will always be a positive number for any nonzero value of x.

JUST FOR FUN

1. (a) Mrs. Kelly offers to give Mr. Nenno $1000.00 a day for a month (30 days) if he will give her 1 cent the first day and double the amount each day for the month. If Mr. Nenno accepts this offer, how much will he make or lose?

(b) Indicate the total amount that Mrs. Kelly will receive as a number in exponential form.

2. Evaluate each expression. Leave answers in exponential form.

(a) $2^2 \cdot 2^3$ **(b)** $3^2 \cdot 3^3$ **(c)** $x^m \cdot x^n$ **(d)** $\dfrac{2^3}{2^2}$

(e) $\dfrac{3^4}{3^2}$ **(f)** $\dfrac{x^m}{x^n}$ **(g)** $(2^3)^2$ **(h)** $(3^3)^2$

(i) $(x^m)^n$ **(j)** $(2x)^2$ **(k)** $(3x)^2$ **(l)** $(ax)^2$

General rules that may be used to solve problems of this type will be discussed in Chapter 4.

1.8

Use of Parentheses and Priority of Operations

1 *Learn the priority of operations.*

2 *Evaluate expressions for given values of the variable.*

---

1   Evaluate $2 + 3 \cdot 4$. Is it 20? Is it 14? To be able to answer questions of this type, we must know the order of operations to follow when evaluating a mathematical expression. You will often have to evaluate expressions containing multiple operations.

To Evaluate Mathematical Expressions, Use the Following Order

1. First, evaluate the information within parentheses, ( ), or brackets, [ ]. If the expression contains nested parentheses (one pair of parentheses within another pair), evaluate the information in the innermost parentheses first.

2. Next, evaluate all exponents.

3. Next, evaluate all multiplications or divisions in the order in which they occur, working from left to right.

4. Finally, evaluate all additions or subtractions in the order in which they occur, working from left to right.

We can now answer the question posed above. Since multiplications are performed before additions,

$$2 + 3 \cdot 4 \quad \text{means} \quad 2 + (3 \cdot 4) = 2 + 12 = 14$$

Parentheses may be used (1) to change the order of operations to be followed in evaluating an algebraic expression or (2) to help clarify the understanding of an expression.

In the example above, $2 + 3 \cdot 4$, if we wished to have the addition performed before the multiplication, we could indicate this by placing parentheses about the $2 + 3$:

$$(2 + 3) \cdot 4 = 5 \cdot 4 = 20$$

Consider the expression $1 \cdot 3 + 2 \cdot 4$. According to the priority, multiplications are to be performed before additions. We can rewrite this expression as $(1 \cdot 3) + (2 \cdot 4)$. Note that the priority of operations was not changed. The parentheses were used only to help clarify the order to be followed.

HELPFUL HINT

To help you remember the order of operations, remember the word

"PEMD AS"

Parentheses, Exponents, Multiplication, Division, Addition, Subtraction

Note the line under M and D, and the line under A and S. *Multiplication and division have the same priority.* That is, when an expression has both multiplication and division, and parentheses are not used, work from left to right. Similarly *addition and subtraction have the same priority,* and are evaluated from left to right.

EXAMPLE 1   Evaluate $2 + 3 \cdot 5^2 - 7$.

Solution:
$$2 + 3 \cdot 5^2 - 7$$
$$= 2 + 3 \cdot 25 - 7$$
$$= 2 + 75 - 7$$
$$= 77 - 7$$
$$= 70 \quad \blacksquare$$

EXAMPLE 2   Evaluate $6 + 3[(12 \div 4) + 5]$.

Solution:
$$6 + 3[(12 \div 4) + 5]$$
$$= 6 + 3[3 + 5]$$
$$= 6 + 3(8)$$
$$= 6 + 24$$
$$= 30 \quad \blacksquare$$

EXAMPLE 3   Evaluate $(4 \div 2) + 4(5 - 2)^2$.

Solution:

$$(4 \div 2) + 4(5 - 2)^2$$
$$= 2 + 4(3)^2$$
$$= 2 + 4 \cdot 9$$
$$= 2 + 36$$
$$= 38 \quad \blacksquare$$

EXAMPLE 4   Evaluate $5 + 2^2 \cdot 3 - 3^2$.

Solution:

$$5 + 2^2 \cdot 3 - 3^2$$
$$= 5 + 4 \cdot 3 - 9$$
$$= 5 + 12 - 9$$
$$= 17 - 9$$
$$= 8 \quad \blacksquare$$

EXAMPLE 5   Evaluate $-8 - 81 \div 9 \cdot 2^2 + 7$.

Solution:

$$-8 - 81 \div 9 \cdot 2^2 + 7$$
$$= -8 - 81 \div 9 \cdot 4 + 7$$
$$= -8 - 9 \cdot 4 + 7$$
$$= -8 - 36 + 7$$
$$= -44 + 7$$
$$= -37 \quad \blacksquare$$

EXAMPLE 6   Evaluate each expression.

(a) $-4^2 + 6 \div 3$ (b) $(-4)^2 + 6 \div 3$

Solution:

(a) $-4^2 + 6 \div 3$ (b) $(-4)^2 + 6 \div 3$
   $= -16 + 6 \div 3$ $= 16 + 6 \div 3$
   $= -16 + 2$ $= 16 + 2$
   $= -14$ $= 18 \quad \blacksquare$

EXAMPLE 7   Evaluate $\dfrac{3}{8} - \dfrac{2}{5} \cdot \dfrac{1}{12}$

Solution: First perform the multiplication.

$$\frac{3}{8} - \frac{2}{5} \cdot \frac{1}{12}$$

$$= \frac{3}{8} - \left(\frac{\overset{1}{2}}{5} \cdot \frac{1}{\underset{6}{12}} \right)$$

$$= \frac{3}{8} - \frac{1}{30}$$

$$= \frac{45}{120} - \frac{4}{120}$$

$$= \frac{41}{120} \quad \blacksquare$$

EXAMPLE 8   Indicate the order of operations using parentheses and brackets, then evaluate. Multiply 5 by 3. To this product add 6. Multiply this sum by 7.

Solution: $5 \cdot 3$ multiply 5 by 3

$(5 \cdot 3) + 6$ add 6

$7[(5 \cdot 3) + 6]$ multiply by 7

Now evaluate to determine the answer.

$$7[(5 \cdot 3) + 6]$$
$$= 7[15 + 6]$$
$$= 7(21)$$
$$= 147 \quad \blacksquare$$

Sometimes brackets are used in place of parentheses to help avoid confusion. The preceding expression could have been written using only parentheses as $7((5 \cdot 3) + 6)$.

EXAMPLE 9   Indicate the order of operations using parentheses and brackets, then evaluate. Subtract 3 from 15. Divide this difference by 2. Multiply this quotient by 4.

Solution: $15 - 3$ subtract 3 from 15

$(15 - 3) \div 2$ divide by 2

$4[(15 - 3) \div 2]$ multiply by 4

Now evaluate.

$$4[(15 - 3) \div 2]$$
$$= 4[12 \div 2]$$
$$= 4(6)$$
$$= 24 \quad \blacksquare$$

2   Now we will evaluate some expressions when given the values of the variables.

EXAMPLE 10   Evaluate $7x - 2$ when $x = 2$.

Solution: Substitute 2 for each x in the expression.

$$7x - 2 = 7(2) - 2$$
$$= 14 - 2$$
$$= 12 \quad \blacksquare$$

EXAMPLE 11   Evaluate $(3x + 1) + 2x^2$ when $x = 4$.

Solution: $(3x + 1) + 2x^2 = [3(4) + 1] + 2(4)^2$
$$= [12 + 1] + 2(4)^2$$
$$= (13) + 2(16)$$
$$= 13 + 32$$
$$= 45 \quad \blacksquare$$

EXAMPLE 12 Evaluate $-y^2 + 3(x + 2) - 5$ when $x = -3$ and $y = -2$.

Solution: $-y^2 + 3(x + 2) - 5 = -(-2)^2 + 3(-3 + 2) - 5$

$$= -(-2)^2 + 3(-1) - 5$$
$$= -(4) + 3(-1) - 5$$
$$= -4 - 3 - 5$$
$$= -7 - 5$$
$$= -12 \quad \blacksquare$$

Calculator Corner

We now know that $2 + 3 \times 4$ means $2 + (3 \times 4)$ and has a value of 14. What will a calculator display if you key in the following?

$$\boxed{c}\ 2\ \boxed{+}\ 3\ \boxed{\times}\ 4\ \boxed{=}$$

The answer depends on your calculator. Many calculators contain a mathematical hierarchy. These calculators will evaluate an expression following the rules stated earlier in this section.

Calculators with hierarchy: $\boxed{c}\ 2\ \boxed{+}\ 3\ \boxed{\times}\ 4\ \boxed{=}\ 14$

Other calculators, generally the less expensive ones, will perform operations in the order they are entered.

Calculators without hierarchy: $\boxed{c}\ 2\ \boxed{+}\ 3\ \boxed{\times}\ 4\ \boxed{=}\ 20$

Remember that in algebra, unless otherwise instructed by parentheses, we always perform multiplications and divisions before additions and subtractions.

Does your calculator use a mathematical hierarchy?

Exercise Set 1.8

Evaluate each expression.

1. $3 + 4 \cdot 5$
2. $2 - 3^2 + 4$
3. $2 - 2 + 5$
4. $(6^2 \div 3) - (6 - 4)$
5. $1 + 3 \cdot 2^2$
6. $4 \cdot 3^2 - 2 \cdot 5$
7. $-3^2 + 5$
8. $(-2)^3 + 8 \div 4$
9. $(4 - 3) \cdot (5 - 1)^2$
10. $20 - 6 - 3 - 2$
11. $3 \cdot 7 + 4 \cdot 2$
12. $6 + 9(3 + 4)$
13. $[1 - (4 \cdot 5)] + 6$
14. $[12 - (4 \div 2)] - 5$
15. $4^2 - 3 \cdot 4 - 6$
16. $5 - 3 + 4^2 - 6$
17. $-3[-4 + (6 - 8)]$
18. $(-3)^2 + (3 - 4)^3 - 5$
19. $(6 \div 3)^3 + 4^2 \div 8$
20. $5^2 - 2^2(4 - 2)^2$
21. $-4^2 + 8 \div 2 \cdot 5 + 3$
22. $-4 - (-12 + 4) \div 2 + 1$

23. $3 + (4^2 - 10)^2 - 3$

24. $[-(1 - 4)^2]^2 + 9$

25. $-[12 - (-4 - 5)]^2$

26. $(-2)^2 + 4^2 \div 2^2 + 3$

27. $(3^2 - 1) \div (3 + 1)^2$

28. $-4(5 - 2)^2 + 5$

29. $2[(36 \div 9) + 1]$

30. $3[(4 + 6) \div 2]$

31. $2[3(8 - 2^2) - 6]$

32. $(13 + 5) - (4 - 2)^2$

33. $10 - [8 - (3 + 4)]^2$

34. $2.5 + 7.56 \div 2.1 + (9.2)^2$

35. $(8.4 + 3.1)^2 - (3.64 - 1.2)$

36. $2[1.63 + 5(4.7)] - 3.15$

37. $(4.3)^2 + 2(5.3) - 3.05$

38. $\frac{2}{3} + \frac{3}{8} \cdot \frac{4}{5}$

39. $(\frac{2}{7} + \frac{3}{8}) - \frac{3}{112}$

40. $(\frac{5}{6} \cdot \frac{4}{5}) + (\frac{2}{3} \cdot \frac{5}{8})$

41. $\frac{3}{4} - 4 \cdot \frac{5}{40}$

42. $\frac{2}{3} + 4 \div 3^2$

43. $2(3 + \frac{2}{5}) \div (\frac{3}{5})^2$

44. $64 \cdot \frac{1}{2} \div 8 + \frac{3}{4}$

Indicate the order of operations using parentheses and brackets, then evaluate.

45. Multiply 6 by 3. From this product, subtract 4. From this difference, subtract 2.

46. Add 4 to 9. Divide this sum by 2. Add 10 to this quotient. $[(9 + 4) \div 2] + 10$,

47. Divide 20 by 5. Add 12 to this quotient. Subtract 8 from this sum. Multiply this difference by 9.

48. Multiply 6 by 3. To this product, add 27. Divide this sum by 8. Multiply this quotient by 10.

49. Add $\frac{4}{5}$ to $\frac{3}{7}$. Multiply this sum by $\frac{2}{3}$.

50. Multiply $\frac{3}{8}$ by $\frac{4}{5}$. To this product, add $\frac{7}{120}$. From this sum, subtract $\frac{1}{60}$.

Evaluate each expression for the values given.

51. $x + 4$, when $x = -2$

52. $2x - 4x + 5$, when $x = 1$

53. $3x - 2$, when $x = 4$

54. $3(x - 2)$, when $x = 5$

55. $x^2 - 6$, when $x = -3$

56. $x^2 + 4$, when $x = 5$

57. $-3x^2 - 4$, when $x = 1$

58. $2x^2 + x$, when $x = 3$

59. $-4x^2 - 2x + 5$, when $x = -3$

60. $-3x^2 + 6x + 5$, when $x = 5$

61. $3(x - 2)^2$, when $x = 7$

62. $4(x + 1)^2 - 6x$, when $x = 5$

63. $2(x - 3)(x + 4)$, when $x = 1$

64. $3x^2(x - 1) + 5$, when $x = -4$

65. $-6x + 3y$, when $x = 2$ and $y = 4$

66. $6x + 3y^2 - 5$, when $x = 1$ and $y = -3$

67. $x^2 - y^2$, when $x = -2$, and $y = -3$

68. $x^2 - y^2$ when $x = 2$ and $y = -4$

69. $4(x + y)^2 + 4x - 3y$, when $x = 2$ and $y = -3$

70. $(4x - 3y)^2 - 5$, when $x = 4$ and $y = -2$

71. $3(a + b)^2 + 4(a + b) - 6$, when $a = 4$ and $b = -1$

72. $4xy - 6x + 3$, when $x = 5$ and $y = 2$

73. $x^2y - 6xy + 3x$, when $x = 2$ and $y = 3$

74. $\dfrac{6x^2}{3} + \dfrac{2x^2}{2}$, when $x = 2$

JUST FOR FUN

Evaluate each expression.

1. $4([3(x - 2)]^2 + 4)$, when $x = 4$

2. $[(3 - 6)^2 + 4]^2 + 3 \cdot 4 - 12 \div 3$

3. $-2[(3x^2 + 4)^2 - (3x^2 - 2)^2]$, when $x = -2$

1.9

**Properties
of the Real
Number System**

□1   *Identify the commutative property.*
□2   *Identify the associative property.*
□3   *Identify the distributive property.*

Here, we introduce various properties of the real number system. We will use these properties throughout the text.

□1   The *commutative property of addition* states that the order in which any two real numbers are added does not matter.

Commutative Property of Addition

If a and b represent any two real numbers, then

$$a + b = b + a$$

For example,

$$4 + 3 = 3 + 4$$
$$7 = 7$$

The *commutative property of multiplication* states that the order in which any two real numbers are multiplied does not matter.

Commutative Property of Multiplication

If a and b represent any two real numbers, then

$$a \cdot b = b \cdot a$$

For example,

$$6 \cdot 3 = 3 \cdot 6$$
$$18 = 18$$

The commutative property does not hold for subtraction or division. For example, $4 - 6 \neq 6 - 4$ and $6 \div 3 \neq 3 \div 6$.

□2   The *associative property of addition* states that in the addition of three or more numbers, parentheses may be placed around any two adjacent numbers without changing the results.

Associative Property of Addition

If a, b, and c represent any three real numbers, then

$$(a + b) + c = a + (b + c)$$

Notice that the associative property involves a change of *grouping*. For example,

$$(3 + 4) + 5 = 3 + (4 + 5)$$
$$7 + 5 = 3 + 9$$
$$12 = 12$$

In this example the 3 and 4 are grouped together on the left, and the 4 and 5 are grouped together on the right.

The *associative property of multiplication* states that in the multiplication of three or more numbers, parentheses may be placed around any two adjacent numbers without changing the results.

Associative Property of Multiplication

If a, b, and c represent any three real numbers, then

$$(a \cdot b) \cdot c = a \cdot (b \cdot c)$$

For example,

$$(6 \cdot 2) \cdot 4 = 6 \cdot (2 \cdot 4)$$
$$12 \cdot 4 = 6 \cdot 8$$
$$48 = 48$$

Notice that the associative property involves a change of contents within the parentheses.

The associative property does not hold for subtraction or division. For example, $(4 - 1) - 3 \neq 4 - (1 - 3)$ and $(8 \div 4) \div 2 \neq 8 \div (4 \div 2)$.

3 A very important property of the real numbers is the *distributive property of multiplication over addition*.

Distributive Property

If a, b, and c represent any three real numbers, then

$$a(b + c) = ab + ac$$

For example, if we let $a = 2$, $b = 3$, and $c = 4$, then

$$2(3 + 4) = (2 \cdot 3) + (2 \cdot 4)$$
$$2 \cdot 7 = 6 + 8$$
$$14 = 14$$

Therefore, one may either add first and then multiply, or multiply first and then add. The distributive property will be discussed in more detail in Chapter 2.

HELPFUL HINT

> *Commutative property:* change in order
>
> *Associative property:* change in grouping
>
> *Distributive property:* two operations, multiplication and addition

If we assume that x represents any real number, then:

$$x + 4 = 4 + x \text{ by the commutative property of addition,}$$
$$x \cdot 4 = 4 \cdot x \text{ by the commutative property of multiplication,}$$
$$(x + 4) + 7 = x + (4 + 7) \text{ by the associative property of addition,}$$
$$(x \cdot 4) \cdot 6 = x \cdot (4 \cdot 6) \text{ by the associative property of multiplication,}$$
$$3(x + 4) = (3 \cdot x) + (3 \cdot 4) \text{ or } 3x + 12 \text{ by the distributive property.}$$

EXAMPLE 1   Name the following properties.

(a) $4 + (-2) = -2 + 4$ (b) $x + y = y + x$
(c) $x \cdot y = y \cdot x$ (d) $(-12 + 3) + 4 = -12 + (3 + 4)$

Solution: (a) Commutative property of addition.
(b) Commutative property of addition.
(c) Commutative property of multiplication.
(d) Associative property of addition.   ■

EXAMPLE 2   Name the following properties.

(a) $2(x + 2) = (2 \cdot x) + (2 \cdot 2) = 2x + 4$
(b) $4(x + y) = (4 \cdot x) + (4 \cdot y) = 4x + 4y$
(c) $3x + 3y = (3 \cdot x) + (3 \cdot y) = 3(x + y)$
(d) $(3 \cdot 6) \cdot 5 = 3 \cdot (6 \cdot 5)$

Solution: (a) Distributive property.
(b) Distributive property.
(c) Distributive property (in reverse order).
(d) Associative property of multiplication.   ■

EXAMPLE 3   Name the following properties.

(a) $(3 + 4) + 5 = (4 + 3) + 5$ (b) $(2 + 3) + (4 + 5) = (4 + 5) + (2 + 3)$
(c) $3(x + 4) = 3(4 + x)$ (d) $3(x + 4) = (x + 4)3$

Solution: (a) Commutative property of addition. $3 + 4$ was changed to $4 + 3$; the same numbers remain within parentheses.
(b) Commutative property of addition. The order of parentheses was changed; however, the same numbers remain within the parentheses.
(c) Commutative property of addition. $x + 4$ was changed to $4 + x$.
(d) Commutative property of multiplication. The information within parentheses is not changed.   ■

EXAMPLE 4   Name the property used to go from one step to the next.

(a) $9 + 4(x + 5)$
(b) $= 9 + 4x + 20$
(c) $= 9 + 20 + 4x$
(d) $= 29 + 4x$ addition facts
(e) $= 4x + 29$

Solution: (a to b) Distributive property.
(b to c) Commutative property of addition; $4x + 20 = 20 + 4x$.
(d to e) Commutative property of addition; $29 + 4x = 4x + 29$.   ∎

The distributive property can be expanded in the following manner:

$$a(b + c + d + \cdots + n) = ab + ac + ad + \cdots + an$$

For example, $3(x + y + 5) = 3x + 3y + 15$.

Exercise Set 1.9

Name the property illustrated.

1. $3(4 + 2) = 3(4) + 3(2)$

2. $3 + y = y + 3$

3. $5 \cdot y = y \cdot 5$

4. $1(x + 3) = (1)(x) + (1)(3) = x + 3$

5. $2(x + 4) = 2x + 8$

6. $3(4 + x) = 12 + 3x$

7. $x \cdot (y \cdot z) = (x \cdot y) \cdot z$

8. $1(x + 4) = x + 4$

9. $1(x + 3) = x + 3$

10. $3 + (4 + x) = (3 + 4) + x$

Complete each exercise using the given property.

11. $3 + 4 =$
commutative property of addition

12. $-3 + 4 =$
commutative property of addition

13. $-6 \cdot (4 \cdot 2) =$
associative property of multiplication

14. $-4 + (5 + 3) =$
associative property of addition.

15. $(6)(y) =$
commutative property of multiplication

16. $4(x + 3) =$
distributive property

17. $1(x + y) =$
distributive property

18. $6(x + y) =$
distributive property

19. $4x + 3y =$
commutative property of addition

20. $3(x + y) =$
distributive property

21. $5x + 5y =$
distributive property (in reverse order)

22. $(3 + x) + y =$
associative property of addition

23. $(x + 2)3 =$
commutative property of multiplication

24. $2x + 2z =$
distributive property (in reverse order)

25. $(3x + 4) + 6 =$
associative property of addition

26. $3(x + y) =$
commutative property of addition

27. $3(x + y) =$
commutative property of multiplication

28. $(3x)y =$
associative property of multiplication

29. $4(x + y + 3) =$
distributive property

30. $3(x + y + 2) =$
distributive property

Name the property illustrated to go from one step to the next. See Example 4.

$(3 + x) + 4$
31. $= (x + 3) + 4$
32. $= x + (3 + 4)$
$= x + 7$ addition facts

$6 + 5(x + 3)$
33. $= 6 + 5x + 15$
34. $= 6 + 15 + 5x$
$= 21 + 5x$ addition facts
35. $= 5x + 21$

$(x + 4)5$
36. $= 5(x + 4)$
37. $= 5x + 20$
38. $= 20 + 5x$

Summary _____

Glossary

Absolute value: The distance between a number and 0 on the number line. The absolute value of any nonzero number will be positive.

Additive inverses or opposites: Two numbers whose sum is 0.

Denominator: The bottom number of a fraction.

Factor: If $a \cdot b = c$ then a and b are factors of c.

Lowest common denominator: The smallest number divisible by two or more denominators.

Numerator: The top number of a fraction.

Reduced to its lowest terms: A fraction is reduced to its lowest terms when its numerator and denominator have no common factor other than 1.

Variable: A letter used to represent a number.

Important Facts

Fractions

$$\frac{a}{c} + \frac{b}{c} = \frac{a + b}{c} \qquad \frac{a}{c} - \frac{b}{c} = \frac{a - b}{c}$$

$$\frac{a}{b} \cdot \frac{c}{d} = \frac{ac}{bd} \qquad \frac{a}{b} \div \frac{c}{d} = \frac{a}{b} \cdot \frac{d}{c} = \frac{ad}{bc}$$

Sets of Numbers

Natural numbers: $\{1, 2, 3, 4, \ldots\}$
Whole numbers: $\{0, 1, 2, 3, 4, \ldots\}$
Integers: $\{\ldots, -3, -2, -1, 0, 1, 2, 3, \ldots\}$
Rational numbers: {quotient of two integers, denominator not 0}
Real numbers: {all numbers that can be represented on the number line}
Irrational numbers: {real numbers that are not rational numbers}

To *add real numbers with the same sign* add their absolute values. The sum has the same sign as the number being added.

To *add real numbers with different signs* find the difference between the larger absolute value and the smaller absolute value. The answer has the sign of the number with the larger absolute value.

To subtract b from a, add the opposite of *b* to *a.*

$$a - b = a + (-b)$$

Multiplication and division:

$$(+)(+) = + \qquad (+)(-) = -$$
$$(-)(-) = + \qquad (-)(+) = -$$

Division involving 0:

$$\frac{0}{a} = 0, \, a \neq 0$$

$$\frac{a}{0} \text{ is undefined,} \qquad a \neq 0$$

$$\frac{0}{0} \text{ is indeterminate}$$

Exponents:

$$b^n = \underbrace{b \cdot b \cdot b \cdot \cdots \cdot b}_{n \text{ factors of } b}$$

Priority of Operations

1. Parentheses
2. Exponents
3. Multiplication or division working left to right
4. Addition or subtraction working left to right.

Properties of the Real Number Systems

Property	Addition	Multiplication
Commutative	$a + b = b + a$	$ab = ba$
Associative	$(a + b) + c = a + (b + c)$	$(ab)c = a(bc)$
Distributive	$a(b + c) = ab + ac$	

Review Exercises

[1.1] Perform the indicated operations. Reduce answers to lowest terms.

1. $\dfrac{3}{5} \cdot \dfrac{5}{6}$ **2.** $\dfrac{2}{5} \div \dfrac{10}{9}$ **3.** $\dfrac{5}{12} \div \dfrac{3}{5}$

4. $\dfrac{5}{6} + \dfrac{1}{3}$ **5.** $\dfrac{3}{8} - \dfrac{1}{9}$ **6.** $2\dfrac{1}{3} - 1\dfrac{1}{5}$

[1.2]

7. List the set of natural numbers.   **8.** List the set of whole numbers.

9. List the set of integers.
10. Describe the set of rational numbers.
11. Describe the set of real numbers.
12. Consider the set of numbers

$$\left\{3, -5, -12, 0, \frac{1}{2}, -0.62, \sqrt{7}, 426, -3\frac{1}{4}\right\}$$

List those that are
(a) Positive integers.
(b) Whole numbers.
(c) Integers.
(d) Rational numbers.
(e) Irrational numbers.
(f) Real numbers.

13. Consider the set of numbers

$$\left\{-2.3, -8, -9, 1\frac{1}{2}, \sqrt{2}, -\sqrt{2}, 1, -\frac{3}{17}\right\}$$

List those that are
(a) Natural numbers.
(b) Whole numbers.
(c) Negative integers.
(d) Integers.
(e) Rational numbers.
(f) Real numbers.

[1.3] Insert either $<$, $>$, or $=$ to make each expression true.

14. $-3 \quad -5$
15. $-2 \quad 1$
16. $0.6 \quad -1.3$
17. $-2.6 \quad -3.6$
18. $0.50 \quad 0.509$
19. $4.6 \quad 4.06$
20. $-3.2 \quad -3.02$
21. $5 \quad |-3|$
22. $-3 \quad |-7|$
23. $|-2.5| \quad \left|\frac{5}{2}\right|$

[1.4–1.5] Evaluate as indicated.

24. $-3 + 6$
25. $-4 + (-5)$
26. $-6 + 6$
27. $4 + (-9)$
28. $0 + (-3)$
29. $-10 + 4$
30. $-8 - (-2)$
31. $-9 - (-4)$
32. $4 - (-4)$
33. $0 - 2$
34. $-8 - 1$
35. $2 - 12$
36. $7 - 2$
37. $2 - 7$
38. $0 - (-4)$
39. $-7 - 5$

Evaluate as indicated.

40. $6 - 4 + 3$
41. $-5 + 7 - 6$
42. $-5 - 4 - 3$
43. $-2 + (-3) - 2$
44. $-(-4) + 5 - (+3)$
45. $7 - (+4) - (-3)$
46. $5 - 2 - 7 + 3$
47. $4 - (-2) + 3$

[1.6] Evaluate as indicated.

48. $-4(7)$
49. $(-9)(-3)$
50. $4(-9)$
51. $-2(3)$
52. $\left(\frac{3}{5}\right)\left(\frac{-2}{7}\right)$
53. $\left(\frac{10}{11}\right)\left(\frac{3}{-5}\right)$
54. $\left(\frac{-5}{8}\right)\left(\frac{-3}{7}\right)$
55. $0 \cdot \frac{4}{9}$
56. $4(-2)(-6)$
57. $(-1)(-3)(4)$
58. $-5(2)(7)$
59. $(-3)(-4)(-5)$
60. $-1(-2)(3)(-4)$
61. $(-4)(-6)(-2)(-3)$

Evaluate as indicated.

62. $15 \div (-3)$
63. $6 \div (-2)$
64. $-20 \div 5$
65. $-36 \div (-2)$
66. $0 \div 4$
67. $0 \div (-4)$
68. $72 \div (-9)$
69. $-40 \div (-8)$

70. $-4 \div \left(\dfrac{-4}{9}\right)$

71. $\dfrac{15}{32} \div (-5)$

72. $\dfrac{3}{8} \div \left(\dfrac{-1}{2}\right)$

73. $\dfrac{28}{-3} \div \dfrac{9}{-2}$

74. $\dfrac{14}{3} \div \left(\dfrac{-6}{5}\right)$

75. $\left(\dfrac{-5}{12}\right) \div \left(\dfrac{-5}{12}\right)$

Indicate whether each of the following is 0, undefined, or indeterminate.

76. $0 \div 4$

77. $0 \div (-6)$

78. $8 \div 0$

79. $-4 \div 0$

80. $0 \div 0$

81. $0 \div 1$

[1.4–1.6] Evaluate as indicated.

82. $-4(2 - 8)$

83. $2(4 - 8)$

84. $(3 - 6) + 4$

85. $(-4 + 3) - (2 - 6)$

86. $[4 + 3(-2)] - 6$

87. $(-4 - 2)(-3)$

88. $[4 + (-4)] + (6 - 8)$

89. $9[3 + (-4)] + 5$

90. $-4(-3) + [4 \div (-2)]$

91. $(-3 \cdot 4) \div (-2 \cdot 6)$

92. $(-3)(-4) + 6 - 3$

93. $[-2(3) + 6] - 4$

[1.7] Evaluate each expression.

94. 4^2

95. 6^2

96. 9^3

97. 1^5

98. 3^4

99. 2^4

100. $(-3)^3$

101. $(-1)^9$

102. $(-2)^5$

103. $\left(\dfrac{2}{7}\right)^2$

104. $\left(\dfrac{-3}{5}\right)^2$

105. $\left(\dfrac{2}{5}\right)^3$

Express in exponential form.

106. xxy

107. xyy

108. $xxyyx$

109. $yyzz$

110. $2 \cdot 2 \cdot 3 \cdot 3 \cdot 3xyy$

111. $5 \cdot 7 \cdot 7 \cdot xxy$

112. $xyxyz$

Express as a product of factors.

113. $x^2 y$

114. xz^3

115. $y^3 z$

116. $2x^3 y^2$

Evaluate each expression for the values given.

117. $-x^2$, when $x = 3$

118. $-x^2$, when $x = -4$

119. $-x^3$, when $x = 3$

120. $-x^4$, when $x = -2$

[1.8] Evaluate each expression.

121. $3 + 5 \cdot 4$

122. $7 - 3^2$

123. $3 \cdot 5 + 4 \cdot 2$

124. $(3 - 7)^2 + 6$

125. $6 + 4 \cdot 5$

126. $8 - 36 \div 4 \cdot 3$

127. $6 - 3^2 \cdot 5$

128. $2 - (8 - 3)$

129. $[6 - (3 \cdot 5)] + 5$

130. $3[9 - (4^2 + 3)] \cdot 2$

131. $(-3^2 + 4^2) + (3^2 \div 3)$

132. $2^3 \div 4 + 6 \cdot 3$

133. $(4 \div 2)^4 + 4^2 \div 2^2$

134. $(15 - 2^2)^2 - 4 \cdot 3 + 10 \div 2$

135. $4^3 \div 4^2 - 5(2 - 7) \div 5$

136. $10 - 15 \div 5 + 10(-3) - 12$

Evaluate each expression for the values given.

137. $4x - 6$, when $x = 5$

138. $8 - 3x$, when $x = 2$

139. $6 - 4x$, when $x = -5$

140. $x^2 - 5x + 3$, when $x = 6$

141. $5y^2 + 3y - 2$, when $y = -1$

142. $-x^2 + 2x - 3$, when $x = 2$

143. $-x^2 + 2x - 3$, when $x = -2$

144. $-3x^2 - 5x + 5$, when $x = 1$

145. $3xy - 5x$, when $x = 3$ and $y = 4$

[1.9] Name the property illustrated.

146. $(4 + 3) + 9 = 4 + (3 + 9)$

147. $6 \cdot x = x \cdot 6$

148. $4(x + 3) = 4x + 12$

149. $(x + 4)3 = 3(x + 4)$

150. $6x + 3x = 3x + 6x$

151. $(x + 7) + 4 = x + (7 + 4)$

152. $-6x + 3 = 3 + (-6x)$

Practice Test

1. Consider the set of numbers

$$\left\{-6,\, 42,\, -3\tfrac{1}{2},\, 0,\, 6.52,\, \sqrt{5},\, \tfrac{5}{9},\, -7,\, -1\right\}$$

  List those that are

  (a) Natural numbers.   **(b)** Whole numbers.

  (c) Integers.   **(d)** Rational numbers.

  (e) Irrational numbers.   **(f)** Real numbers.

Insert either $<$, $>$, or $=$ in the space provided to make each expression true.

2. $-6 \quad -3$

3. $|-3| \quad |-2|$

Evaluate each expression.

4. $-4 + (-8)$

5. $-6 - 5$

6. $4 - (-12)$

7. $5 - 12 - 7$

8. $(-4 + 6) - 3(-2)$

9. $(-4)(-3)(2)(-1)$

10. $\left(\dfrac{-2}{9}\right) \div \left(\dfrac{-7}{8}\right)$

11. $\left(-12 \cdot \dfrac{1}{2}\right) \div 3$

12. $3 \cdot 5^2 - 4 \cdot 6^2$

13. $(4 - 6^2) \div [4(2 + 3) - 4]$

14. $-6[(-2 - 3)] \div 5 \cdot 2$

15. $(-3)^4$

16. $\left(\dfrac{3}{5}\right)^3$

17. Write $2 \cdot 2 \cdot 5 \cdot 5 \cdot yyzzz$ in exponential form.

18. Write $2^2 3^3 x^4 y^2$ as a product of factors.

Evaluate each expression for the values given.

19. $2x^2 - 6$, when $x = -4$

20. $6x - 3y^2 + 4$, when $x = 3$ and $y = -2$

21. $-x^2 - 6x + 3$, when $x = -2$

Name the property illustrated.

22. $x + 3 = 3 + x$

23. $4(x + 9) = 4x + 36$

24. $(2 + x) + 4 = 2 + (x + 4)$

25. $5(x + y) = (x + y)5$

2

Solving Linear Equations

See Section 2.6, Exercise 29

1. Identify terms.
2. Identify like terms.
3. Combine like terms.
4. Use the distributive property to remove parentheses.
5. Remove parentheses when they are preceded by a plus or minus sign.
6. Simplify an expression.

---

1 In Section 1.1 and other sections of the text, we indicated that letters called **variables** (or **literal numbers**) are used to represent numbers.

An **algebraic expression** is a collection of numbers, variables, grouping symbols, and operation symbols. Examples of algebraic expressions are

$$5, \quad x^2 - 6, \quad 4x - 3, \quad 2(x + 5) + 6, \quad \frac{x + 3}{4}$$

When an algebraic expression consists of several parts, the parts that are added are called the **terms** of the expression. The expression $2x - 3y - 5$, which means $2x + (-3y) + (-5)$, has three terms: $2x$, $-3y$, and -5. The expression $3x + 2xy + 5(x + y)$ also has 3 terms: $3x$, $2xy$, and $5(x + y)$.

The $+$ and $-$ signs that break the expression into terms are a part of the term. However, when listing the terms of an expression, it is not necessary to list the $+$ sign at the beginning of a term.

Expression	*Terms*
$-2x + 3y - 8$	$-2x, \quad 3y, \quad -8$
$3y - 2x + \dfrac{1}{2}$	$3y, \quad -2x, \quad \dfrac{1}{2}$
$7 + x + 4 - 5x$	$7, \quad x, \quad 4, \quad -5x$
$3(x - 1) - 4x + 2$	$3(x - 1), \quad -4x, \quad 2$
$\dfrac{x + 4}{3} - 5x + 3$	$\dfrac{x + 4}{3}, \quad -5x, \quad 3$

The numerical part (constant) of a term is called its **numerical coefficient** or simply its **coefficient.** In the term $6x$, the 6 is the numerical coefficient. Note that $6x$ means the variable x is multiplied by 6.

Term	*Numerical coefficient*
$3x$	3
$-\dfrac{1}{2}x$	$-\dfrac{1}{2}$
$4(x - 3)$	4
$\dfrac{2x}{3}$	$\dfrac{2}{3}, \quad \dfrac{2x}{3}$ means $\dfrac{2}{3}x$
$\dfrac{x + 4}{3}$	$\dfrac{1}{3}, \quad \dfrac{x + 4}{3}$ means $\dfrac{1}{3}(x + 4)$

Whenever a term appears without a numerical coefficient we assume that the numerical coefficient is 1.

Examples

x means $1x$	$-x$ means $-1x$
x^2 means $1x^2$	$-x^2$ means $-1x^2$
xy means $1xy$	$-xy$ means $-1xy$
$(x + 2)$ means $1(x + 2)$	$-(x + 2)$ means $-1(x + 2)$

2   **Like terms** are terms that have the same variables with the same exponents.

Like terms	*Unlike terms*	
$3x,\ -4x$	$3x,\ \ 2$	(variables differ)
$4y,\ 6y$	$3x,\ \ 4y$	(variables differ)
$5,\ -6$	$x,\ \ 3$	(variables differ)
$3(x + 1),\ -2(x + 1)$	$2x,\ \ 3xy$	(variables differ)
$3x^2,\ 4x^2$	$3x,\ \ 4x^2$	(exponents differ)

EXAMPLE 1   Determine if there are any like terms in each algebraic expression.

(a) $2x + 3x + 4$ (b) $2x + 3y + 2$ (c) $x + 3 + y - \frac{1}{2}$

Solution: (a) $2x$ and $3x$ are like terms.
(b) No like terms.
(c) 3 and $-\frac{1}{2}$ are like terms.   ■

EXAMPLE 2   Determine if there are any like terms in each algebraic expression.

(a) $5x - x + 6$ (b) $3 - 2x + 4x - 6$ (c) $12 + x + 7$

Solution: (a) $5x$ and $-x$ (or $-1x$) are like terms.
(b) 3 and -6 are like terms; and $-2x$ and $4x$ are like terms.
(c) 12 and 7 are like terms.   ■

3   Often, we would like to simplify expressions by combining like terms. **To combine like terms** means to add or subtract the like terms in an expression. To combine like terms we can use the procedure that follows.

To Combine Like Terms

1. Determine which terms are like terms.
2. Add or subtract the coefficients of the like terms.
3. Multiply the number found in (2) by the common variables.

Examples 3 through 9 illustrate this procedure.

EXAMPLE 3   Combine like terms: $4x + 3x$.

Solution: $4x$ and $3x$ are like terms with the common variable x.
Since $4 + 3 = 7$, $4x + 3x = 7x$.   ■

EXAMPLE 4   Combine like terms: $4x - 3x$.

Solution: Since $4 - 3 = 1$, $4x - 3x = x$ (or $1x$).   ■

EXAMPLE 5   Combine like terms: $5a - 7a$.

Solution: Since $5 - 7 = -2$, $5a - 7a = -2a$.   ■

EXAMPLE 6   Combine like terms: $3x + x + 5$

Solution: The $3x$ and x are like terms.

$$3x + x + 5$$
$$\text{means} \quad 3x + 1x + 5$$
$$\text{which equals} \quad 4x + 5 \quad ■$$

EXAMPLE 7   Combine like terms: $12 + x + 7$.

Solution: The 12 and 7 are like terms. We can rearrange the terms to get

$$x + 12 + 7 \quad \text{or} \quad x + 19 \quad ■$$

EXAMPLE 8   Combine like terms: $3y + 4x - 3 - 2x$.

Solution: $4x$ and $-2x$ are the only like terms.

$$\text{Rearranging terms:} \quad 4x - 2x + 3y - 3$$
$$\text{Combining like terms:} \quad 2x + 3y - 3 \quad ■$$

EXAMPLE 9   Combine like terms: $-2x + 3y - 4x + 3 - y + 5$.

Solution: $-2x$ and $-4x$ are like terms.
$3y$ and $-y$ are like terms.
3 and 5 are like terms.

Grouping the like terms together gives

$$-2x - 4x + 3y - y + 3 + 5$$
$$-6x \quad + \quad 2y \quad + \quad 8 \quad ■$$

The commutative and associative properties were used to rearrange the terms in Examples 7, 8 and 9. The order of the terms in the answer is not critical. Thus $2y - 6x + 8$ is also an acceptable answer to Example 9.

COMMON STUDENT ERROR

Students often misinterpret the meaning of a term like $3x$. What does $3x$ mean?

Correct	*Wrong*
$3x = x + x + x$	~~$3x = x \cdot x \cdot x$~~

Just as $2 + 2 + 2$ can be expressed as $3 \cdot 2$, $x + x + x$ can be expressed as $3 \cdot x$ or $3x$. Note that when we combine like terms in $x + x + x$ we get $3x$. Also note that $x \cdot x \cdot x = x^3$, not $3x$.

◢ We introduced the distributive property in Section 1.9. Because this property is so important, we will study it again. But before we do, let us go back briefly to the subtraction of real numbers. Recall from Section 1.5 that

$$6 - 3 = 6 + (-3)$$

For any real numbers a and b,

$$a - b = a + (-b)$$

We will use the fact that $a + (-b)$ means $a - b$ in discussing the distributive property.

Distributive Property

For any real numbers a, b, and c,

$$a(b + c) = ab + ac$$

EXAMPLE 10   Use the distributive property to remove parentheses.

(a) $2(x + 4)$ (b) $-2(x + 4)$

Solution: (a) $2(x + 4) = 2x + 2 \cdot 4 = 2x + 8$
(b) $-2(x + 4) = -2x + (-2)(4) = -2x + (-8) = -2x - 8$

Note in part b that instead of leaving the answer $-2x + (-8)$ we wrote it as $-2x - 8$. ■

EXAMPLE 11   Use the distributive property to remove parentheses.

(a) $3(x - 2)$ (b) $-2(4x - 3)$

Solution: (a) By the definition of subtraction we may write $x - 2$ as $x + (-2)$.

$$3(x - 2) = 3[x + (-2)] = 3x + 3(-2)$$
$$= 3x + (-6)$$
$$= 3x - 6$$

(b) $-2(4x - 3) = -2[4x + (-3)] = -2(4x) + (-2)(-3) = -8x + 6$ ■

HELPFUL HINT

With a little practice, you will be able to eliminate some of the intermediate steps when you use the distributive property to remove parentheses. When using the distributive property, there are 8 possibilities with regard to signs. The 8 possibilities follow: study and learn them.

Positive Coefficient   **Negative Coefficient**

$2(x) = 2x$ $(-2)(x) = -2x$

(a) $2(x + 3) = 2x + 6$ (e) $-2(x + 3) = -2x - 6$

$2(+3) = +6$ $(-2)(+3) = -6$

$2(x) = 2x$ $(-2)(x) = -2x$

(b) $2(x - 3) = 2x - 6$ (f) $-2(x - 3) = -2x + 6$

$2(-3) = -6$ $(-2)(-3) = +6$

$2(-x) = -2x$ $(-2)(-x) = 2x$

(c) $2(-x + 3) = -2x + 6$ (g) $-2(-x + 3) = 2x - 6$

$2(+3) = +6$ $(-2)(+3) = -6$

$2(-x) = -2x$ $(-2)(-x) = 2x$

(d) $2(-x - 3) = -2x - 6$ (h) $-2(-x - 3) = 2x + 6$

$2(-3) = -6$ $(-2)(-3) = +6$

The distributive property can be expanded as follows:

$$a(b + c + d + \cdots + n) = ab + ac + ad + \cdots + an$$

Examples of the expanded distributive property are

$$3(x + y + z) = 3x + 3y + 3z$$
$$2(x + y - 3) = 2x + 2y - 6$$

EXAMPLE 12 Use the distributive property to remove parentheses.

(a) $4(x - 3)$ (b) $-2(2x - 4)$ (c) $-3(x + \frac{1}{6})$ (d) $-2(3x - 2y + 4z)$

Solution: (a) $4(x - 3) = 4x - 12$ (b) $-2(2x - 4) = -4x + 8$

(c) $-3(x + \frac{1}{6}) = -3x - \frac{1}{2}$ (d) $-2(3x - 2y + 4z) = -6x + 4y - 8z$ ∎

5   Consider the expression $(4x + 3)$. How do we remove parentheses? Recall that the coefficient of a term is assumed to be 1 if none is present. Therefore we may write

$$(4x + 3) = 1(4x + 3)$$
$$= 1(4x) + (1)(3)$$
$$= 4x + 3$$

Note that $(4x + 3) = 4x + 3$. **When no sign or a plus sign precedes parentheses, the parentheses may be removed without having to change the expression inside the parentheses.**

Examples

$$(x + 3) = x + 3$$
$$(2x - 3) = 2x - 3$$
$$+(2x - 5) = 2x - 5$$
$$+(x + 2y - 6) = x + 2y - 6$$

Now consider the expression $-(4x + 3)$. How do we remove parentheses? Again we assume the coefficient is 1 and write

$$-(4x + 3) = -1(4x + 3)$$
$$= -1(4x) + (-1)(3)$$
$$= -4x + (-3)$$
$$= -4x - 3$$

Note that $-(4x + 3) = -4x - 3$. **When a minus sign precedes parentheses, the signs of all the terms within the parentheses are changed when the parentheses are removed.**

Examples

$$-(x + 4) = -x - 4$$
$$-(-2x + 3) = 2x - 3$$
$$-(5x - y + 3) = -5x + y - 3$$
$$-(-2x - 3y - 5) = 2x + 3y + 5$$

6

To Simplify an Expression Means to

1. Remove parentheses when present by using the distributive law.
2. Combine like terms.

EXAMPLE 13 Simplify $6 - (2x + 3)$.

Solution: $6 - (2x + 3) = 6 - 2x - 3$ distributive property
$\qquad\qquad\qquad = -2x + 3$ combine like terms ∎

Note: $3 - 2x$ is the same as $-2x + 3$, however we generally write the term containing the variable first.

EXAMPLE 14 Simplify $6x + 4(2x + 3)$.

Solution: $6x + 4(2x + 3) = 6x + 8x + 12$ distributive property
$\qquad\qquad\qquad\quad = 14x + 12$ combine like terms ∎

EXAMPLE 15   Simplify $2(x - 1) + 9$.

Solution: $2(x - 1) + 9 = 2x - 2 + 9$ distributive property
$= 2x + 7$ combine like terms ∎

EXAMPLE 16   Simplify $2(x + 3) - 3(x - 2) - 4$.

Solution: $2(x + 3) - 3(x - 2) - 4 = 2x + 6 - 3x + 6 - 4$ distributive property
$= 2x - 3x + 6 + 6 - 4$ rearrange terms
$= -x + 8$ combine like terms ∎

Exercise Set 2.1

Combine like terms when possible.

1. $3x + 5x$
2. $4x + 3$
3. $2x - 3x$
4. $4x + 3y$
5. $12 + x - 3$
6. $-2x - 3x$
7. $-4x + 7x$
8. $4x - 7x + 4$
9. $x + 3x - 7$
10. $3 + 2x - 5$
11. $6 - 3 + 2x$
12. $2 + 2x + 3x$
13. $-7 + 5x + 12$
14. $-2x - 3x - 2 - 3$
15. $5x + 2y + 3 + y$
16. $-x + 2 - x - 2$
17. $4x - 2x + 3 - 7$
18. $x - 4x + 3$
19. $4 + x + 3$
20. $x + 2x + y + 2$
21. $-3x + 2 - 5x$
22. $x + 4 - 6$
23. $5 + 2x - 4x + 6$
24. $3x + 4x - 2 + 5$
25. $x - 2 - 4 + 2x$
26. $2x + 4 - 3 + x$
27. $2 - 3x - 2x + 1$
28. $3x - x + 4 - 6$
29. $2y + 4y + 6$
30. $6 - x - x$
31. $x - 6 + 3x - 4$
32. $-2x + 4x - 3$
33. $4 - x + 4x - 8$
34. $x + 4 + 5$
35. $x + 7 - 2$
36. $2x - 7 + 3x$
37. $2x - x + 7$
38. $2x + 3y + 1$
39. $x + 2y - 3y$
40. $2x + 3 + 4x + 5$
41. $-4x - 3 - 5$
42. $-x + 2x + y$
43. $1 + x + 6$
44. $2x - 7 - 5x$
45. $3x - 7 - 9 + 4x$
46. $x - y - 2y + 3$
47. $4x + 6 + 3x - 7$
48. $-y - 6 - 3y - y$
49. $-4 + x - 6 + 2$
50. $x - 3y + 2x + 4$
51. $-6x + 3x + 2x - 1$
52. $6x - 2 - 6x - 2$
53. $5x - 3 - 7x - 2$
54. $5y + 3x - 2x - 2y$

Use the distributive property to remove parentheses.

55. $2(x + 4)$
56. $3(x - 2)$
57. $4(x + 5)$
58. $-2(x + 3)$
59. $-2(x - 4)$
60. $3(-x + 5)$
61. $-\frac{1}{2}(2x - 4)$
62. $-5(x + 6)$
63. $1(-4 + x)$
64. $3(y + 3)$
65. $\frac{1}{4}(x - 12)$
66. $5(x + y + 4)$
67. $-6(3x - 5)$
68. $-(x - 3)$
69. $7(-2x + 6)$
70. $-2(x + y - z)$
71. $(x + 6)$
72. $-(x + 4y)$
73. $-(-x + y)$
74. $(3x + 4y - 6)$
75. $-(2x - 6y + 8)$
76. $-(-2x + 6 - y)$
77. $3(4 - 2x + y)$
78. $-2(-x + 3y + 5)$
79. $-(-x + 4 - y)$
80. $2(3 - 2x + 4y)$
81. $(x + 3y - 9)$
82. $(-x + 5 - 2y)$
83. $-(-x + 4 + 2y)$

Simplify when possible.

84. $2(x + 3) + 4$

85. $4(x - 2) - x$

86. $6 - (x + 3)$

87. $-2(3 - x) + 1$

88. $-(2x + 3) + 5$

89. $6x + 2(4x + 9)$

90. $3(x + y) + 2y$

91. $2(x - y) + 2x + 3$

92. $6 + (x - 8) + 2x$

93. $(x + y) - 2x + 3$

94. $4 - (2x + 3) + 5$

95. $8x - (x - 3)$

96. $-(x - 5) - 3x + 4$

97. $2(x - 3) - (x + 3)$

98. $3y - (2x + 2y) - 6x$

99. $4(x - 3) + 2(x - 2) + 4$

100. $4(x + 3) - 2x$

101. $2(x - 4) - 3x + 6$

102. $6 - 2(x + 3) + 5x$

103. $-3(x - 4) + 2x - 6$

104. $-(x + 2) + 3x - 6$

105. $4(x - 3) + 4x - 7$

106. $-3(x + 2y) + 3y + 4$

107. $4 + (x + 2) - 6 + 2$

108. $4 - (2 - x) + 3x$

109. $9 - (-3x + 4) - 5$

110. $2y - 6(y - 2) + 3$

111. $4(x + 2) - 3(x - 4) - 5$

112. $4 - (y - 5) + 2x + 3$

113. $-2(2 - x) + 4(y + 2) + 8$

114. $-5(-y + 2) + 3(2 - x) - 4$

115. $-6x + 3y - (6 + x) + (x + 3)$

116. $(x + 3) + (x - 4) - 6x$

117. $-(x + 3) + (2x + 4) - 6$

2.2

The Addition Property

1 *Identify equations.*

2 *Identify and define equivalent equations.*

3 *Use the addition property to solve equations.*

---

1   When two algebraic expressions are joined by an equal sign, $=$, the result is called an **equation.** For example, $4x + 3 = 2x - 4$ is an equation. In this chapter we learn procedures used to solve *linear equations in one variable.*

> A **linear equation** in one variable is an equation of the form
> $$ax + b = c$$
> for real numbers a, b, and c, $a \neq 0$.

Examples of linear equations in one variable

$$x + 4 = 7$$
$$2x - 4 = 6$$

The **solution of an equation** is the number or numbers that make the equation a true statement. For example, the solution to $x + 4 = 7$ is 3. We can check the solution by inserting it back into the original equation.

Check: $x = 3$ $x + 4 = 7$

                                        $3 + 4 = 7$

                                                $7 = 7$ true

Since the check results in a true statement, 3 is a solution to the equation $x + 4 = 7$.

EXAMPLE 1   Consider the equation $2x - 4 = 6$. Determine whether

(a) 3 is a solution   (b) 5 is a solution.

Solution: (a) To determine whether 3 is a solution to the equation substitute 3 for x.

$$\text{Check:} \quad x = 3 \qquad 2x - 4 = 6$$
$$2(3) - 4 = 6$$
$$6 - 4 = 6$$
$$2 = 6 \qquad \text{not a true statement}$$

Since the value 3 does not check, 3 is not a solution.

(b) *Check:* $x = 5 \qquad 2x - 4 = 6$
$$2(5) - 4 = 6$$
$$10 - 4 = 6$$
$$6 = 6 \qquad \text{a true statement}$$

Since the value 5 checks, 5 is a solution to the equation.   ■

2   Complete procedures for solving an equation will be given shortly. For now, you need to understand that **to solve an equation, it is necessary to get the variable all by itself on one side of the equal sign. We say that we isolate the variable.** To isolate the variable, we make use of two properties: addition and multiplication. Look first at Fig. 2.1.

Figure 2.1

Think of an equation as a balanced statement whose left side is balanced by its right side. When solving an equation, we must make sure that the equation remains balanced at all times. That is, both sides must always remain equal. **We ensure that an equation always remains equal by doing the same thing to both sides of the equation.** For example, if we decide to add something to the left side of the equation, we must add exactly the same amount to the right side. If we decide to multiply the right side of the equation by some number, we must multiply the left side by the same number.

When we add the same number to both sides of an equation, or multiply both sides of an equation by the same nonzero number, we do not change the solution to the equation, just the form. Two or more equations with the same solution are called **equivalent equations.** The equations $2x - 4 = 2$, $2x = 6$ and $x = 3$ are equivalent, since the solution to each is 3.

Check: $x = 3 \qquad 2x - 4 = 2 \qquad\qquad 2x = 6 \qquad\qquad x = 3$
$$2(3) - 4 = 2 \qquad\qquad 2(3) = 6 \qquad\qquad 3 = 3 \qquad \text{true}$$
$$6 - 4 = 2 \qquad\qquad 6 = 6 \qquad \text{true}$$
$$2 = 2 \qquad \text{true}$$

When solving an equation, we use the addition and multiplication properties to express a given equation as simpler equivalent equations until we obtain the solution.

3 In this section we use the addition property to solve equations. In Section 2.3 we use the multiplication property to solve equations.

Addition Property

   If $a = b$, then $a + c = b + c$ for any real numbers a, b, and c.

This property implies that the same number can be added to both sides of an equation without changing the solution. **The addition property is used to solve equations of the form $x + a = b$.** To isolate the variable x in equations of this form add the opposite, or additive inverse of a, $-a$, to both sides of the equation.

   To isolate the variable when solving equations of the form $x + a = b$ we use the addition property to eliminate the number **on the same side of the equal sign as the variable.** Study the following examples carefully.

Equation	*To solve, use the addition property to eliminate the number*
$x + 8 = 10$	8
$x - 7 = 12$	-7
$5 = x - 12$	-12
$-4 = x + 9$	9

Now let us work some problems.

EXAMPLE 2   Solve the equation $x - 4 = 3$.

Solution: To isolate the variable, x, we must eliminate the -4 from the left side of the equation. To do this we add 4, the opposite of -4, to both sides of the equation.

$$x - 4 = 3$$
$$x - 4 + 4 = 3 + 4 \qquad \text{add 4 to both sides of equation}$$
$$x + 0 = 7$$
$$x = 7$$

Note how the process helps to isolate x.

Check: $x - 4 = 3$
$$7 - 4 = 3$$
$$3 = 3 \qquad \text{true} \qquad ■$$

EXAMPLE 3   Solve the equation $y - 3 = -5$.

Solution: To solve this equation we must isolate the variable, y. To eliminate the -3 from the left side of the equation we add its opposite, $+3$, to both sides of the equation.

$$y - 3 = -5$$
$$y - 3 + 3 = -5 + 3 \qquad \text{add 3 to both sides of equation}$$
$$y + 0 = -2$$
$$y = -2 \qquad ■$$

EXAMPLE 4   Solve the equation $x + 5 = 9$.

Solution: To solve this equation we must isolate the x. Therefore, we must eliminate the $+5$ from the left side of the equation. To do this we add the opposite of $+5$, -5, to both sides of the equation.

$$x + 5 = 9$$
$$x + 5 + (-5) = 9 + (-5) \qquad \text{add } -5 \text{ to both sides of equation}$$
$$x + 0 = 4$$
$$x = 4 \qquad \blacksquare$$

In Example 4 we added -5 to both sides of the equation. From Section 1.5 we know that $5 + (-5) = 5 - 5$. Thus we can see that adding a negative 5 to both sides of the equation is equivalent to subtracting a 5 from both sides of an equation. The addition property says the same number may be *added* to both sides of an equation. **Since subtraction is defined in terms of addition, the addition property also allows us to *subtract* the same number from both sides of the equation.** Thus Example 4 could have also been worked as follows:

$$x + 5 = 9$$
$$x + 5 - 5 = 9 - 5 \qquad \text{subtract 5 from both sides of equation}$$
$$x + 0 = 4$$
$$x = 4$$

In this text, unless there is a specific reason to do otherwise, rather than adding a negative number to both sides of the equation we will subtract a number from both sides of the equation.

EXAMPLE 5   Solve the equation $x + 7 = -3$.

Solution:
$$x + 7 = -3$$
$$x + 7 - 7 = -3 - 7 \qquad \text{subtract 7 from both sides of equation}$$
$$x + 0 = -10$$
$$x = -10$$

Check:
$$x + 7 = -3$$
$$-10 + 7 = -3$$
$$-3 = -3 \qquad \text{true} \qquad \blacksquare$$

EXAMPLE 6   Solve the equation $4 = x - 5$.

Solution: In this example the variable x is on the right side of the equation. To isolate the x we must eliminate the -5 from the right side of the equation. This can be accomplished by adding 5 to both sides of the equation.

$$4 = x - 5$$
$$4 + 5 = x - 5 + 5 \qquad \text{add 5 to both sides of equation}$$
$$9 = x + 0$$
$$9 = x$$

Thus the solution is 9.   ■

EXAMPLE 7   Solve the equation $-6 = x + 12$.

Solution: The variable is on the right side of the equation. Subtract 12 from both sides of the equation to isolate the variable.

$$-6 = x + 12$$
$$-6 \; -12 \; = x + 12 \; -12 \qquad \text{subtract 12 from both sides of equation}$$
$$-18 = x + 0$$
$$-18 = x$$

The solution is -18.   ∎

COMMON STUDENT ERROR

When solving equations our goal is to get the variable all by itself on one side of the equal sign. Consider the equation $x + 3 = -4$: How do we solve it?

Correct

$$x + 3 = -4$$
$$x + 3 - 3 = -4 - 3$$
$$x = -7$$

variable is now isolated

Wrong

$$x + 3 = -4$$
$$x + 3 + 4 = -4 + 4$$
$$\underbrace{x + 7}_{\uparrow} = 0$$

variable is **not** isolated

Use the addition property to **remove the number that is on the same side of the equal sign as the variable.**

Consider the following two problems:

$$x - 5 = 12 \qquad\qquad 15 = x + 3$$
$$x - 5 + 5 = 12 + 5 \qquad 15 - 3 = x + 3 - 3$$
$$x + 0 = 12 + 5 \qquad 15 - 3 = x + 0$$
$$x = 17 \qquad\qquad 12 = x$$

Note in these problems how the number on the same side of the equal sign as the variable is transferred to the opposite side of the equal sign when the addition property is used. Also note that the sign of the number changes when transferred from one side of the equal sign to the other.

When you feel comfortable using the addition property you may wish to do some of the steps mentally to reduce some of the written work. For example the two problems above may be shortened as follows.

Shortened form

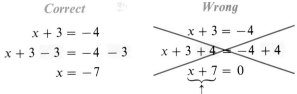

$$x - 5 = 12 \qquad\qquad\qquad x - 5 = 12$$
$$x - 5 + 5 = 12 + 5 \longleftarrow \begin{array}{l}\text{do this} \\ \text{step mentally}\end{array} \qquad x = 12 + 5$$
$$x = 12 + 5 \qquad\qquad\qquad x = 17$$
$$x = 17$$

Shortened form

$$15 = x + 3$$
$$15 - 3 = x + 3 - 3 \longleftarrow \text{do this}$$
$$\qquad\qquad\qquad\text{step mentally}$$
$$15 - 3 = x$$
$$12 = x$$

$$15 = x + 3$$
$$15 - 3 = x$$
$$12 = x$$

Exercise Set 2.2

Solve each equation, and check your solution.

1. $x + 3 = 8$ **2.** $x - 4 = 9$ **3.** $x + 7 = -3$

4. $x - 4 = -8$ **5.** $x + 3 = -4$ **6.** $x - 16 = 36$

7. $x + 43 = -18$ **8.** $6 + x = 9$ **9.** $-6 + x = 12$

10. $13 = 5 + x$ **11.** $27 = x - 16$ **12.** $-9 = x - 25$

13. $-13 = x - 1$ **14.** $4 = 11 + x$ **15.** $29 = -43 + x$

16. $-18 = -14 + x$ **17.** $7 + x = -19$ **18.** $9 + x = 9$

19. $x + 29 = -29$ **20.** $4 + x = -9$ **21.** $6 = x - 4$

22. $5 + x = 12$ **23.** $x + 7 = -5$ **24.** $6 = 4 + x$

25. $9 + x = 12$ **26.** $-4 = x - 3$ **27.** $-5 = 4 + x$

28. $12 = 16 + x$ **29.** $30 = x - 19$ **30.** $15 + x = -5$

31. $x - 12 = -9$ **32.** $x + 6 = -12$ **33.** $4 + x = 9$

34. $-6 = 9 + x$ **35.** $-8 = -9 + x$ **36.** $-12 = 8 + x$

37. $2 = x - 9$ **38.** $2 = x + 9$ **39.** $9 + x = 24$

40. $-9 + x = -24$ **41.** $16 + x = -20$ **42.** $-25 = 18 + x$

43. $40 + x = -7$ **44.** $-40 + x = -12$ **45.** $-37 + x = 9$

46. $15 + x = 7$ **47.** $x - 9 = -30$ **48.** $6 + x = 12$

49. $14 = 17 + x$ **50.** $39 = x - 17$ **51.** $60 = x - 12$

52. $-4 = x + 5$

53. Explain what is meant by equivalent equations.

54. When solving the equation $x - 4 = 6$, would you add 4 to both sides of the equation or subtract 6 from both sides of the equation? Explain your answer.

55. When solving the equation $5 = x + 3$, would you subtract 5 from both sides of the equation or subtract 3 from both sides of the equation? Explain your answer.

2.3
The Multiplication Property

1   *Identify reciprocals.*

2   *Use the multiplication property to solve equations.*

3   *Solve equations of the form* $-x = a$.

1   Before we discuss the multiplication property, let us discuss what is meant by the **reciprocal** of a number. Two numbers are reciprocals of each other when their product is 1. Some examples of numbers and their reciprocals follow.

Number	Reciprocal	Product
3	$\dfrac{1}{3}$	$(3)\left(\dfrac{1}{3}\right) = 1$
-2	$-\dfrac{1}{2}$	$(-2)\left(-\dfrac{1}{2}\right) = 1$
$\dfrac{1}{4}$	4	$\left(\dfrac{1}{4}\right)(4) = 1$
$\dfrac{-3}{5}$	$\dfrac{-5}{3}$	$\left(\dfrac{-3}{5}\right)\left(\dfrac{-5}{3}\right) = 1$
-1	-1	$(-1)(-1) = 1$

Note that the reciprocal of a positive number is a positive number and the reciprocal of a negative number is a negative number.

In general, if a represents any number, its reciprocal is $1/a$. For example, the reciprocal of 3 is $\frac{1}{3}$, the reciprocal of -2 is $\frac{1}{-2}$ or $-\frac{1}{2}$, and the reciprocal of $-\frac{3}{5}$ is $1 \div \left(\frac{-3}{5}\right)$ or $-\frac{5}{3}$.

2 In Section 2.2 we used the addition property to solve equations of the form $x + a = b$. In this section we use the multiplication property to solve equations of the form $ax = b$. Notice the difference in the two forms of equations.

Multiplication Property

If $a = b$, then $a \cdot c = b \cdot c$ for any numbers a, b, and c.

The multiplication property implies that both sides of an equation can be multiplied by the same number without changing the solution. **The multiplication property can be used to solve equations of the form $ax = b$.** We can isolate the variable in equations of this form by multiplying both sides of the equation by the reciprocal of a, $\frac{1}{a}$. By doing so the number on the same side of the equal sign as the variable, x, is eliminated.

Equation	To solve, use the multiplication property to eliminate the number
$4x = 9$	4
$-5x = 20$	-5
$15 = \dfrac{1}{2}x$	$\dfrac{1}{2}$
$7 = -9x$	-9

Now let us work some problems.

EXAMPLE 1   Solve the equation $3x = 6$.

Solution: To isolate the variable, x, we must eliminate the 3 from the left side of the equation. To do this we multiply both sides of the equation by the reciprocal of 3, $\frac{1}{3}$.

$$3x = 6$$

$$\frac{1}{3} \cdot 3x = \frac{1}{3} \cdot 6 \qquad \text{multiply both sides of equation by } \frac{1}{3}$$

$$\frac{1}{3} \cdot 3x = \frac{1}{3} \cdot 6 \qquad \text{divide out common factors}$$

$$x = 2 \qquad \blacksquare$$

EXAMPLE 2   Solve the equation $\frac{x}{2} = 4$.

Solution: Since dividing by 2 is the same as multiplying by $\frac{1}{2}$, $\frac{x}{2} = 4$ is the same as $\frac{1}{2}x = 4$. We will therefore multiply both sides of the equation by the reciprocal of $\frac{1}{2}$, which is 2.

$$\frac{x}{2} = 4$$

$$2\left(\frac{x}{2}\right) = 2 \cdot 4 \qquad \text{multiply both sides of equation by 2}$$

$$1x = 2 \cdot 4$$

$$x = 8$$

Check: $\dfrac{x}{2} = 4$

$$\frac{8}{2} = 4$$

$$4 = 4 \qquad \text{true} \qquad \blacksquare$$

EXAMPLE 3   Solve the equation $\frac{2}{3}x = 6$.

Solution: The reciprocal of $\frac{2}{3}$ is $\frac{3}{2}$. Multiply both sides of the equation by $\frac{3}{2}$.

$$\frac{2}{3}x = 6$$

$$\frac{3}{2} \cdot \frac{2}{3}x = \frac{3}{2} \cdot 6$$

$$1x = 9$$

$$x = 9 \qquad \blacksquare$$

In Example 1, $3x = 6$, we multiplied both sides of the equation by $\frac{1}{3}$ to isolate the variable. We could have also isolated the variable by dividing both sides of the equation by 3 as follows.

$$3x = 6$$

$$\frac{\overset{1}{\cancel{3}}x}{\underset{1}{\cancel{3}}} = \frac{\overset{2}{\cancel{6}}}{\underset{1}{\cancel{3}}} \qquad \text{divide both sides of equation by 3}$$

$$x = 2$$

We can do this because dividing by 3 is equivalent to multiplying by $\frac{1}{3}$. **Since division can be defined in terms of multiplication ($\frac{a}{b}$ means $a \cdot \frac{1}{b}$) the multiplication property also allows us to divide both sides of an equation by the same nonzero number.**

EXAMPLE 4 Solve the equation $8p = 5$.

Solution: $8p = 5$

$$\frac{8p}{8} = \frac{5}{8} \qquad \text{divide both sides of equation by 8}$$

$$p = \frac{5}{8} \quad \blacksquare$$

EXAMPLE 5 Solve the equation $-12 = -3x$.

Solution: In this equation the variable, x, is on the right side of the equal sign. To isolate x, we divide both sides of the equation by -3.

$$-12 = -3x$$

$$\frac{\overset{4}{-\cancel{12}}}{\underset{1}{-\cancel{3}}} = \frac{\overset{1}{-\cancel{3}}x}{\underset{1}{-\cancel{3}}} \qquad \text{divide both sides of equation by } -3$$

$$4 = x \quad \blacksquare$$

HELPFUL HINT

When solving an equation of the form $ax = b$ we can isolate the variable by:

(a) multiplying both sides of the equation by the reciprocal of a, $\frac{1}{a}$, as was done in Examples 1, 2, and 3, or

(b) dividing both sides of the equation by a, as was done in Examples 4, and 5.

Either method may be used to isolate the variable. However if the equation contains a fraction, or fractions, you will arrive at a solution more quickly by multiplying by the reciprocal of a. This is illustrated in Example 6.

EXAMPLE 6   Solve the equation $-2x = \dfrac{3}{5}$.

Solution: Since this equation contains a fraction we will isolate the variable by multiplying both sides of the equation by $-\frac{1}{2}$, the reciprocal of -2.

$$-2x = \frac{3}{5}$$

$$\left(-\frac{1}{2}\right)(-2x) = \left(-\frac{1}{2}\right)\left(\frac{3}{5}\right) \qquad \text{multiply both sides of equation by } -\frac{1}{2}$$

$$1x = \left(-\frac{1}{2}\right)\left(\frac{3}{5}\right)$$

$$x = -\frac{3}{10}$$

Check: $$-2x = \frac{3}{5}$$

$$-\overset{1}{2}\left(\frac{-3}{\underset{5}{10}}\right) = \frac{3}{5}$$

$$\frac{3}{5} = \frac{3}{5} \qquad \text{true} \qquad \blacksquare$$

In Example 6, if you wished to solve the equation by dividing both sides of the equation by -2, you would have to divide the fraction $\frac{3}{5}$ by -2.

3   When solving an equation in the following sections we may obtain an equation like $-x = 7$. This is *not* a solution. The solution to an equation will be of the form $x = $ some number. When given an equation of the form $-x = 7$ we can solve for x by multiplying both sides of the equation by -1, as illustrated in the following example.

EXAMPLE 7   Solve the equation $-x = 7$.

Solution: $-x = 7$ means that $-1x = 7$. To obtain a positive x we can multiply both sides of the equation by -1.

$$-x = 7$$

$$-1x = 7$$

$$(-1)(-1x) = (-1)(7) \qquad \text{multiply both sides of equation by } -1$$

$$1x = -7$$

$$x = -7$$

Check: $$-x = 7$$

$$-(-7) = 7$$

$$7 = 7 \qquad \text{true} \qquad \blacksquare$$

Whenever we have the negative of a variable equal to a quantity as in Example 7, we can solve for the variable by multiplying both sides of the equation by -1.

EXAMPLE 8   Solve the equation $-x = -5$

Solution:
$$-x = -5$$
$$-1x = -5$$
$$(-1)(-1x) = (-5)(-1)$$
$$1x = 5$$
$$x = 5 \quad \blacksquare$$

HELPFUL HINT

For any real number a, $a \neq 0$

$$\text{If} \quad -x = a$$
$$\text{then} \quad x = -a$$

Examples:

$$-x = 7 \qquad\qquad -x = -2$$
$$x = -7 \qquad\qquad x = -(-2)$$
$$\qquad\qquad\qquad x = 2$$

When you feel comfortable using the multiplication property you may wish to do some of the steps mentally to reduce some of the written work. Below are two examples worked out in detail, along with their shortened form.

EXAMPLE 9   Solve the equation $-3x = -21$.

Solution:

$$-3x = -21$$
$$\frac{-3x}{-3} = \frac{-21}{-3} \quad \longleftarrow \quad \text{do this step mentally}$$
$$x = \frac{-21}{-3}$$
$$x = 7$$

Shortened Form

$$-3x = -21$$
$$x = \frac{-21}{-3}$$
$$x = 7 \quad \blacksquare$$

EXAMPLE 10   Solve the equation $\frac{1}{3}x = 9$.

Solution:

$$\frac{1}{3}x = 9$$
$$3\left(\frac{1}{3}x\right) = 3(9) \quad \longleftarrow \quad \text{do this step mentally}$$
$$x = 3(9)$$
$$x = 27$$

Shortened Form

$$\frac{1}{3}x = 9$$
$$x = 3(9)$$
$$x = 27 \quad \blacksquare$$

In the previous section we discussed the addition property and in this section we discussed the multiplication property. It is important that you understand the difference between the two. The helpful hint below should be studied carefully.

HELPFUL HINT

The **addition property** is used to solve equations of the form $x + a = b$.

$$x + 3 = -6 \qquad\qquad x - 5 = -2$$
$$x + 3 - 3 = -6 - 3 \qquad x - 5 + 5 = -2 + 5$$
$$x = -9 \qquad\qquad x = 3$$

The **multiplication property** is used to solve equations of the form $ax = b$.

$$3x = 6 \qquad \frac{x}{2} = 4 \qquad\qquad \frac{2}{5}x = 12$$
$$\frac{3x}{3} = \frac{6}{3} \qquad 2\left(\frac{x}{2}\right) = 2(4) \qquad \left(\frac{5}{2}\right)\left(\frac{2}{5}x\right) = \left(\frac{5}{2}\right)(12)$$
$$x = 2 \qquad\quad x = 8 \qquad\qquad x = 30$$

Note: The *addition property* is used when a number is *added to or subtracted from* a variable. The *multiplication property* is used when a variable is *multiplied or divided by a number.*

Exercise Set 2.3

Solve each equation, and check your solution.

1. $3x = 9$

2. $4x = 16$

3. $\dfrac{x}{2} = 4$

4. $\dfrac{x}{3} = 12$

5. $-4x = 8$

6. $8 = 16y$

7. $\dfrac{x}{6} = -2$

8. $\dfrac{x}{3} = -2$

9. $\dfrac{x}{5} = 1$

10. $-2x = 12$

11. $-32x = -96$

12. $16 = -4y$

13. $-6 = 4z$

14. $\dfrac{x}{8} = -3$

15. $-x = -4$

16. $-x = 9$

17. $-2 = -y$

18. $-3 = \dfrac{x}{5}$

19. $-\dfrac{x}{7} = -7$

20. $4 = \dfrac{x}{9}$

21. $9 = -18x$

22. $12y = -15$

23. $-\dfrac{x}{3} = -2$

24. $-\dfrac{a}{8} = -7$

25. $19x = 35$

26. $-24x = -18$

27. $-9 = 42x$

28. $-9 = -45y$

29. $7x = -7$

30. $3x = \dfrac{3}{5}$

31. $5x = -\dfrac{3}{8}$

32. $-2b = -\dfrac{4}{5}$

33. $15 = -\dfrac{x}{5}$

34. $\dfrac{x}{16} = -4$

35. $-\dfrac{x}{5} = -25$

36. $-x = -\dfrac{5}{9}$

37. $\dfrac{x}{5} = -7$

38. $-3x = -18$

39. $6 = \dfrac{x}{4}$

40. $-3 = \dfrac{x}{-5}$

41. $6c = -30$

42. $\dfrac{2}{7}x = 7$

43. $\dfrac{y}{-2} = -6$

44. $-2x = \dfrac{3}{5}$

45. $\dfrac{-3}{8}x = 6$

46. $-x = \dfrac{4}{7}$

47. $\dfrac{1}{3}x = -12$

48. $6 = \dfrac{3}{5}x$

49. $-4 = -\dfrac{2}{3}z$

50. $-8 = \dfrac{-4}{5}x$

51. $4x = -9$

52. $-4x = -13$

53. $2x = -\dfrac{5}{2}$

54. $6x = \dfrac{8}{3}$

55. $\dfrac{2}{3}x = 6$

56. $-\dfrac{1}{2}x = \dfrac{2}{3}$

57. When solving the equation $3x = 5$ would you divide both sides of the equation by 3 or by 5? Explain why.

58. When solving the equation $-2x = 6$ would you divide both sides of the equation by -2 or by 6? Explain why.

59. Consider the equation $\frac{2}{3}x = 4$. This equation could be solved by multiplying both sides of the equation by $\frac{3}{2}$, the reciprocal of $\frac{2}{3}$, or by dividing both sides of the equation by $\frac{2}{3}$. Which method do you feel would be easier? Explain your answer. Find the solution to the equation.

60. Consider the equation $4x = \frac{3}{5}$. Would it be easier to solve this equation by dividing both sides of the equation by 4, or by multiplying both sides of the equation by $\frac{1}{4}$, the reciprocal of 4? Explain your answer. Find the solution to the problem.

61. Consider the equation $\frac{3}{7}x = \frac{4}{5}$. Would it be easier to solve this equation by dividing both sides of the equation by $\frac{3}{7}$, or by multiplying both sides of the equation by $\frac{7}{3}$, the reciprocal of $\frac{3}{7}$? Explain your answer. Find the solution to the equation.

2.4

Solving Linear Equations with a Variable on Only One Side of the Equation

1 *Solve linear equations that contain a variable on only one side of the equal sign.*

1 In this section we discuss how to solve linear equations when the variable appears on only one side of the equal sign. In Section 2.5 we discuss how to solve linear equations when the variable appears on both sides of the equal sign.

No one method is the "best" to solve all linear equations. Following is a general procedure that can be used to solve linear equations when the variable appears on only one side of the equation and the equation does not contain fractions.

**To Solve Linear Equations
with a Variable on Only One Side of the Equal Sign**

1. Use the distributive property to remove parentheses.
2. Combine like terms on the same side of the equal sign.
3. Use the addition property to obtain an equation with the term containing the variable on one side of the equal sign and a constant on the other side. This will result in an equation of the form $ax = b$.
4. Use the multiplication property to isolate the variable. This will give an answer of the form $x = \dfrac{b}{a}$ $\left(\text{or } 1x = \dfrac{b}{a}\right)$.
5. Check the solution in the *original* equation.

Equations containing fractions will be solved using a different procedure. We will discuss how to solve equations containing fractions in Section 6.6.

When solving an equation remember that our goal is to obtain the variable all by itself on one side of the equation.

EXAMPLE 1   Solve the equation $2x - 5 = 9$.

Solution: We will follow the procedure just outlined for solving equations. Since the equation contains no parentheses and since there are no like terms to be combined, we start with Step 3.

Step 3

$$2x - 5 = 9$$

$$2x - 5 + 5 = 9 + 5 \qquad \text{add 5 to both sides of equation}$$

$$2x = 14$$

Step 4

$$\frac{2x}{2} = \frac{14}{2} \qquad \text{divide both sides of equation by 2}$$

$$x = 7$$

Step 5   *Check:*

$$2x - 5 = 9$$

$$2(7) - 5 = 9$$

$$14 - 5 = 9$$

$$9 = 9 \qquad \text{true}$$

Since the check is true, the solution is 7. Note that after completing Step 3 we obtain an equation of the form $ax = b$. And after completing Step 4 we obtain the answer in the form $x = $ some number.   ■

EXAMPLE 2   Solve the equation $-2x - 6 = -3$.

Solution:

$$-2x - 6 = -3$$

Step 3

$$-2x - 6 + 6 = -3 + 6 \qquad \text{add 6 to both sides of equation}$$

$$-2x = 3$$

Step 4

$$\frac{-2x}{-2} = \frac{3}{-2} \qquad \text{divide both sides of equation by } -2$$

$$x = -\frac{3}{2}$$

Step 5   *Check:*

$$-2x - 6 = -3$$

$$-2\left(-\frac{3}{2}\right) - 6 = -3$$

$$3 - 6 = -3$$

$$-3 = -3 \qquad \text{true}$$

The solution is $-\dfrac{3}{2}$.   ■

Note that checks are always made with the original equation. In some of the following examples the check will be omitted to save space.

EXAMPLE 3   Solve the equation $16 = 4x + 6 - 2x$.

Solution: Again we must isolate the variable x. Since the right side of the equation has two like terms containing the variable x, we will first combine these like terms.

Step 2 $16 = 4x + 6 - 2x$ combine like terms to obtain

$16 = 2x + 6$

Step 3 $16 - 6 = 2x + 6 - 6$ subtract 6 from both sides of equation

$10 = 2x$

Step 4 $\dfrac{10}{2} = \dfrac{2x}{2}$ divide both sides of equation by 2

$5 = x$ ∎

The solution above can be condensed as follows.

$16 = 4x + 6 - 2x$

$16 = 2x + 6$ (like terms were combined)

$10 = 2x$ (6 was subtracted from both sides of equation)

$5 = x$ (both sides of equation were divided by 2)

EXAMPLE 4   Solve the equation $2(x + 4) - 5x = -3$.

Solution: $2(x + 4) - 5x = -3$ use the distributive property to obtain

$2x + 8 - 5x = -3$ now combine like terms to obtain

$-3x + 8 = -3$

$-3x + 8 - 8 = -3 - 8$ subtract 8 from both sides of equation

$-3x = -11$

$\dfrac{-3x}{-3} = \dfrac{-11}{-3}$ divide both sides of equation by -3

$x = \dfrac{11}{3}$ ∎

The solution above can be condensed as follows:

$2(x + 4) - 5x = -3$

$2x + 8 - 5x = -3$ (the distributive property was used)

$-3x + 8 = -3$ (like terms were combined)

$-3x = -11$ (8 was subtracted from both sides of equation)

$x = \dfrac{11}{3}$ (both sides of equations were divided by -3)

EXAMPLE 5   Solve the equation $2x - (x + 2) = 6$.

Solution: $2x - (x + 2) = 6$

$$2x - x - 2 = 6 \quad \text{(the distributive property was used)}$$

$$x - 2 = 6 \quad \text{(like terms were combined)}$$

$$x = 8 \quad \text{(2 was added to both sides of equation)}$$

Check: $2x - (x + 2) = 6$

$$2(8) - (8 + 2) = 6$$

$$16 - 10 = 6$$

$$6 = 6 \quad \text{true} \quad \blacksquare$$

Exercise Set 2.4 _____

Solve each equation.

1. $2x + 3 = 7$

2. $2x - 4 = 8$

3. $-2x - 5 = 7$

4. $-4x + 5 = -3$

5. $5x - 6 = 19$

6. $6 - 3x = 18$

7. $5x - 2 = 10$

8. $-9x + 3 = 15$

9. $-x - 4 = 8$

10. $6 = 2x - 3$

11. $12 - x = 9$

12. $-3x - 3 = -12$

13. $9 + 2x = 24$

14. $-7x + 3 = -12$

15. $32x + 9 = -12$

16. $14 = 18 + 7x$

17. $-44 = 9x + 12$

18. $-18 + 18x = -18$

19. $6x - 9 = 21$

20. $-x + 4 = -8$

21. $12 = -6x + 5$

22. $15 = 7x + 1$

23. $-2x - 7 = -13$

24. $-2 - x = -12$

25. $4x + 7 = -9$

26. $15 - 4x = -5$

27. $9x - 4 = 5$

28. $12 = -3x + 9$

29. $-23 = -4x + 7$

30. $4 = 7x - 20$

31. $2(x + 1) = 6$

32. $3(x - 2) = 12$

33. $4(3 - x) = 12$

34. $-2(x + 3) = -9$

35. $-4 = -(x + 5)$

36. $-3(2 - 3x) = 9$

37. $12 = 4(x + 3)$

38. $-2(x + 4) + 5 = 1$

39. $5 = 2(3x + 6)$

40. $-2 = 5(3x + 1) - 12x$

41. $2x + 3(x + 2) = 11$

42. $4 = -2(x + 3)$

43. $x - 3(2x + 3) = 11$

44. $3(4 - x) + 5x = 9$

45. $5x + 3x - 4x - 7 = 9$

46. $-(x + 2) = 4$

47. $3 - (2x - 3) = 9$

48. $12 + (x + 9) = 7$

49. $2(2 - 2x) + 3x = 9$

50. $4(x + 3) - 4 = -6$

51. $3 - 2(x + 3) + 2 = 1$

52. $2(3x - 4) - 4x = 12$

53. $1 - (x + 3) + 2x = 4$

54. $5x - 2x - 7x = -20$

55. $4 - 6x + 9 - 3 = -8$

56. $-4(x + 2) - 3x = 20$

57. When solving equations that do not contain fractions do we normally use the addition or multiplication property first in the process of isolating the variable? Explain your answer.

2.5

Solving Linear Equations with Variables on Both Sides of the Equation

1 *Solve equations when the variable appears on both sides of the equal sign.*

2 *Identify and define identities.*

1   The equation $4x + 6 = 2x + 4$ contains the variable x on both sides of the equal sign. To solve equations of this type, we must use the appropriate properties to rewrite the equation with all terms containing the variable on only one side of the equal sign and all terms not containing the variable on the other side of the equal sign. This will allow us to isolate the variable, which is our goal. Following is a general procedure, similar to the one outlined in Section 2.4, that can be used to solve linear equations with variables on both sides of the equal sign.

To Solve Linear Equations with Variables on Both Sides of the Equal Sign

1. Use the distributive property to remove parentheses.
2. Combine like terms on the same side of the equal sign.
3. Use the addition property to rewrite the equation with all terms containing the variable on one side of the equal sign and all terms not containing the variable on the other side of the equal sign. It may be necessary to use the addition property a number of times to accomplish this. Repeated use of the addition property will eventually result in an equation of the form $ax = b$.
4. Use the multiplication property to isolate the variable. This will give an answer of the form $x =$ some number.
5. Check the solution in the original equation.

Whenever possible, you should check your work. We will not show all the checks to save space.

EXAMPLE 1   Solve the equation $4x + 6 = 2x + 4$.

Solution: Remember that our goal is always to get all terms with the variable on one side of the equal sign and all terms without the variable on the other side of the equal sign. Many methods can be used to isolate the variable. We will illustrate two. In method 1 we will isolate the variable on the left side of the equation. In method 2 we will isolate the variable on the right side of the equation.

Method 1:

$$4x + 6 = 2x + 4$$

Step 3 $4x - 2x + 6 = 2x - 2x + 4$ subtract $2x$ from both sides of equation

$$2x + 6 = 4$$

Step 3 $2x + 6 - 6 = 4 - 6$ subtract 6 from both sides of equation

$$2x = -2$$

Step 4 $\dfrac{2x}{2} = \dfrac{-2}{2}$ divide both sides of equation by 2

$$x = -1$$

Method 2:

$$4x + 6 = 2x + 4$$

Step 3 $4x - 4x + 6 = 2x - 4x + 4$ subtract $4x$ from both sides of equation

$$6 = -2x + 4$$

Step 3 $6 - 4 = -2x + 4 - 4$ subtract 4 from both sides of equation

$$2 = -2x$$

Step 4 $$\frac{2}{-2} = \frac{-2x}{-2}$$ divide both sides of equation by -2

$$-1 = x$$

The same answer is obtained using both methods.

Step 5   *Check:* $4x + 6 = 2x + 4$

$$4(-1) + 6 = 2(-1) + 4$$

$$-4 + 6 = -2 + 4$$

$$2 = 2 \quad \text{true} \quad \blacksquare$$

EXAMPLE 2   Solve the equation $2x - 3 - 5x = 13 + 4x - 2$.

Solution: Since there are like terms *on the same side of the equal sign*, we will begin by combining these like terms.

Step 2 $2x - 3 - 5x = 13 + 4x - 2$ combine like terms to obtain

$$-3x - 3 = 4x + 11$$

Step 3 $-3x + 3x - 3 = 4x + 3x + 11$ add $3x$ to both sides of equation

$$-3 = 7x + 11$$

Step 3 $-3 - 11 = 7x + 11 - 11$ subtract 11 from both sides of equation

$$-14 = 7x$$

Step 4 $$\frac{-14}{7} = \frac{7x}{7}$$ divide both sides of equation by 7

$$-2 = x$$

Step 5   *Check:* $2x - 3 - 5x = 13 + 4x - 2$

$$2(-2) - 3 - 5(-2) = 13 + 4(-2) - 2$$

$$-4 - 3 + 10 = 13 - 8 - 2$$

$$-7 + 10 = 5 - 2$$

$$3 = 3 \quad \text{true}$$

Since the check is true, the solution is -2. $\blacksquare$

The solution to Example 2 could be condensed as follows:

$$2x - 3 - 5x = 13 + 4x - 2$$
$$-3x - 3 = 4x + 11 \qquad \text{(like terms were combined)}$$
$$-3 = 7x + 11 \qquad \text{(3x was added to both sides of equation)}$$
$$-14 = 7x \qquad \text{(11 was subtracted from both sides of equation)}$$
$$-2 = x \qquad \text{(both sides of equation were divided by 7)}$$

EXAMPLE 3   Solve the equation $2(p + 3) = -3p + 10$.

Solution:
$$2(p + 3) = -3p + 10$$

Step 1 $\qquad 2p + 6 = -3p + 10 \qquad$ distributive property was used

Step 3 $\qquad 2p + 3p + 6 = -3p + 3p + 10 \qquad$ add 3p to both sides of equation
$$5p + 6 = 10$$

Step 3 $\qquad 5p + 6 - 6 = 10 - 6 \qquad$ subtract 6 from both sides of equation
$$5p = 4$$

Step 4 $\qquad \dfrac{5p}{5} = \dfrac{4}{5} \qquad$ divide both sides of equation by 5

$$p = \dfrac{4}{5} \quad \blacksquare$$

The solution above could be condensed as follows:

$$2(p + 3) = -3p + 10$$
$$2p + 6 = -3p + 10 \qquad \text{(distributive property was used)}$$
$$5p + 6 = 10 \qquad \text{(3p was added to both sides of equation)}$$
$$5p = 4 \qquad \text{(6 was subtracted from both sides of equation)}$$
$$p = \dfrac{4}{5} \qquad \text{(both sides of equation were divided by 5)}$$

HELPFUL HINT

After the distributive property was used in Step 1, Example 3, we obtained $2p + 6 = -3p + 10$. At this point we had to decide whether to collect terms with the variable on the left or the right side of the equal sign. If we wish the sum of the variable terms to be positive, we use the addition property to eliminate the variable, with the *smaller* numerical coefficient from one side of the equation. Since -3 is smaller than 2, we added 3p to both sides of the equation. This eliminated the $-3p$ from the right side of the equation and resulted in the sum of the variable terms on the left side of the equation, 5p, being positive.

EXAMPLE 4   Solve the equation $2(x - 5) + 3 = 3x + 9$.

Solution: $2(x - 5) + 3 = 3x + 9$

Step 1 $2x - 10 + 3 = 3x + 9$ (distributive property was used)

Step 2 $2x - 7 = 3x + 9$ (like terms were combined)

Step 3 $-7 = x + 9$ (2x was subtracted from both sides of equation)

Step 3 $-16 = x$ (9 was subtracted from both sides of equation)   ∎

EXAMPLE 5   Solve the equation $7 - 2x + 5x = -2(-3x + 4)$.

Solution: $7 - 2x + 5x = -2(-3x + 4)$

Step 1 $7 - 2x + 5x = 6x - 8$ (distributive property was used)

Step 2 $7 + 3x = 6x - 8$ (like terms were combined)

Step 3 $7 = 3x - 8$ (3x was subtracted from both sides of equation)

Step 3 $15 = 3x$ (8 was added to both sides of equation)

Step 4 $5 = x$ (both sides of equation were divided by 3)

The solution is 5.   ∎

❷   Thus far all the equations we have solved have had a single value for a solution. Equations of this type are called **conditional equations,** for they are only true under specific conditions. Some equations, as in Example 6, are true for all values of x. Equations that are true for all values of x are called **identities.** A third type of equation, as in Example 7, has no solution.

EXAMPLE 6   Solve the equation $2x + 6 = 2(x + 3)$.

Solution: $2x + 6 = 2(x + 3)$
$2x + 6 = 2x + 6$

Since the same expression appears on both sides of the equal sign, the statement is true for all values of x. If we proceeded to solve this equation further, we might obtain

$2x = 2x$ (6 was subtracted from both sides of equation)
$0 = 0$ (2x was subtracted from both sides of equation)

Note: The solution process could have been terminated at $2x + 6 = 2x + 6$. Since one side is identical to the other side, the equation is true for all values of x. **Therefore, the answer to this problem is all real numbers.**   ∎

EXAMPLE 7   Solve the equation $-3x + 4 + 5x = 4x - 2x + 5$.

Solution:

$$-3x + 4 + 5x = 4x - 2x + 5$$

$$2x + 4 = 2x + 5 \qquad \text{like terms were combined}$$

$$2x - 2x + 4 = 2x - 2x + 5 \qquad \text{subtract } 2x \text{ from both sides of equation}$$

$$4 = 5 \qquad \text{false}$$

When solving an equation, if you obtain an obviously false statement as above, the equation has no solution. No value of x will make the equation above a true statement. **Therefore, the answer to this problem is no solution.**   ∎

COMMON STUDENT ERROR

At this point some students will begin to confuse combining like terms with using the addition property. Remember that *when combining terms, you work on only one side of the equal sign at a time,* as in

$$3x + 4 - x = 4x - 8$$

$$2x + 4 = 4x - 8 \qquad \text{(the } 3x \text{ and } -x \text{ were combined)}$$

When using the addition property, you add (or subtract) the same quantity to (from) **both sides of the equation.**

Correct

$$2x + 4 = 4x - 8$$

$$2x - 2x + 4 = 4x - 2x - 8 \qquad \text{(}2x \text{ was subtracted from both sides of equation)}$$

$$4 = 2x - 8$$

$$4 + 8 = 2x - 8 + 8 \qquad \text{(8 was added to both sides of equation)}$$

$$12 = 2x$$

$$x = 6$$

Wrong

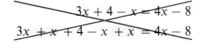

$$3x + 4 - x = 4x - 8$$

$$3x + x + 4 - x + x = 4x - 8 \qquad \text{(wrong use of addition property;}$$
note x was not added to *both* sides of equation)

Ordinarily when solving an equation, combining like terms is done before the use of the addition property.

Exercise Set 2.5

Solve each equation.

1. $2x + 4 = 3x$

2. $x + 5 = 2x - 1$

3. $-4x + 10 = 6x$

4. $6x = 4x + 8$

5. $5x + 3 = 6$

6. $-6x = 2x + 16$

7. $15 - 3x = 4x - 2x$
8. $9 - 12x = 4x + 15$
9. $x - 3 = 2x + 18$
10. $-5x = -4x + 9$
11. $3 - 2x = 9 - 8x$
12. $6x = 3x + 9 - 5x$
13. $3 + 8x = 6 - 2x$
14. $8 + y = 2y - 6 + y$
15. $5x = 2(x + 6)$
16. $8x - 4 = 3(x - 2)$
17. $x - 25 = 12x + 9 + 3x$
18. $5y + 6 = 2y + 3 - y$
19. $2(x + 2) = 4x + 1 - 2x$
20. $4x = 10 - 2(x - 4)$
21. $-(x + 2) = -6x + 32$
22. $15(4 - x) = 5(10 + 2x)$
23. $4 - (2x + 5) = 6x + 31$
24. $4(2x - 3) = -2(3x + 16)$
25. $26 - 10x + 3 = 4(-3x + 4)$
26. $3y - 6y + 2 = 8y + 6 - 5y$
27. $2(x + 4) = 4x + 3 - 2x + 5$
28. $25(y - 6) = 18(y + 10) - 2$
29. $9(-y + 3) = -6y + 15 - 3y + 12$
30. $-4(-y + 3) = 12y + 8 - 2y$
31. $-(3 - x) = -(2x + 3)$
32. $12 - 2x - 3(x + 2) = 4x + 6 - x$
33. $-(x + 4) + 5 = 4x + 1 - 5x$
34. $19x + 3(4x + 9) = -6x - 38$
35. $35(2x + 12) = 7(x - 4) + 3x$
36. $10(x - 10) + 5 = 5(2x - 20)$
37. $4x = 3(x + 6) - x$
38. $3(x - 4) = 2(x - 8) + 5x$
39. $-(x - 5) + 2 = 3(4 - x) + 5x$
40. $-2(3x + 6) + 12 = -6(4x + 1) + 10x$
41. $2(x - 6) + 3(x + 1) = 4x + 3$
42. $-2(-3x + 5) + 6 = 4(x - 2)$
43. $5 + 2x = 6(x + 1) - 5(x - 3)$
44. $4 - (6x + 3) = -(-2x + 3)$
45. $5 - (x - 5) = 2(x + 3) - 6(x + 1)$
46. $12 - 6x + 3(2x + 3) = 2x + 5$

47. When solving an equation, how will you know if the equation is an identity?

48. When solving an equation, how will you know if the equation has no real solution?

JUST FOR FUN

1. Solve the equation $-2(x + 3) + 5x = -3(5 - 2x) + 3(x + 2) + 6x$.
2. Solve the equation $4(2x - 3) - (x + 7) - 4x + 6 = 5(x - 2) - 3x + 7(2x + 2)$

2.6

Ratios and Proportions

☐ *Solve ratio problems.*

☐ *Solve proportions using cross-multiplication.*

☐ *Solve practical application problems.*

☐ A **ratio** is another name for a fraction. Ratios are often used to compare two or more numbers. The ratio of a to b is written a/b or $a:b$. The a and b are the **terms** of the ratio.

EXAMPLE 1   An algebra class consists of 11 males and 15 females.

(a) Find the ratio of males to females.
(b) Find the ratio of females to the entire class.

Solution: (a) $\dfrac{11}{15}$ or 11:15   (b) $\dfrac{15}{26}$ or 15:26   ∎

Many everyday problems involve ratios. For example, a client and lawyer may agree to split their settlement in a ratio of 2:1, respectively. Or an insect spray may have to be mixed with an 8:1 ratio of water to insecticide. To set up and solve a word problem where you are given specific ratios, use the following procedure.

To Solve Ratio Problems

1. Multiply each term of the ratio by x
2. Set the sum of each of the terms in (1) equal to the total amount.
3. Solve the resulting equation.
4. Answer the questions asked.

EXAMPLE 2 Mr. James left his estate to be divided between his wife, son, and favorite charity in the ratio 5:3:2, respectively. If Mr. James' estate was valued at $150,000, how much did each receive?

Solution: The ratio 5:3:2 means that for each $5 the wife receives, the son receives $3 and the favorite charity receives $2.

Step 1   Let $5x$ = amount wife receives

$3x$ = amount son receives

$2x$ = amount charity receives

$$\underset{\text{receives}}{\text{amount wife}} + \underset{\text{receives}}{\text{amount son}} + \underset{\text{receives}}{\text{amount charity}} = \text{total amount}$$

Step 2 $5x$ + $3x$ + $2x$ =   150,000

Step 3 $10x = 150,000$

$x = 15,000$

Step 4   The wife receives $5x$ or $5(15,000)$ = $\ 75,000$
The son receives $3x$ or $3(15,000)$ =   45,000
The charity receives $2x$ or $2(15,000)$ =   30,000
total = $150,000   ■

2 A **proportion** is a special type of equation. It is a statement of equality between two ratios. One way of denoting a proportion is $a:b = c:d$, which is read "a is to b as c is to d." In this text we write proportions as

$$\frac{a}{b} = \frac{c}{d}$$

The a and d are referred to as the **extremes** and the b and c are referred to as the **means** of the proportion. One method that can be used in evaluating proportions is **cross-multiplication:**

Cross-Multiplication

$$\text{If } \frac{a}{b} = \frac{c}{d} \quad \text{then} \quad ad = bc.$$

Note that the product of the means is equal to the product of the extremes.

If any three of the four quantities of a proportion are known, the fourth quantity can easily be found.

EXAMPLE 3 Solve for x by cross-multiplying $\dfrac{x}{3} = \dfrac{25}{15}$.

Solution:
$$\frac{x}{3} = \frac{25}{15}$$
$$x \cdot 15 = 3 \cdot 25$$
$$15x = 75$$
$$x = \frac{75}{15} = 5$$

Check:
$$\frac{x}{3} = \frac{25}{15}$$
$$\frac{5}{3} = \frac{25}{15}$$
$$\frac{5}{3} = \frac{5}{3} \quad \text{true} \quad \blacksquare$$

EXAMPLE 4 Solve for x by cross-multiplying $\dfrac{-8}{3} = \dfrac{64}{x}$.

Solution:
$$-\frac{8}{3} = \frac{64}{x}$$
$$-8x = 3 \cdot 64$$
$$-8x = 192$$
$$\frac{-8x}{-8} = \frac{192}{-8}$$
$$x = -24$$

Check:
$$\frac{-8}{3} = \frac{64}{x}$$
$$\frac{-8}{3} = \frac{\overset{8}{\cancel{64}}}{\underset{3}{\cancel{-24}}}$$
$$\frac{-8}{3} = \frac{8}{-3}$$
$$\frac{-8}{3} = \frac{-8}{3} \quad \text{true} \quad \blacksquare$$

3 Often, practical applications can be solved using proportions. To solve such problems using proportions, use the following procedure.

To Solve Problems Using Proportions

1. Represent the unknown quantity by a letter.
2. Set up the proportion by listing the given ratio on the left side of the equal sign, and the unknown and other given quantity on the right side of the equal sign. When setting up the right side of the proportion the same respective quantities should occupy the same respective positions on the left and right. For example, an acceptable proportion might be

$$\text{given ratio} \left\{ \frac{\text{miles}}{\text{hour}} = \frac{\text{miles}}{\text{hour}} \right.$$

3. Once the proportion is correctly written, drop the units and cross-multiply.
4. Solve the resulting equation.
5. Answer the questions asked.

Note that the ratios must have the same units. For example, if one ratio is given in miles/hour and the second ratio is given in feet/hour, one of the ratios must be changed before setting up the proportion.

EXAMPLE 5   A 30-pound bag of fertilizer will cover an area of 2500 square feet.

(a) How many pounds of fertilizer are needed to cover an area of 16,000 square feet?
(b) How many bags of fertilizer are needed?

Solution: (a) The given ratio is 30 pounds per 2500 square feet. The unknown quantity is the number of pounds necessary to cover 16,000 square feet.

Step 1   Let x = number of pounds

Step 2   given ratio $\begin{cases} \dfrac{30 \text{ pounds}}{2500 \text{ square feet}} = \dfrac{x \text{ pounds}}{16,000 \text{ square feet}} \end{cases}$ ← unknown
$\quad$ ← given quantity

Note how the pounds and the area are given in the same relative positions.

Step 3
$$\frac{30}{2500} = \frac{x}{16,000}$$
$$30(16,000) = 2500x$$

Step 4
$$480,000 = 2500x$$
$$\frac{480,000}{2500} = x$$
$$192 = x$$

Step 5   One hundred ninety-two pounds of fertilizer are needed.

(b) Since each bag weighs 30 pounds, the number of bags is found by division.

$$192 \div 30 = 6.4 \text{ bags}$$

The number of bags needed is therefore 7, since you can purchase only whole bags.   ■

EXAMPLE 6   In Washington County the tax rate is $80.65 per $1000 of assessed value. If a house and its property have been assessed at $12,400, find the tax the owner will have to pay.

Solution: The unknown quantity is the tax the property owner must pay. Let us call this unknown x.

$$\frac{\text{tax}}{\text{assessed value}} = \frac{\text{tax}}{\text{assessed value}}$$

given tax rate $\begin{cases} \dfrac{80.65}{1000} = \dfrac{x}{12,400} \end{cases}$

$$(80.65)(12,400) = 1000x$$
$$1,000,060 = 1000x$$
$$\$1000.06 = x$$

The owner will have to pay $1000.06 tax.   ■

COMMON STUDENT ERROR _____

> Note that when you plan the proportion, the same units should not be multiplied by themselves during cross-multiplication. *The following is wrong.*
>
> *Wrong*
>
> $$\frac{\text{miles}}{\text{hour}} \diagdown \frac{\text{hour}}{\text{miles}}$$

Exercise Set 2.6 _____

1. The results of a mathematics examination are 5 A's, 6 B's, 8 C's, 4 D's, and 2 F's.
   (a) Write the ratio of A's to C's.
   (b) Write the ratio of A's to total grades.
   (c) Write the ratio of D's to F's.
   (d) Write the ratio of grades better than C's to total grades.
   (e) Write the ratio of total grades to D's.

2. Two business partners agree to divide the profits from the business in the ratio 5:3. If the profits for the year are $120,000, how much will each partner receive?

3. A chain saw runs on a mixture of gas and oil in the ratio 15:1 respectively. How much of each ingredient is needed to make 4 gallons of this mixture?

4. Three partners share the purchase price of a piece of property in the ratio 3:2:2. If oil is found on their land and they lease their land for $140,000, how much should each of the three partners receive?

5. Brass is an alloy of copper and zinc in the ratio 3:2 respectively. How many tons of each are needed to make 33 tons of brass?

6. Mr. Carpenter left his estate to be divided among his wife, two children, and favorite charity in the ratio 5:2:2:1, respectively. If his estate is valued at $180,000, how much will each receive?

7. The angles of a triangle are in the ratio 4:3:2. How large is each angle if the sum of the angles of a triangle is 180 degrees?

8. A small car dealership employs three salespeople, *A*, *B*, and *C*. The yearly car sales were in the ratio 5:3:2, respectively. At the end of the year the owner offers a bonus of $4000 to be divided among the three salespeople on the basis of their sales ratios. How much should each receive?

9. In many legal agreements, the ratio of the amount the client receives to the amount the lawyer receives is 2:1 respectively. Find the amount each receives if the settlement is $48,000.

Solve for the variable by cross-multiplying.

10. $\dfrac{4}{x} = \dfrac{5}{20}$

11. $\dfrac{x}{4} = \dfrac{12}{48}$

12. $\dfrac{5}{3} = \dfrac{75}{x}$

13. $\dfrac{x}{32} = \dfrac{-5}{4}$

14. $\dfrac{90}{x} = \dfrac{-9}{10}$

15. $\dfrac{-3}{8} = \dfrac{x}{40}$

16. $\dfrac{1}{9} = \dfrac{x}{45}$

17. $\dfrac{y}{6} = \dfrac{7}{42}$

18. $\dfrac{3}{z} = \dfrac{2}{-20}$

19. $\dfrac{5}{12} = \dfrac{-40}{z}$

20. $\dfrac{15}{20} = \dfrac{x}{8}$

21. $\dfrac{12}{3} = \dfrac{x}{-100}$

Write a proportion that can be used to solve the problem. Solve the problem and find the desired value.

22. A car can travel 32 miles on 1 gallon of gasoline. How far can it travel on 12 gallons of gasoline?

23. A car can travel 23 miles on 1 gallon of gasoline. How far can it travel on 297 gallons?

24. If 100 feet of wire has an electrical resistance of 7.3 ohms, find the electrical resistance of 40 feet of wire.

25. The property tax in the town of Waverly is $27.445 per $1000 of assessed value. If the Litton's house is assessed at $40,600, how much property tax will they pay?

26. A quality control worker can check 12 units in 2.5 minutes. How long will it take her to check 60 units?

27. A blueprint of a shopping mall is in the scale of 1:150. Thus 1 foot on a blueprint represents 150 feet of actual length. One part of the mall is to be 190 feet long. How long will this part be on the blueprint?

28. A model railroad is made in the scale 1:87. A boxcar measures 12.2 meters. How large will the boxcar be in the model railroad set?

29. A photograph shows a boy is standing next to a tall cactus. If the boy, who is actually 48 inches tall, measures 0.6 inches in the photograph, how tall is the cactus that measures 3.25 inches in the photo?

See Exercise 26.

30. If a 40-pound bag of fertilizer covers 5000 square feet of area, how many pounds of fertilizer are needed to cover an area of 26,000 square feet?

31. The instructions on a bottle of liquid insecticide say "use 3 teaspoons of insecticide per gallon of water." If your sprayer has an 8-gallon capacity, how much insecticide should be used to fill the sprayer?

32. A recipe for McGillicutty Stew calls for $4\frac{1}{2}$ pounds of beef. If the recipe is for 20 servings, how much beef is needed to make 12 servings?

JUST FOR FUN

1. The recipe to make a deep-dish apple pie includes:

12 cups sliced apples	$\frac{1}{4}$ teaspoon salt
$\frac{1}{2}$ cup flour	2 tablespoons butter or margarine
1 teaspoon nutmeg	$1\frac{1}{2}$ cups sugar
1 teaspoon cinnamon	

Determine the amount of each of the other ingredients that should be used if only 8 cups of apples are available.

2. Insulin comes in 10-cubic centimeter (cc) vials labeled in the number of units of insulin per cubic centimeter. Thus a vial labeled U40 means there are 40 units of insulin per cubic centimeter of fluid. If a patient needs 25 units of insulin, how many cubic centimeters of fluid should be drawn up into a syringe from the U40 vial?

2.7

Inequalities in One Variable

❶ *Solve inequalities.*

❶ The greater-than symbol, $>$, and less-than symbol, $<$, were introduced in Section 1.3. The symbol $\geq$ means greater than or equal to and $\leq$ means less than or equal to. A mathematical expression containing one or more of these symbols is called an **inequality.** The direction of the symbol is sometimes called the **sense** of the inequality.

Examples of inequalities in one variable

$$x + 3 < 5, \qquad x + 4 \geq 2x - 6, \qquad 4 > -x + 3$$

To solve an inequality, we must get the variable by itself on one side of the inequality sign. To do this, we make use of properties very similar to those used to solve equations. Here are 4 properties used to solve inequalities. Later in this section we will introduce 2 additional properties.

Properties Used to Solve Inequalities

For real numbers a, b, and c:

1. If $a > b$, then $a + c > b + c$.

2. If $a > b$, then $a - c > b - c$.

3. If $a > b$ **and** $c > 0$, **then** $ac > bc$.

4. If $a > b$ **and** $c > 0$, **then** $\dfrac{a}{c} > \dfrac{b}{c}$.

Property 1 says the same number may be added to both sides of an inequality. Property 2 says the same number may be subtracted from both sides of an inequality. Property 3 says the same *positive* number may be used to multiply both sides of an inequality. Property 4 says the same *positive* number may be used to divide both sides of an inequality. When any of these four properties is used, the solution to the inequality does not change.

EXAMPLE 1   Solve the inequality $x - 4 > 7$, and graph the solution on the real number line.

Solution: To solve this inequality we need to isolate the variable, x. Therefore we must eliminate the -4 from the left side of the inequality. To do this we add 4 to both sides of the inequality.

$$x - 4 > 7$$
$$x - 4 + 4 > 7 + 4 \qquad \text{add 4 to both sides of inequality}$$
$$x > 11$$

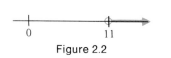

Figure 2.2

The solution is all real numbers greater than 11. We can illustrate the solution on the number line by placing an open circle at 11 on the number line and drawing an arrow to the right, see Fig. 2.2.

The open circle at the 11 indicates that the 11 is *not* part of the solution. The arrow going to the right indicates that all the values greater than 11 are solutions to the inequality. ■

EXAMPLE 2   Solve the inequality $2x + 6 \leq -2$, and graph the solution on the real number line.

Solution: To isolate the variable we must eliminate the $+6$ from the left side of the inequality. We do this by subtracting 6 from both sides of the inequality.

$$2x + 6 \leq -2$$
$$2x + 6 - 6 \leq -2 - 6 \qquad \text{subtract 6 from both sides of inequality}$$
$$2x \leq -8$$
$$\frac{2x}{2} \leq -\frac{8}{2} \qquad \text{divide both sides of equation by 2}$$
$$x \leq -4$$

Figure 2.3

The solution is all real numbers less than or equal to -4. We can illustrate the solution on the number line by placing a closed, or darkened, circle at -4 and drawing an arrow to the left, see Fig. 2.3.

The darkened circle at -4 indicates that -4 *is* a part of the solution. The arrow going to the left indicates that all the values less than -4 are also solutions to the inequality. ■

Notice that in Properties 3 and 4 we specified that $c > 0$. What happens when an inequality is multiplied or divided by a negative number? Examples 3 and 4 will illustrate this.

EXAMPLE 3   Multiply both sides of the inequality $8 > 4$ by -2.

Solution: $$8 > 4$$
$$-2 \cdot 8 < -2 \cdot 4$$
$$-16 < -8 \quad ■$$

EXAMPLE 4   Divide both sides of the inequality $8 > 4$ by -2.

Solution: $$8 > 4$$
$$\frac{8}{-2} < \frac{4}{-2}$$
$$-4 < -2 \quad ■$$

Examples 3 and 4 illustrate that **when an inequality is multiplied or divided by a negative number, the sense (direction) of the inequality changes.**

Additional Properties Used to Solve Inequalities

5. If $a > b$ and $c < 0$, then $ac < bc$.

6. If $a > b$ and $c < 0$, then $\dfrac{a}{c} < \dfrac{b}{c}$.

EXAMPLE 5   Solve the inequality $-2x > 6$, and graph the solution on the real number line.

Solution: To isolate the variable we must eliminate the -2 on the left side of the inequality. To do this we can divide both sides of the inequality by -2. When we do this, however, we must remember to change the sense of the inequality.

$$-2x > 6$$

$$\frac{-2x}{-2} < \frac{6}{-2} \qquad \text{divide both sides of inequality by } -2$$
$$\text{and change sense of inequality}$$

$$x < -3$$

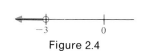

Figure 2.4

The solution is all real numbers less than 3. The solution is graphed on the number line in Fig. 2.4. ■

EXAMPLE 6   Solve the inequality $4 \geq -5 - x$, and graph the solution on the real number line.

Solution: *Method 1:*

$$4 \geq -5 - x$$

$$4 + 5 \geq -5 + 5 - x \qquad \text{add 5 to both sides of inequality}$$

$$9 \geq -x$$

$$-1(9) \leq -1(-x) \qquad \text{multiply both sides of inequality by } -1 \text{ and change}$$
$$\text{sense of inequality}$$

$$-9 \leq x$$

The inequality $-9 \leq x$ can also be written $x \geq -9$.

Method 2:

$$4 \geq -5 - x$$

$$4 + x \geq -5 - x + x \qquad \text{add } x \text{ to both sides of inequality}$$

$$4 + x \geq -5$$

$$4 - 4 + x \geq -5 - 4 \qquad \text{subtract 4 from both sides of inequality}$$

$$x \geq -9$$

Figure 2.5

The number line is shown in Fig. 2.5. Other methods could also be used to solve this problem. ■

Notice in Example 6, method 1, we wrote $-9 \leq x$ as $x \geq -9$. Although the solution $-9 \leq x$ is correct, it is customary to write the solution to an inequality with the variable on the left. One reason we write the variable on the left is that it often makes it easier to graph the solution on the number line. How would you graph $-3 > x$? How would you graph $-5 \leq x$? If you rewrite these inequalities with the variable on the left side, the answer becomes clearer.

$$-3 > x \qquad \text{means} \qquad x < -3$$

and

$$-5 \leq x \qquad \text{means} \qquad x \geq -5$$

Notice that you can change an answer from a greater than statement to a less than statement, or from a less than statement to a greater than statement. Remember that when you change the answer from one form to the other, the inequality symbol must point to the symbol to which it was pointing originally.

HELPFUL HINT

$a > x$ means $x < a$ (note that both inequality symbols point to x)

$a < x$ means $x > a$ (note that both inequality symbols point to a)

Examples: $-3 > x$ means $x < -3$

$-5 \leq x$ means $x \geq -5$

EXAMPLE 7   Solve the inequality $2x + 4 < -x + 12$, and graph the solution on the real number line.

Solution:
$$2x + 4 < -x + 12$$
$$2x + x + 4 < -x + x + 12$$
$$3x + 4 < 12$$
$$3x + 4 - 4 < 12 - 4$$
$$3x < 8$$
$$\frac{3x}{3} < \frac{8}{3}$$
$$x < \frac{8}{3}$$

Figure 2.6

The number line is shown in Fig. 2.6.   ■

EXAMPLE 8   Solve the inequality $-5x + 9 < -2x + 6$, and graph the solution on the real number line.

Solution:

$-5x + 9 < -2x + 6$	subtract 9 from both sides of inequality to obtain
$-5x < -2x - 3$	now add $2x$ to both sides of inequality
$-3x < -3$	now divide both sides of inequality by -3
$x > 1$	and change the sense of inequality

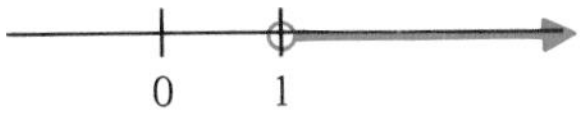

Figure 2.7

The number line is shown in Fig. 2.7.   ■

EXAMPLE 9   Solve the inequality $2(x + 3) \leq 5x - 3x + 8$, and graph the solution on the real number line.

Solution:

$2(x + 3) \leq 5x - 3x + 8$	use distributive property to obtain
$2x + 6 \leq 5x - 3x + 8$	now combine like terms
$2x + 6 \leq 2x + 8$	
$2x - 2x + 6 \leq 2x - 2x + 8$	subtract $2x$ from both sides of inequality
$6 \leq 8$	

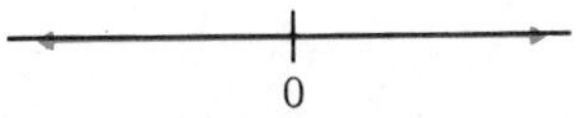

Figure 2.8

Since 6 is always less than or equal to 8, the solution is **"all real numbers"** (Fig. 2.8).
   ■

EXAMPLE 10   Solve the inequality $4(x + 1) > x + 5 + 3x$ and graph the solution on the real number line.

Solution:

$$4(x + 1) > x + 5 + 3x$$
$$4x + 4 > x + 5 + 3x \qquad \text{distributive property}$$
$$4x + 4 > 4x + 5$$
$$4x - 4x + 4 > 4x - 4x + 5 \qquad \text{subtract } 4x \text{ from both sides of inequality}$$
$$4 > 5$$

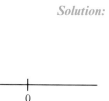

Figure 2.9

Since 4 is not greater than 5, the answer is "**no solution**" (Fig. 2.9).   ∎

Exercise Set 2.7

Solve each inequality, and graph the solution on the real number line.

1. $x + 3 > 7$

2. $x - 4 > -3$

3. $x + 5 \geq 3$

4. $4 - x \geq 3$

5. $-x + 3 < 8$

6. $4 < 3 + x$

7. $6 > x - 4$

8. $-4 \leq -x - 3$

9. $8 \leq 4 - x$

10. $2x < 4$

11. $-2x < 3$

12. $6 \geq -3x$

13. $2x + 3 \leq 5$

14. $-4x - 3 > 5$

15. $12x + 24 < -12$

16. $3x - 4 \leq 9$

17. $4 - 6x > -5$

18. $8 < 4 - 2x$

19. $15 > -9x + 50$

20. $3x - 4 < 5$

21. $4 < 3x + 12$

22. $-4x > 2x + 12$

23. $6x + 2 \leq 3x - 9$

24. $-2x - 4 \leq -5x + 12$

25. $x - 4 \leq 3x + 8$

26. $-3x - 5 \geq 4x - 29$

27. $-x + 4 < -3x + 6$

28. $2(x - 3) < 4x + 10$

29. $-3(2x - 4) > 2(6x - 12)$

30. $-(x + 3) \leq 4x + 5$

31. $x + 3 < x + 4$

32. $x + 5 \geq x - 2$

33. $6(3 - x) < 2x + 12$

34. $2(3 - x) + 4x < -6$

35. $-21(2 - x) + 3x > 4x + 4$

36. $-(x + 3) \geq 2x + 6$

37. $4x - 4 < 4(x - 5)$

38. $-2(-5 - x) > 3(x + 2) + 4 - x$

39. $5(2x + 3) \geq 6 + (x + 2) - 2x$

40. $-3(-2x + 12) < -4(x + 2) - 6$

41. When solving an inequality, under what conditions will it be necessary to change the sense of the inequality?

42. When solving an inequality, if you obtain the results $3 < 5$, what is the solution?

43. When solving an inequality, if you obtain the result $4 \geq 2$, what is the solution?

JUST FOR FUN

1. Solve the inequality

$$3(2 - x) - 4(2x - 3) \le 6 + 2x - 6(x - 5) + 2x$$

2. Solve the inequality

$$-(x + 4) + 6x - 5 > -4(x + 3) + 2(x + 6) - 5x$$

Summary

Glossary

Algebraic expression: A collection of numbers, variables, grouping symbols, and operation symbols.

Coefficient or numerical coefficient: The numerical part of a term.

Equation: Two algebraic expressions joined by an equal sign.

Identity: An equation that is true for all values of the variable.

Inequality: A mathematical statement containing one or more inequality symbols ($>$, $\ge$, $<$, $\le$).

Like terms: Terms that have the same variables with the same exponents.

Linear equation in one variable: Equation of the form $ax + b = c$, $a \ne 0$.

Proportion: A statement of equality between two ratios.

Reciprocal of a: $\dfrac{1}{a}$, $a \ne 0$

Term: The parts that are added in an algebraic expression.

Variable: A letter used to represent a number.

Important Facts

Distributive property: $a(b + c) = ab + ac$

Addition property: If $a = b$, then $a + c = b + c$.

Multiplication property: If $a = b$, then $a \cdot c = b \cdot c$.

Cross-multiplication: If $\dfrac{a}{b} = \dfrac{c}{d}$, then $ad = bc$.

Properties used to solve inequalities

1. If $a > b$, then $a + c > b + c$.

2. If $a > b$, then $a - c > b - c$.

3. If $a > b$ and $c > 0$, then $ac > bc$.

4. If $a > b$ and $c > 0$, then $\dfrac{a}{c} > \dfrac{b}{c}$.

5. If $a > b$ and $c < 0$, then $ac < bc$.

6. If $a > b$ and $c < 0$, then $\dfrac{a}{c} < \dfrac{b}{c}$.

Review Exercises

[2.1] Use the distributive property to remove parentheses.

1. $2(x + 4)$

2. $3(x - 2)$

3. $2(4x - 3)$

4. $-2(x + 4)$

5. $-(x + 2)$

6. $-(x - 2)$

7. $-4(4 - x)$

8. $3(6 - 2x)$

9. $4(5x - 6)$

10. $-3(2x - 5)$

11. $6(6x - 6)$

12. $4(-x + 3)$

13. $-3(x + y)$

14. $-2(3x - 2)$

15. $-(3 + 2y)$

16. $-(x + 2y - z)$

17. $3(x + 3y - 2z)$

18. $-2(2x - 3y + 7)$

Simplify where possible.

19. $2x + 3x$

20. $4y + 3y + 2$

21. $4 - 2y + 3$

22. $1 + 3x + 2x$

23. $6x + 2y + y$

24. $-2x - x + 3y$

25. $2x + 3y + 4x + 5y$

26. $6x + 3y + 2$

27. $2x - 3x - 1$

28. $5x - 2x + 3y + 6$

29. $x + 8x - 9x + 3$

30. $-4x - 8x + 3$

31. $3(x + 2) + 2x$

32. $-2(x + 3) + 6$

33. $2x + 3(x + 4) - 5$

34. $4(3 - 2x) - 2x$

35. $6 - (-x + 3) + 4x$

36. $2(2x + 5) - 10 - 4$

37. $-6(4 - 3x) - 18 + 4x$

38. $6 - 3(x + y) + 6x$

39. $3(x + y) - 2(2x - y)$

40. $3x - 6y + 2(4y + 8)$

41. $3 - (x - y) + (x - y)$

42. $(x + y) - (2x + 3y) + 4$

[2.2–2.5] Solve each equation.

43. $2x = 4$

44. $x + 3 = -5$

45. $x - 4 = 7$

46. $\dfrac{x}{3} = -9$

47. $2x + 4 = 8$

48. $14 = 3 + 2x$

49. $8x - 3 = -19$

50. $6 - x = 9$

51. $-x = -12$

52. $2(x + 2) = 6$

53. $-3(2x - 8) = -12$

54. $4(6 + 2x) = 0$

55. $3x + 2x + 6 = -15$

56. $4 = -2(x + 3)$

57. $27 = 46 + 2x - x$

58. $4x + 6 - 7x + 9 = 18$

59. $4 + 3(x + 2) = 12$

60. $-3 + 3x = -2(x + 1)$

61. $3x - 6 = -5x + 30$

62. $-(x + 2) = 2(3x - 6)$

63. $2x + 6 = 3x + 9$

64. $-5x + 3 = 2x + 10$

65. $3x - 12x = 24 - 6x$

66. $2(x + 4) = -3(x + 5)$

67. $4(2x - 3) + 4 = 9x + 2$

68. $6x + 11 = -(6x + 5)$

69. $2(x + 7) = 6x + 9 - 4x$

70. $-6(4 - 2x) = -11 + 12x - 13$

71. $4(x - 3) - (x + 5) = 0$

72. $-2(4 - x) = 6(x + 2) + 3x$

73. $-3(2x - 5) + 5x = 4x - 7$

[2.6] Solve each proportion.

74. $\dfrac{x}{9} = \dfrac{6}{18}$

75. $\dfrac{15}{100} = \dfrac{x}{20}$

76. $\dfrac{3}{x} = \dfrac{15}{45}$

77. $\dfrac{20}{45} = \dfrac{15}{x}$

78. $\dfrac{6}{2} = \dfrac{-12}{x}$

79. $\dfrac{x}{9} = \dfrac{8}{-3}$

80. $\dfrac{25}{x} = \dfrac{-5}{1}$

81. $\dfrac{200}{4} = \dfrac{x}{12}$

82. $\dfrac{-4}{9} = \dfrac{-16}{x}$

83. $\dfrac{x}{-15} = \dfrac{30}{-5}$

[2.7] Solve each inequality, and graph the solution on the real number line.

84. $2x + 4 \geq 8$

85. $6 - 2x > 4x - 12$

86. $6 - 3x \leq 2x + 18$

87. $2(x + 4) \leq 2x - 5$

88. $2(x + 3) > 6x - 4x + 4$

89. $x + 6 > 9x + 30$

90. $x - 2 \leq -4x + 7$

91. $-(x + 2) < -2(-2x + 5)$

92. $2(x + 3) < -(x + 3) + 4$

93. $-6x - 3 \geq 2(x - 4) + 3x$

94. $-2(x - 4) \leq 3x + 6 - 5x$

95. $2(2x + 4) > 4(x + 2) - 6$

96. $4(x - 2) + 3 \leq 2(x + 3)$

[2.6] Set up an equation that can be used to solve each problem. Solve the equation, and find the value desired.

97. Mr. St. James left his estate to be divided between his wife and child in the ratio 3:2, respectively. If his estate is valued at $55,000, how much does each receive?

98. Two business partners, Alice and Celia, agree to share the profits in the ratio 4:3. If the profits for the year are $63,000, how much does each receive?

99. The New York State Conservation Department agrees to cut any unwanted dead trees from Mrs. Pollinger's land. The Conservation Department makes arrangements such that the cut firewood is divided up in the ratio 4:1:1. Mrs. Pollinger receives 4 shares; the Conservation Department, 1 share; and the individual contracted to do the cutting, 1 share. If a total of 96 face cords of wood are cut, how much will each receive?

Set up an equation that can be used to solve each problem. Solve the equation, and find the value desired.

100. If a car traveling at a specified speed can travel 45 miles in 60 minutes, how many miles can it travel in 90 minutes?

101. If a water pump can remove 8 gallons of water in a minute, how long will it take the pump to empty a basement filled with 126 gallons of water?

102. If the scale of a map is 1 inch to 60 miles, what distance on the map represents 380 miles?

103. Bryce builds a model car to a scale of 1 inch to 0.9 feet. If the completed model is 10.5 inches, what is the size of the actual car?

Practice Test

Use the distributive property to remove parentheses.

1. $-2(4 - 2x)$

2. $-(x + 3y - 4)$

Simplify where possible.

3. $3x - x + 4$

4. $4 + 2x - 3x + 6$

5. $y - 2x - 4x - 6$

6. $x - 4y + 6x - y + 3$

7. $2x + 3 + 2(3x - 2)$

Solve each equation.

8. $2x + 4 = 12$

9. $-x - 3x + 4 = 12$

10. $4x - 2 = x + 4$

11. $3(x - 2) = -(5 - 4x)$

12. $2x - 3(-2x + 4) = -13 + x$

13. $3x - 4 - x = 2(x + 5)$

14. $-3(2x + 3) = -2(3x + 1) - 7$

15. Solve the proportion $\dfrac{9}{x} = \dfrac{3}{-15}$

Solve each inequality, and graph the solution on the real number line.

16. $2x - 4 < 4x + 10$

17. $3(x + 4) \geq 5x - 12$

18. $4(x + 3) + 2x < 6x - 3$

19. Two business partners decide to divide their company's profits in the ratio 2:1. If the profit for the year is $96,000, how much will each receive?

20. If 6 gallons of insecticide can treat 3 acres of land, how many gallons of insecticide are needed to treat 75 acres?

3 Formulas and Applications of Algebra

See Review Exercise 48.

Formulas

1. Evaluate formulas.
2. Solve for a variable in a formula.

---

1 A **formula** is an equation commonly used to express a specific physical concept mathematically. For example, the formula for the area of a rectangle is

$$\text{area} = \text{length} \cdot \text{width} \qquad \text{or} \qquad A = lw$$

To evaluate a formula, substitute the appropriate numerical values for the variables and perform the indicated operations.

The **perimeter,** P, is the sum of the lengths of the sides of a figure. Perimeters are measured in the same common unit as the sides. The **area,** A, is the total surface within the figure's boundaries. Areas are measured in square units. Table 3.1 gives the formulas for finding the areas and perimeters of some common shapes.

Table 3.1 FORMULAS FOR AREAS AND PERIMETERS

Figure	Sketch	Area	Perimeter
Square		$A = s^2$	$P = 4s$
Rectangle		$A = lw$	$P = 2l + 2w$
Parallelogram		$A = lh$	$P = 2l + 2w$
Trapezoid		$A = \frac{1}{2}h(b + d)$	$P = a + b + c + d$
Triangle		$A = \frac{1}{2}bh$	$P = a + b + c$

EXAMPLE 1 Find the perimeter of a rectangle when $l = 6$ feet and $w = 2$ feet.

Solution: Substitute 6 for l and 2 for w in the formula for the perimeter of a rectangle.

$$P = 2l + 2w$$
$$= 2(6) + 2(2)$$
$$= 12 + 4 = 16 \text{ feet} \quad \blacksquare$$

EXAMPLE 2   Find the length of a rectangle when the perimeter is 22 inches and the width is 3 inches.

Solution: Substitute 22 for P and 3 for w in the formula $P = 2l + 2w$, then solve for the length, l.

$$P = 2l + 2w$$
$$22 = 2l + 2(3)$$
$$22 = 2l + 6$$
$$22 - 6 = 2l + 6 - 6$$
$$16 = 2l$$
$$\frac{16}{2} = \frac{2l}{2}$$
$$8 = l$$

The length is 8 inches.   ■

EXAMPLE 3   Find the height of a triangle if it has an area of 30 square feet and a base of 12 feet.

Solution: $A = \dfrac{1}{2}bh$

$$30 = \frac{1}{2}(12)h$$
$$30 = 6h$$
$$\frac{30}{6} = \frac{6h}{6}$$
$$5 = h$$

The height of the triangle is 5 feet.   ■

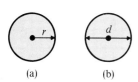

(a)   (b)

Figure 3.1

Another figure that we see and use daily is the circle.

The **circumference,** C, is the length (or perimeter) of the curve that forms a circle. The **radius,** r, is the line segment from the center of the circle to any point on the circle (see Fig. 3.1a). A **diameter** of a circle is a line segment through the center with both end points on the circle (see Fig. 3.1b). Note that the diameter is twice the radius.

The formulas for the area and circumference of a circle are given in Table 3.2.

Table 3.2

Circle	Area	Circumference
	$A = \pi r^2$	$C = 2\pi r$

Pi, symbolized by the Greek lowercase letter π, has a value of *approximately* 3.14.

EXAMPLE 4 Determine the area and circumference of a circle with a diameter of 16 inches.

Solution: The radius is half the diameter, $r = \frac{16}{2} = 8$ inches.

$$A = \pi r^2 \qquad\qquad\qquad C = 2\pi r$$
$$A = 3.14(8)^2 \qquad\qquad C = 2(3.14)(8)$$
$$A = 3.14(64) \qquad\qquad C = 50.24 \text{ inches}$$
$$A = 200.96 \text{ square inches} \qquad\qquad \blacksquare$$

Now let us look at some other uses of formulas.

EXAMPLE 5 The formula to find *simple interest* (i) in dollars is $i = prt$, where p is the principal or amount invested, r is the interest rate in decimal form, and t is the amount of time of the investment. Find i when $p = \$2000$, $r = 12\%$, and $t = 3$ years.

Solution: Change 12% to 0.12 and then substitute the given values into the formula.

$$i = prt$$
$$i = 2000(0.12)3$$
$$i = 720.$$

The simple interest is $720.   ■

EXAMPLE 6 The number of diagonals, d, in a polygon of n sides is given by the formula $d = \frac{1}{2}n^2 - \frac{3}{2}n$.
(a) How many diagonals has a quadrilateral (four sides)?
(b) How many diagonals has an octagon (8 sides)?

Solution: (a) $n = 4$,   *Check:* for (a)

$$d = \frac{1}{2}(4)^2 - \frac{3}{2}(4)$$

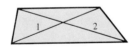

$$= \frac{1}{2}(16) - 6$$

$$= 8 - 6 = 2$$

(b) $n = 8$,

$$d = \frac{1}{2}(8)^2 - \frac{3}{2}(8)$$

$$= \frac{1}{2}(64) - 12$$

$$= 32 - 12 = 20 \quad ■$$

EXAMPLE 7 The average, A, of 3 test grades a, b, and c is found by the formula $A = \dfrac{a + b + c}{3}$.

If Juan's first two test grades are 74 and 88, find the grade that he must get on the third test to obtain an average of 80.

Solution: We are given test grades a and b and the desired average, A. We are asked to find test grade c.

$$A = \frac{a + b + c}{3}$$

$$80 = \frac{74 + 88 + c}{3} \qquad \text{add the two test grades}$$

$$80 = \frac{162 + c}{3}$$

$$3(80) = \cancel{3}\left(\frac{162 + c}{\cancel{3}}\right) \qquad \text{multiply both sides of equation by 3}$$

$$240 = 162 + c$$

$$240 - 162 = 162 - 162 + c$$

$$78 = c$$

If Juan obtains a grade of 78 on his third test he will have an 80 average.   ■

Solving for a Variable in a Formula

2   Often in a science, mathematics, or other course you will be given an equation or formula solved for one variable and be asked to solve it for a different variable. To do this, treat each of the quantities, except the one you are solving for, as if they were constants. Then solve for the desired variable by isolating it on one side of the equation, using the properties discussed previously.

EXAMPLE 8   Solve the equation $A = lw$ for w.

Solution: We must get l all by itself on one side of the equation. We begin by removing the $2w$ from the right side of the equation to isolate the term containing the l.

$$A = lw$$

$$\frac{A}{l} = \frac{\cancel{l}w}{\cancel{l}} \qquad \text{divide both sides of equation by } l$$

$$\frac{A}{l} = w \qquad ■$$

EXAMPLE 9   Solve the equation $P = 2l + 2w$ for l.

Solution: We must get l all by itself on one side of the equation. We begin by removing the $2w$ from the right side of the equation to isolate the term containing the l.

$$P = 2l + 2w$$

$$P - 2w = 2l + 2w - 2w \qquad \text{subtract } 2w \text{ from both sides of equation}$$

$$P - 2w = 2l$$

$$\frac{P - 2w}{2} = \frac{\cancel{2}l}{\cancel{2}} \qquad \text{divide both sides of equation by 2}$$

$$\frac{P - 2w}{2} = l \qquad \left(\text{or } l = \frac{P}{2} - w\right) \qquad ■$$

EXAMPLE 10   An equation we will use in Chapter 7 is $y = mx + b$. Solve for m.

Solution: We must get the m all by itself on one side of the equal sign.

$$y = mx + b$$
$$y - b = mx + b - b \qquad \text{subtract } b \text{ from both sides of equation}$$
$$y - b = mx$$
$$\frac{y - b}{x} = \frac{m\cancel{x}}{\cancel{x}} \qquad \text{divide both sides of equation by } x$$
$$\frac{y - b}{x} = m \qquad \left(\text{or } m = \frac{y}{x} - \frac{b}{x} \right) \qquad ■$$

EXAMPLE 11   Solve the equation $2x + 3y = 12$ for y, then find y when $x = 6$.

Solution:
$$2x + 3y = 12$$
$$2x - 2x + 3y = 12 - 2x \qquad \text{subtract } 2x \text{ from both sides of equation}$$
$$3y = 12 - 2x$$
$$\frac{\cancel{3}y}{\cancel{3}} = \frac{12 - 2x}{3} \qquad \text{divide both sides of equation by } 3$$
$$y = \frac{12 - 2x}{3}$$

Now we substitute $x = 6$ and determine the value of y.

$$y = \frac{12 - 2x}{3}$$
$$y = \frac{12 - 2(6)}{3}$$
$$y = \frac{12 - 12}{3}$$
$$y = \frac{0}{3}$$
$$y = 0$$

We see that when $x = 6$, $y = 0$.   ■

EXAMPLE 12   Solve the formula $i = prt$ for p.

Solution: We must isolate the p. Since p is multiplied by both r and t we divide both sides of the equation by rt.

$$i = prt$$
$$\frac{i}{rt} = \frac{p\cancel{r}\cancel{t}}{\cancel{r}\cancel{t}}$$
$$\frac{i}{rt} = p. \qquad ■$$

Some formulas contain fractions. When a formula contains a fraction we can eliminate the fraction by multiplying both sides of the equation by the denominator as illustrated in Example 13.

EXAMPLE 13   Solve the formula $A = \frac{1}{2}h(b + d)$ for b.

Solution:

$$A = \frac{1}{2}h(b + d)$$

$$2A = 2\left(\frac{1}{2}\right)h(b + d) \qquad \text{multiply both sides of equation by 2}$$

$$2A = h(b + d)$$

$$\frac{2A}{h} = \frac{h(b + d)}{h} \qquad \text{divide both sides of equation by } h$$

$$\frac{2A}{h} = b + d$$

$$\frac{2A}{h} - d = b + d - d \qquad \text{subtract } d \text{ from both sides of equation}$$

$$\frac{2A}{h} - d = b \qquad\qquad \left(\text{or } b = \frac{2A - hd}{h}\right) \quad \blacksquare$$

Exercise Set 3.1

Use the formula to find the value of the indicated variable for the values given.

1. $A = s^2$, find A when $s = 4$.
2. $A = 2l + 2w$, find A when $l = 8$ and $w = 5$.
3. $A = \frac{1}{2}bh$, find A when $b = 12$ and $h = 8$.
4. $P = \frac{1}{2}h(b + d)$, find A when $h = 6$, $b = 18$ and $d = 24$.
5. $P = a + b + c$, find P when $a = 4$, $b = 3$ and $c = 5$.
6. $A = \pi r^2$, find A when $r = 6$ and $\pi = 3.14$.
7. $C = 2\pi r$, find C when $\pi = 3.14$ and $r = 2$.
8. $P = 2l + 2w$, find w when $P = 28$ and $l = 12$.
9. $A = \frac{1}{2}bh$, find h when $A = 20$ and $b = 4$.
10. $V = \frac{1}{3}Bh$, find h when $V = 40$ and $B = 12$.
11. $i = prt$, find p when $i = 120$, $r = 0.12$ and $t = 1$.
12. $i = prt$, find t when $i = 480$, $p = 2000$ and $r = 0.08$.
13. $d = rt$, find r when $d = 1500$ and $t = 150$.
14. $V = lwh$, find l when $V = 18$, $w = 1$ and $h = 3$.
15. $V = lwh$, find h when $V = 80$, $l = 20$ and $w = 2$.

16. $T = \frac{RS}{R + S}$, find T when $R = 50$ and $S = 50$.
17. $A = P(1 + rt)$, find A when $P = 1000$, $r = 0.08$ and $t = 1$.
18. $A = P(1 + rt)$, find r when $A = 1500$, $t = 1$ and $P = 1000$.
19. $A = \pi r^2$, find A when $r = 4$ and $\pi = 3.14$.
20. $M = \frac{a + b}{2}$, find b when $M = 36$ and $a = 16$.
21. $M = \frac{a + b}{2}$, find a when $M = 30$ and $b = 6$.
22. $C = \frac{5}{9}(F - 32)$, find C when $F = 41$.
23. $C = \frac{5}{9}(F - 32)$, find C when $F = 59$.
24. $F = \frac{9}{5}C + 32$, find F when $C = 10$.
25. $F = \frac{9}{5}C + 32$, find F when $C = 20$.

26. $z = \dfrac{x - m}{s}$, find z when $x = 115$, $m = 100$ and $s = 15$.

27. $z = \dfrac{x - m}{s}$, find x when $z = 2$, $m = 50$ and $s = 5$.

28. $z = \dfrac{x - m}{s}$, find s when $z = 3$, $x = 80$ and $m = 59$.

Solve each equation for y, then find the value of y for the given value of x.

29. $2x + y = 8$, when $x = 2$.
30. $6x + 2y = -12$, when $x = -3$.
31. $2x = 6y - 4$, when $x = 10$.
32. $-3x - 5y = -10$, when $x = 0$.
33. $2y = 6 - 3x$ when $x = 2$.
34. $15 = 3y - x$ when $x = 3$.

35. $-4x + 5y = -20$ when $x = 4$.
36. $3x - 2y = -18$ when $x = -1$.
37. $-3x = 18 - 6y$ when $x = 0$.
38. $-12 = -2x - 3y$ when $x = -2$.
39. $-8 = -x - 2y$ when $x = -4$.
40. $2x + 5y = 20$ when $x = -5$.

Solve for the variable indicated.

41. $d = rt$, for t
42. $d = rt$ for r
43. $i = prt$, for p
44. $i = prt$, for r
45. $C = \pi d$, for d
46. $v = lwh$, for w
47. $A = \dfrac{1}{2} bh$, for b
48. $E = IR$, for I
49. $P = 2l + 2w$, for w
50. $PV = KT$, for T
51. $4n + 3 = m$, for n
52. $3t - 4r = 25$, for t
53. $y = mx + b$, for b
54. $y = mx + b$, for x
55. $I = P + Prt$, for r
56. $A = \dfrac{m + d}{2}$, for m

57. $A = \dfrac{m + 2d}{3}$, for d
58. $R = \dfrac{l + 3w}{2}$, for w
59. $d = a + b + c$, for b
60. $A = \dfrac{a + b + c}{3}$, for b
61. $I = \dfrac{E + L}{2}$, for E
62. $W = 3L + 2D$, for D
63. $ax + by = c$, for y
64. $ax + by + c = 0$, for y
65. $C = \dfrac{5}{9}(F - 32)$, for F
66. $F = \dfrac{9}{5}C + 32$, for C

67. Mr. Wicker borrowed $4000 for 3 years at 12% simple interest. How much interest did he pay?
68. Ms. Bianco lent her brother $4000 for a period of 2 years. At the end of the 2 years her brother repaid the $4000 plus $640 interest. What interest rate did her brother pay?

In Exercises 69 and 70 use the formula in Example 6 to find the number of diagonals in a figure with the given number of sides.

69. 10 sides

70. 6 sides

In Exercises 71 and 72 use the formula given in Example 7.

71. If the first two test grades are 75 and 77, what third grade is needed to obtain an average of 80?

72. If the first two test grades are 83 and 97, what third grade is needed to obtain an average of 90?

In Exercises 73 and 74 use the formula $C = \frac{5}{9}(F - 32)$ *to find the Celsius temperature (C) for the given Fahrenheit temperature (F).*

73. $50°F$ **74.** $86°F$

In Exercises 75 and 76 use the formula $F = \frac{9}{5}C + 32$ *to find the Fahrenheit temperature (F) for the given Celsius temperature (C).*

75. $35°C$ **76.** $10°C$

A formula in the study of chemistry is $P = \dfrac{KT}{V}$, *where P is pressure, T is temperature, V is volume, and K is a constant. In Exercises 77 through 80 find the missing quantity.*

77. $T = 10$, $K = 1$, $V = 1$ **78.** $T = 30$, $P = 3$, $K = 0.5$
79. $P = 80$, $T = 100$, $V = 5$ **80.** $P = 100$, $K = 2$, $V = 6$

The sum of the first n *even numbers can be found by the formula* $S = n^2 + n$. *In Exercises 81 through 83 find the sum of the numbers indicated.*

81. First 3 even numbers.   **82.** First 5 even numbers.   **83.** First 10 even numbers.

JUST FOR FUN

1. (a) Using the formulas presented in this section, write an equation in d that can be used to find the shaded area in the figure shown.

(b) Find the shaded area when $d = 4$ feet.
(c) Find the shaded area when $d = 6$ feet.

2. A cereal box is to be made by folding the cardboard along the dashed lines as shown in the figure.

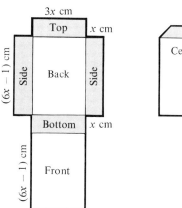
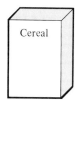

(a) Using the formula, volume = length · width · height, write an equation for the volume of the box.
(b) Find the volume of the box when $x = 7$ cm.

    (c) Write an equation for the surface area of the box.

    (d) Find the surface area when $x = 7$ cm.

3. The total pressure, P, in pounds per square inch, exerted on an object x feet be-low sea level is given by the formula $P = 14.70 + 0.43x$. As shown in the accom-panying diagram, the 14.70 represents the weight in pounds of the column of air (from sea level to the top of the atmosphere) standing over a 1-inch by 1-inch square of seawater. The $0.43x$ represents the weight, in pounds, of a column of water 1 inch by 1 inch by x feet.

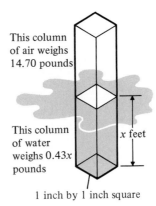

This column of air weighs 14.70 pounds

This column of water weighs $0.43x$ pounds

x feet

1 inch by 1 inch square

    (a) A submarine can withstand a total pressure of 162 pounds per square inch. How deep can the submarine go?

    (b) If the pressure gauge in the submarine registers a total pressure of 97.26 pounds per square inch, how deep is the submarine?

3.2

Changing Word Problems into Equations

  1 *Translate phrases into mathematical expressions.*

  2 *Translate word problems into equations.*

---

  1 One practical advantage of knowing algebra is that you can use it to solve everyday problems involving mathematics. For algebra to be useful to you in solving everyday problems, you must first be able to transform the verbal, or word, problem into mathematical language. The purpose of this section is to help you take a verbal problem and write it as a mathematical expression.

    Often the most difficult part of solving a word problem is translating it into an equation. Here are examples of statements represented as algebraic expressions.

Verbal	Algebraic
5 more than a number	$x + 5$
a number increased by 3	$x + 3$
7 less than a number	$x - 7$
a number decreased by 12	$x - 12$

Verbal	Algebraic
twice a number	$2x$
the product of 6 and a number	$6x$
one-eighth of a number	$\frac{1}{8}x$ or $\frac{x}{8}$
a number divided by 3	$\frac{1}{3}x$ or $\frac{x}{3}$
4 more than twice a number	$2x + 4$
5 less than three times a number	$3x - 5$
3 times the sum of a number and 8	$3(x + 8)$

Sometimes in a problem, two numbers are related to each other in a certain way. We often represent the simplest, or most basic number that needs to be expressed, as a variable, and the other as an expression containing that variable. Some examples are illustrated below.

Verbal	One number	Second number
two numbers differ by 3	x	$x + 3$
John's age now and John's age in 6 years	x	$x + 6$
one number is six times the other number	x	$6x$
two consecutive integers	x	$x + 1$
two consecutive odd (or even) integers	x	$x + 2$
the sum of two numbers is 10	x	$10 - x$
a 25-foot tree cut in two pieces	x	$25 - x$
one number is 12% less than the other	x	$x - 0.12x$

Note that often there are more than one pair of expressions that can be used to represent the two numbers. For example, two numbers differ by 3 can also be expressed as x and $x - 3$.

EXAMPLE 1   Express each phrase as an algebraic expression.
(a) The distance, d, increased by 10 miles.
(b) 6 less than twice the area.
(c) 3 pounds more than four times the weight.
(d) Twice the sum of the height plus 3 feet.
(e) The cost increased by 6%.

Solution: (a) $d + 10$
(b) $2a - 6$
(c) $4w + 3$
(d) $2(h + 3)$
(e) $c + 0.06c$; note that 6% is written as a decimal. If the cost is c, 6% of the cost is 0.06c.   ■

In Example 1, the letter x (or any other letter) could have been used in place of those selected.

EXAMPLE 2   For each relationship, select a variable to represent one quantity and express the other quantity in terms of the first.

(a) A boy is 15 years older than his brother.
(b) The speed of the second car is 1.4 times the speed of the first.
(c) $75 is divided between two people.
(d) John has $5 more than three times the amount of money that Dee has.
(e) The length of a rectangle is 3 units less than four times its width.

Solution: (a) x, $x + 15$
(b) x, $1.4x$·
(c) x, $75 - x$
(d) x, $3x + 5$
(e) x, $4x - 3$ ■

EXAMPLE 3   Write each of the following as an algebraic expression.

(a) The cost of purchasing x items at $2 each.
(b) Five percent commission on x dollars in sales.
(c) The number of calories in x potato chips, where each potato chip has 8 calories.
(d) The increase in population in n years for a city growing at a rate of 300 per year.
(e) The distance traveled in t hours when 55 miles are traveled each hour.

Solution: (a) We can reason like this: one item would cost 1(2) dollars, two items would cost 2(2) dollars, three items 3(2), four items 4(2), and so on. Continuing this reasoning process, we can see that x items would cost $x(2)$ or $2x$ dollars.

(b) A 5% commission on $1 sales would be 0.05(1), on $2 sales 0.05(2), on $3 sales 0.05(3), on $4 sales 0.05(4), and so on. Therefore, the commission on x dollar sales would be $0.05(x)$ or $0.05x$.

(c) $8x$
(d) $300n$
(e) $55t$ ■

EXAMPLE 4   A slice of white bread contains 65 calories and a slice of whole-wheat bread contains 55 calories. Write an algebraic expression to represent the total number of calories in x slices of white and y slices of whole-wheat bread.

Solution: x slices of white bread contain $65x$ calories.

y slices of whole-wheat bread contain $55y$ calories.

Together they contain $65x + 55y$ calories.   ■

EXAMPLE 5   Write an algebraic expression for each phrase.

(a) The number of ounces in x pounds.
(b) The number of cents in a dimes and b nickels.
(c) The number of seconds in x hours, y minutes, and z seconds.

Solution: (a) $16x$ (16 ounces = 1 pound)
(b) $10a + 5b$
(c) $3600x + 60y + z$ (3600 seconds = 1 hour)   ■

2   The word *is* in a verbal problem means *is equal to* and is represented by an equal sign. Below are some examples of verbal problems written as equations.

Verbal Problem	Equation
6 more than twice a number *is* 4	$2x + 6 = 4$
a number decreased by 4 *is* 3 more than twice the number	$x - 4 = 2x + 3$
the product of two consecutive integers *is* 56	$x(x + 1) = 56$
one number is 4 more than three times the other number; their sum *is* 60	$x + (3x + 4) = 60$
a number increased by 15% *is* 120	$x + 0.15x = 120$
the sum of two consecutive odd integers *is* 24	$x + (x + 2) = 24$

EXAMPLE 6   Write each problem as an equation.

(a) One number is 4 less than twice the other. Their sum is 14.
(b) For two consecutive integers, the sum of the smaller and three times the larger is 23.

Solution: (a)   Let x = one number

then $2x - 4$ = second number

First number + second number = 14

$$x + (2x - 4) = 14$$

(b)   Let x = smaller consecutive integer

then $x + 1$ = larger consecutive integer

Smaller + three times the larger = 23

$$x + 3(x + 1) = 23$$ ■

EXAMPLE 7   Write the following problem as an equation. One train travels 3 miles more than twice the distance another train travels. The total distance traveled by both trains is 800 miles.

Solution: Let x = distance traveled by one train

then $2x + 3$ = distance traveled by second train

Distance of train 1 + distance of train 2 = total distance

$$x + (2x + 3) = 800$$ ■

EXAMPLE 8   Express each of the following as an equation.

(a) The cost of renting a snow blower for x days at $12 per day is $60.
(b) Parktown is increasing at a rate of 500 people per year. The increase in population in t years is 2500.
(c) The distance Dawn and Jack traveled for x days at 600 miles per day is 1500 miles.
(d) The number of cents in d dimes is 120.

Solution: (a) $12x = 60$ (b) $500t = 2500$
(c) $600x = 1500$ (d) $10d = 120$ ■

Exercise Set 3.2

Express as an algebraic expression.

1. Five more than a number.
2. Seven less than a number.
3. Four times a number.
4. The product of a number and eight.
5. Three less than six times a number.
6. The product of six and three less than a number.
7. Seven plus three-fourths of a number.
8. Four times a number decreased by two.

9. Twice the sum of a number and 8.
10. Seventeen decreased by x.
11. The number of cents in x quarters.
12. The number of cents in x quarters and y dimes.
13. The number of inches in x feet.
14. The number of inches in x feet and y inches.
15. The number of ounces in a pounds and b ounces.

Select a variable to represent one quantity, and express the second quantity in terms of the first.

16. Eileen's salary is $45 more than Martin's salary.
17. A boy is 12 years older than his brother.
18. A number is one-third of another.
19. Two consecutive integers.
20. Two consecutive even integers.
21. One hundred dollars divided between two people.
22. Two numbers differ by 12.

23. A number is 5 less than four times another number.
24. A number is 3 more than one-half another number.
25. A Cadillac costs 1.7 times as much as a Ford.
26. A number is 4 less than three times another number.
27. An 80-foot tree cut into two pieces.
28. Two consecutive odd integers.

Write as an algebraic expression.

29. The cost of purchasing x items at $4 each.
30. The rental fee for subscribing to Home Box Office for x months at a fee of $12 per month.
31. The cost of traveling x miles at 23 cents per mile.
32. The cost of paying a consultant $75 per day for t days.
33. The cost of paying a $15 per hour tennis court fee for x hours.
34. The distance traveled in t hours when traveling 30 miles per hour.
35. The number of employees hired when 10 employees are hired per day for x days.
36. The cost of renting a telephone for b months at a cost of $3.12 per month.

37. The population growth of a city in n years if the city is growing at a rate of 300 per year.
38. The population decline of a city in n years if the city is losing 400 residents a year.
39. The sales tax on x dollars if the sales tax rate is 7.5%.
40. The number of ounces in y pounds.
41. The number of cents in a dimes.
42. The number of seconds in m minutes.
43. The number of dollars in p $5 bills.
44. The number of pennies in a dimes and b quarters.

Express as an equation.

45. One number is five times another. The sum of the two numbers is 18.
46. Marie is 6 years older than Denise. The sum of their ages is 48.
47. The sum of two consecutive integers is 47.
48. The product of two consecutive even integers is 48.
49. Twice a number decreased by 8 is 12.
50. For two consecutive integers the sum of the smaller and twice the larger is 27.

51. One-fifth of the sum of a number and 10 is 150.
52. One train travels six times as far as another. The total distance traveled by both trains is 700 miles.
53. One train travels 8 miles less than twice the other. The total distance traveled by both trains is 1000 miles.
54. One number is 3 greater than six times the other. Their product is 408.
55. A number increased by 8% is 92.

56. The product of a number and the number plus 5% is 120.

57. One number is 3 less than twice another number. Their sum is 21.

58. The cost of renting a phone at a cost of $2.37 per month for x months is $27.

59. The distance traveled by a car going 40 miles per hour for t hours is 180 miles.

60. The cost of traveling x miles at 23 cents per mile is $12.80.

61. The number of calories in y french fried potatoes at 15 calories per french fry is 215.

62. Milltown is increasing at a rate of 200 per year. The increase in population in t years is 2400.

63. The number of cents in q quarters is 150.

64. The number of ounces in p pounds is 64.

3.3

Solving Word Problems

1   *Set up and solve word problems.*

---

1   Transforming verbal problems into mathematical terms is something we do all the time without realizing it. For example, if you need 3 cups of milk for a recipe and the measuring cup holds only 2 cups, you reason that you need 1 additional cup of milk after the initial 2 cups. You may not realize it, but when you do this simple operation, you are using algebra.

Let x = number of additional cups of total milk needed

Thought process: initial 2 cups + number of additional cups = total milk needed

Equation to represent problem: $2 + x = 3$

When we solve for x we get 1 cup of milk.

You probably said to yourself: Why do I have to go through all this when I know that the answer is $3 - 2$ or 1 cup? When you perform this subtraction, you have mentally solved the equation $2 + x = 3$.

$$2 + x = 3$$
$$2 - 2 + x = 3 - 2$$
$$x = 3 - 2$$
$$x = 1$$

Let's look at another example.

EXAMPLE 1   Suppose that you are at a supermarket, and your purchases so far total $13.20. In addition to groceries, you wish to purchase as many packages of gum as possible, but you have a total of only $18. If a package of gum cost $1.15, how many can you purchase?

Solution: How can we represent this problem as an equation? We might reason as follows. We need to find the number of packages of gum. Let us call this unknown quantity x.

Let x = number of packages of gum

Thought process: cost of groceries + cost of gum = total cost

Now substitute $13.20 for the cost of groceries and $18 for the total cost to get

$$13.20 + \text{cost of gum} = 18$$

At this point you might be tempted to replace the cost of gum with the letter x. But look at what x represents. The variable x represents the *number* of packages of gum, not the cost of the gum. In the preceding section we learned that the cost of x packages of gum at $1.15 per package is $1.15x$. Now substitute the cost of the x packages of gum, $1.15x$, into the equation to obtain

Equation to represent problem: $13.20 + 1.15x = 18$

When we solve this equation we obtain $x = 4.2$ packages (to the nearest tenth). Since you cannot purchase a part of a pack of gum, we reason that only 4 packages of gum can be purchased.   ■

Now let us look at the procedure for setting up and solving a word problem.

To Solve a Word Problem

1. Read the question carefully.
2. If possible, draw a sketch to help visualize the problem.
3. Determine which quantity you are being asked to find. Choose a letter to represent this unknown quantity. Write down exactly what this letter represents.
4. Write the word problem as an equation.
5. Solve the equation for the unknown quantity.
6. Answer the question or questions asked.
7. Check the solution.

Let us now set up and solve some word problems using this procedure.

EXAMPLE 2   Two subtracted from four times a number is 10. Find the number.

Solution: We are asked to find the number. Let us designate the unknown number by the letter x. Let $x = $ unknown number.

$$\underbrace{\text{2 subtracted from 4 times a number}}\ \text{is 10}$$

Write the equation: $4x - 2\ = 10$
Solve the equation: $4x = 12$
Answer the question: $x = 3$

Check: $4x - 2 = 10$
$4(3) - 2 = 10$
$10 = 10$ true

Since the solution checks, the unknown number is 3.   ■

EXAMPLE 3   The sum of two numbers is 17. Find the two numbers if the larger is five more than twice the smaller number.

Solution: We are asked to find *two* numbers. We will call the smaller number x.

$$\text{Let } x = \text{smaller number}$$
$$\text{then } 2x + 5 = \text{larger number}$$

The sum of the two numbers is 17. Therefore we write the equation.

$$x + (2x + 5) = 17$$

Now solve the equation: $$3x + 5 = 17$$
$$3x = 12$$
$$x = 4$$

Answer the questions: smaller number $= 4$
$$\text{larger number} = 2x + 5$$
$$= 2(4) + 5 = 13$$

Check: sum of two numbers $= 17$
$$4 + 13 = 17$$
$$17 = 17 \qquad \text{true} \qquad \blacksquare$$

EXAMPLE 4   The population of a growing city is 40,000. If the population is increasing by 300 a year, in how many years will the population reach 44,500 people?

Solution: We are asked to find the number of years.

$$\text{Let } n = \text{number of years.}$$
$$\text{then } 300n = \text{increase in population over } n \text{ years}$$

$$\text{Present population} + \left(\begin{array}{c}\text{increase in population} \\ \text{over } n \text{ years}\end{array}\right) = \text{future population}$$
$$40{,}000 + 300n = 44{,}500$$
$$300n = 4500$$
$$n = \frac{4500}{300}$$
$$n = 15 \text{ years}$$

Check: $$40{,}000 + 300n = 44{,}500$$
$$40{,}000 + 300(15) = 44{,}500$$
$$40{,}000 + 4500 = 44{,}500$$
$$44{,}500 = 44{,}500 \qquad \text{true} \qquad \blacksquare$$

EXAMPLE 5   The monthly rental fee for a telephone from the Rochester Telephone Company is $2.63. Radio Shack is selling new telephones for $34.99. In how many months will the rental fee equal the cost of the new telephone?

Solution: We are asked to find the number of months.

$$\text{Let } x = \text{number of months}$$

$$\text{then } 2.63x = \text{cost of renting phone for } x \text{ months}$$

$$\text{Cost of renting phone for } x \text{ months} = \text{cost of new phone}$$

$$2.63x = 34.99$$

$$x = \frac{34.99}{2.63} = 13.3 \text{ months}$$

The rental fee will equal the cost of a new telephone in 13.3 months.   ■

EXAMPLE 6   The cost of renting an automobile is $25 a day plus 20 cents a mile. Find the maximum mileage Janet can drive in 1 day if she has only $56.

Solution: We are asked to find the number of miles Janet can drive.

$$\text{Let } x = \text{number of miles Janet can drive}$$

$$\text{then } 0.20x = \text{cost of driving } x \text{ miles}$$

$$\text{Total cost} = \text{daily fee} + \text{mileage cost}$$

$$56 = 25 + 0.20x$$

$$31 = 0.20x$$

$$\frac{31}{0.20} = x$$

$$155 = x$$

Janet can drive a maximum of 155 miles in 1 day.   ■

EXAMPLE 7   United Airlines wishes to keep its airfare, including a 7% tax, between Dallas, Texas, and Los Angeles, California, at exactly $160. Find the cost of the ticket before tax.

Solution: We are asked to find the cost of the ticket before tax.

$$\text{Let } x = \text{cost of the ticket before tax}$$

$$\text{then } 0.07x = \text{tax on the ticket}$$

$$\frac{\text{Cost of ticket}}{\text{before tax}} + \frac{\text{tax on}}{\text{the ticket}} = 160$$

$$x + 0.07x = 160$$

$$1.07x = 160$$

$$x = \frac{160}{1.07}$$

$$x = 149.53$$

Thus if United prices the ticket at $149.53, the total cost including a 7% tax will be $160.   ■

EXAMPLE 8   Four brothers must divide a total of $6000. How much will each receive if the two older brothers are each to receive twice as much as the two younger brothers?

Solution: We are asked to find how much each of the four brothers receives.

Let x = amount each of the two younger brothers receives
then $2x$ = amount each of the two older brothers receives

The total received by the four brothers is $6000. Thus the equation we use is

2 younger brothers   2 older brothers
each receive x each receive $2x$

$$\overbrace{x + x} \quad + \quad \overbrace{2x + 2x} \quad = 6000$$
$$6x = 6000$$
$$x = 1000$$

The two younger brothers each receive $1000 and the two older brothers each receive $2(1000) = $2000.   ■

Exercise Set 3.3

Set up an algebraic equation that can be used to solve each problem. Solve the equation, and find the values desired.

1. The sum of two consecutive integers is 97. Find the numbers.
2. The sum of two consecutive integers is 225. Find the numbers.
3. The sum of two consecutive even integers is 154. Find the numbers.
4. The sum of two consecutive odd numbers is 76. Find the numbers.
5. One number is 3 more than twice a second number. Their sum is 27. Find the numbers.
6. One number is 5 less than three times a second number. Their sum is 43. Find the numbers.
7. The sum of three consecutive integers is 39. Find the three integers.
8. The sum of three consecutive odd integers is 87. Find the three integers.
9. The sum of three integers is 29. Find the three numbers if one number is twice the smallest and the third number is 4 more than twice the smallest.
10. The larger of two integers is 4 less than three times the smallest. When the smaller number is subtracted from the larger, the difference is 12. Find the two numbers.
11. The larger of two integers is 8 less than twice the smaller. When the smaller number is subtracted from the larger, the difference is 17. Find the two numbers.
12. A village of 2500 is growing by 200 each year. In how many years will the population reach 4900?

13. A city with a population of 50,000 is decreasing by 300 per year. In how many years will the population drop to 44,000?
14. The monthly rental fee for a telephone is $2.90. The telephone store is selling telephones for $37.70. How long will it take for the monthly rental fee to equal the cost of a new phone?
15. It cost the Weismans $4.50 a week to wash and dry their clothing at the corner laundry. If a washer and dryer cost a total of $612, how many weeks will it take for the laundry cost to equal the cost of purchasing a washer and dryer?
16. The cost of renting an automobile is $28 a day plus 12 cents a mile. How much will it cost Colette to rent the car for 1 day and drive a total distance of 200 miles?
17. The cost of renting a car is $15 a day plus 20 cents per mile. How far can Milt drive in one day if he has only $55?
18. Ace Company reimburses its employees $20 a day plus 13 cents a mile when they use their own vehicles on company business. If Mrs Kohn takes a 1-day trip and is reimbursed $72, how far did she travel?
19. Murgatronics Company reimburses its employees $25 a day and 10 cents a mile when they use their own vehicles on company business. How far will Mr. Abbott have to travel in one day to be reimbursed $42?

20. Mr. Connally receives a weekly salary of $210. He also receives a 6% commission on the total dollar volume of all sales he makes. What must his dollar volume be in a week if he is to make a total of $450?

21. Jamie, a hot dog vendor, wishes to price her hot dogs such that the total cost of the hot dog, including a 5% tax, is $1.00. What will be the price of a hot dog before tax?

22. Monroe County has a 7% sales tax. How much does Robert's car cost before tax, if the total cost of the car plus its sales tax is $7200?

23. Clark is on a diet and can have only 500 calories for lunch. He orders a hamburger without a roll and french fries. How many french fries can he eat if the hamburger has 300 calories and each french fry has 20 calories?

24. Janice's investment of $7500 on solar heating equipment results in a yearly savings of $500 in the Jackson's heating bills. How many years will it take for the savings to equal the investment?

25. Eighty-four hours of overtime must be split among four workers. The two younger workers are to be assigned the same number of hours. The third worker is to be assigned twice as much as each of the younger workers. The fourth worker is to be assigned three times as much as each of the younger workers. How much overtime should be assigned to each worker?

26. Sam Fulkerson worked a 55-hour week last week.

He is not sure of his hourly rate, but knows that he is paid $1\frac{1}{2}$ times his regular hourly rate for all hours over a 40-hour week. His pay last week was $400. What has his hourly rate?

27. After Mrs Linton is seated in a restaurant, she realizes that she has only $15.00. If from this $15 she must pay a 7% tax and she wishes to leave a 15% tip, what is the maximum price for a meal that she can afford to pay?

28. The fine for speeding in a certain community is $3 per mile per hour over the speed limit plus a $15 administrative charge. Michelle received a speeding ticket and had to pay a fine of $45. How many miles per hour in excess of the speed limit was she traveling?

29. The Winton Racquetball Club has two payment plans for its members. One plan is a flat $35 per month. The second is to pay $7.25 per hour for court time. If court time can be rented only in 1-hour intervals, how many hours would you have to play per month to make it advantageous to select the flat fee?

30. The Midtown Tennis Club has two payment plans for its members. Plan 1 is a monthly fee of $20 per month plus $8 per hour court rental time. Plan 2 is no monthly fee, but court time is $16.25 per hour. If court time can be rented only in 1-hour intervals how many hours would one have to play per month so that plan 1 becomes advantageous?

JUST FOR FUN

1. Mr. and Mrs. Burnham are considering buying a Volkswagen Rabbit. The 1988 Rabbit with its standard 105-horsepower four-cylinder engine has a list price of $6615 and has a rated fuel economy of 32 miles per gallon. The same Rabbit with an optional 97-horsepower four-cylinder diesel engine has a list price of $7140 and has a rated fuel economy of 45 miles per gallon. How many miles would the Burnhams have to drive in the diesel before they regain the extra cost of the car? Assume that diesel fuel costs $1.17 per gallon and unleaded gasoline costs $1.28 per gallon.

2. Pick any number: say, 9

Multiply the number by 4: $9 \cdot 4 = 36$

Add 6 to the product: $36 + 6 = 42$

Divide the sum by 2: $42 \div 2 = 21$

Subtract 3 from the quotient: $21 - 3 = 18$

The solution is twice the number you started with. Show that when you select n to represent the given number, the solution will always be $2n$.

3.4 _____

Motion Problems

■   *Set up and solve motion (or rate) problems.*

_____

■   A motion (or rate) problem is one in which an item is moving at a specified rate for a specified period of time. A car traveling at a constant speed, a swimming pool being filled or drained (the water is moving at a specified rate), or spaghetti being cut on a conveyor belt (conveyor belt moving at a specified speed) are all motion problems.

The formula often used to solve motion problems is

Motion Problems
amount = rate · time

The amount can be a measure of many different quantities, depending on the rate. For example, if the rate is measuring distance per unit time, the amount will be distance. If the rate is measuring volume per unit time, the amount will be volume; and so on.

To find the distance traveled by a car going at 30 miles per hour for 3 hours, we use: distance = rate · time.

$$d = r \cdot t$$
$$= 30 \cdot 3 = 90 \text{ miles}$$

When filling a swimming pool, the amount of water placed in the pool can be found by: volume = rate · time.

EXAMPLE 1   A swimming pool is being filled at a rate of 10 gallons per minute. How many gallons have been added after 25 minutes?

Solution: volume = rate · time
$$= 10 \cdot 25 = 250 \text{ gallons} \quad ■$$

When multiplying rate by time, the time units must be the same as in the rate. For example, if rate is 60 feet per second, the time must be in seconds.

EXAMPLE 2   An oil-well-drilling device can drill 3 feet per hour. How long will it take to drill to a depth of 1870 feet?

Solution: Let t = time needed to drill 1870 feet.

$$d = rt$$
$$1870 = 3t$$
$$\frac{1870}{3} = t$$
$$623.33 = t$$

or $$t = 623.33 \text{ hours} \quad ■$$

EXAMPLE 3   A patient is to receive 1200 cubic centimeters of fluid intravenously over an 8-hour period. What should be the average intravenous flow rate?

Solution: Here we are given the total volume and the length of time the fluid is to be administered. We are asked to find the rate.

$$\text{Let } r = \text{rate}$$

$$\text{Volume} = \text{rate} \cdot \text{time}$$

$$1200 = r \cdot 8$$

$$\frac{1200}{8} = r$$

$$150 = r$$

The fluid should be administered at a rate of 150 cubic centimeters per hour.   ■

EXAMPLE 4   A conveyor belt transporting uncut spaghetti moves at a rate of 1.5 feet per second. A cutting blade is activated at regular intervals to cut the spaghetti into proper lengths. At what time intervals should the blade be activated if the spaghetti is to be in 9-inch lengths?

Solution: Since the rate is given in feet per second and the length is given in inches, one of these units must be changed. One foot equals 12 inches; thus, to change from feet per second to inches per second, we multiply the rate by 12.

$$1.5 \text{ feet per second} = (1.5)(12) = 18 \text{ inches per second}$$

We are asked to find the time.

$$\text{Let } t = \text{time}$$

$$\text{length} = \text{rate} \cdot \text{time}$$

$$9 = 18 \cdot t$$

$$\frac{9}{18} = t$$

$$\frac{1}{2} = t$$

The blade should cut at $\frac{1}{2}$-second intervals.   ■

EXAMPLE 5   A mother and daughter plan to go cross-country skiing on a straight path. The mother skis at 4 miles per hour, the daughter at 5 miles per hour. If the daughter begins skiing $\frac{1}{2}$ hour after the mother, and they ski the same path:

(a) How long will it take for the mother and daughter to meet?
(b) How far from the starting point will they be when they meet?

Solution: (a) Since the daughter is the faster skier, she will cover the same distance in less time. When they meet, they have both traveled the same distance. Since the rate is given in miles per hour, the time will be in hours. The daughter begins $\frac{1}{2}$ hour later, therefore her time will be $\frac{1}{2}$ hour less than her mother's. We are asked to find the time for the mother and daughter to meet. Sometimes when a problem involves two different rates, it is helpful to set up a table to analyze it.

Let t = time mother is skiing

then $t - \dfrac{1}{2}$ = time daughter is skiing

mother, 4 mph →

daughter, 5 mph →

Skier	Rate	Time	Distance
Mother	4	t	$4t$
Daughter	5	$t - \frac{1}{2}$	$5(t - \frac{1}{2})$

distance mother = distance daughter

$$4t = 5\left(t - \frac{1}{2}\right)$$

$$4t = 5t - \frac{5}{2}$$

$$4t - 5t = 5t - 5t - \frac{5}{2}$$

$$-t = -\frac{5}{2}$$

$$t = \frac{5}{2}$$

Thus they will meet in $\frac{5}{2}$ or $2\frac{1}{2}$ hours.

(b) The distance can be found using either the mother or daughter. We will use the mother.

$$d = r \cdot t$$

$$= 4 \cdot \frac{5}{2} = \frac{20}{2} = 10 \text{ miles}$$

The mother and daughter will meet 10 miles from the starting point. ■

EXAMPLE 6   Two trains leave the same station along parallel tracks going in opposite directions. The train traveling east has a speed of 40 miles per hour. The train traveling west has a speed of 60 miles per hour. In how many hours will they be 500 miles apart?

Solution: When the trains are 500 miles apart, each train has traveled the same number of hours. We are asked to find the number of hours. Let t = the unknown number of hours. We set up another table to help analyze the problem.

← West | East →

60 mph | 40 mph

500 miles

Train	Rate	Time	Distance
East	40	t	$40t$
West	60	t	$60t$

Since the trains are traveling in opposite directions, the sum of their distances must be 500 miles.

$$40t + 60t = 500$$
$$100t = 500$$
$$t = 5$$

The two trains will be 500 miles apart in 5 hours.   ∎

Exercise Set 3.4

Set up an algebraic equation that can be used to solve each problem. Solve the equation, and find the values desired.

1. Royce jogs at an average of 5 miles per hour. How far will he jog in 1.2 hours?
2. How fast must a car travel to cover 150 miles in $2\frac{1}{2}$ hours?
3. An airplane travels 650 miles per hour. How long will it take to travel a distance of 3000 miles?
4. Apollo 11 took approximately 87 hours to reach the moon, a distance of about 238,000 miles. Find the average rate of speed of the Apollo.
5. Alice's small aboveground swimming pool has a capacity of 13,500 gallons of water. If a hose can supply 2.2 gallons per minute, how long will it take Alice's son to fill the pool?
6. Marty can lay 42 bricks per hour. How long will it take him to lay 572 bricks?
7. A tree grows 4.5 feet per year. How many feet will it grow in $7\frac{1}{2}$ years?
8. At what rate must a photocopying machine copy to make 100 copies in 2.5 minutes?
9. A certain laser can cut through steel at a rate of 0.2 centimeter per minute. How thick is the steel door if it requires 12 minutes to cut through it?
10. Morton can check 12 parts per minute on the assembly line. How many parts will he be able to check in an 8-hour day?
11. A patient is to receive 1500 cubic centimeters of an intravenous fluid over a period of 6 hours. What should be the intravenous flow rate?
12. How long will it take a submarine traveling at 24 knots per hour to travel 600 knots?
13. Two planes leave an airport at the same time. One plane flies north at 500 miles per hour. The other flies south at 650 miles per hour. In how many hours will they be 4025 miles apart?
14. Two runners start from the same point and run in the same direction. One runs at 8 miles per hour, and the other runs at 11 miles per hour. In how many hours will the runners be 9 miles apart?
15. Two trains are 804 miles apart. Both start at the same time and travel toward each other. They meet 6 hours later. If the speed of the faster train is 30 miles per hour faster than the slower train, find the speed of each train.
16. Two trains leave South Station at the same time traveling in the same direction along parallel tracks. Train 1 travels 80 miles per hour, and train 2 travels 62 miles per hour. In how many hours will they be 144 miles apart?
17. Two rockets are to be launched in the same direction one hour apart. The first rocket is launched at noon and travels at 12,000 miles per hour. If the second rocket travels at 14,400 miles per hour, at what time will the rockets be the same distance from Earth?
18. Two rockets are to be launched from the same location. The first rocket travels at 8000 miles per hour, the second at 9500 miles per hour. How long after the first should the second rocket be launched if the second rocket is to meet the first at a distance of 38,000 miles from Earth?
19. Two trains that are 460 miles apart travel toward each other on different tracks. One travels at 60 miles per hour while the other travels at 55 miles per hour. When will they pass each other?

JUST FOR FUN

JUST FOR FUN

1. Sixty-six lines of type will fit on a standard $8\frac{1}{2} \times 11$-inch sheet of paper. A Centronics 739 printer types 10 characters per inch at a speed of 100 characters per second. How long will it take the printer to type a full page of 66 lines if the page is set for a 1-inch margin on the left and a $\frac{1}{2}$-inch margin on the right (disregard the carriage return time).
2. An automatic garage door opener is designed to begin to open when a car is 100 feet from the garage. At what rate will the garage door have to open if it is to raise 6 feet by the time a car traveling at 4 miles per hour reaches it? (1 mile per hour = 1.47 feet per second.)

3.5
Mixture and Geometric Problems

1   Set up and solve mixture problems.

2   Set up and solve geometric problems.

1   Any problem in which two or more quantities are combined to produce a different quantity, or a single quantity is separated into two or more different quantities, may be considered a mixture problem. Mixture problems are familiar to everyone, as we can see in the everyday examples that follow.

EXAMPLE 1   General Motors stock is selling at $37 a share. Eastman Kodak stock is selling at $75 a share. Mr. Abelard has a maximum of $8000 to invest. He wishes to purchase five times as many shares of General Motors as of Kodak.

(a) How many shares of each will he purchase?
(b) How much money will be left over if stocks can be purchased only in whole shares?

Solution: (a) We are asked to find the number of shares of stock for Eastman Kodak and General Motors.

$$\text{Let } x = \text{number of shares of Kodak}$$
$$\text{then } 5x = \text{number of shares of GM}$$

The cost of purchasing a stock is found by multiplying the price per share times the number of shares purchased. Let us make a table to help analyze the problem.

Stock	Price	Number of Shares	Cost of Stock
Kodak	75	x	$75x$
GM	37	$5x$	$37(5x)$

$$\text{cost of Kodak} + \text{cost of GM} = \text{total cost}$$
$$75x + 37(5x) = 8000$$
$$75x + 185x = 8000$$
$$260x = 8000$$
$$x = \frac{8000}{260} = 30.8$$

Thus 30 shares of Kodak and $5(30) = 150$ shares of GM can be purchased.

(b) Total spent = cost of GM + cost of Kodak

Total spent = $37(150) + 75(30)$

$$= 5550 + 2250 = 7800$$

The money left over = $8000 - $7800 = $200.   ■

EXAMPLE 2   Deborah wishes to mix coffee worth $6 per pound with 12 pounds of coffee worth $3 per pound.
(a) How many pounds of coffee worth $6 per pound must be mixed to obtain a mixture worth $5 per pound?
(b) How much of the mixture will be produced?

Solution: (a) We are asked to find the number of pounds of $6 coffee.

Let x = number of pounds of $6 coffee

The value of the coffee is found by multiplying the number of pounds by the price per pound.

Coffee	Price	Number of Pounds	Value of Coffee
More expensive	6	x	$6x$
Less expensive	3	12	$3(12)$
Mixture	5	$x + 12$	$5(x + 12)$

Value of $6 coffee + value of $3 coffee = value of $5 mixture

$$6x + 3(12) = 5(x + 12)$$
$$6x + 36 = 5x + 60$$
$$x + 36 = 60$$
$$x = 24 \text{ pounds}$$

Thus 24 pounds of the $6 coffee must be mixed with 12 pounds of the $3 coffee to obtain a mixture worth $5 per pound.

(b) The number of pounds of the mixture is

$$x + 12 = 24 + 12 = 36 \text{ pounds}   ■$$

EXAMPLE 3   Jerry's Juice Company sells apple juice for 5 cents an ounce and apple drink for 2 cents an ounce. They wish to market and sell for 4 cents an ounce cans of juice-drink that are part juice and part drink. How many ounces of each will be used if the juice-drink is to be sold in 8-ounce cans?

Solution: We are asked to find the number of ounces of juice and drink needed.

Let x = the number of ounces of apple juice

then $8 - x$ = number of ounces of apple drink

The value of the juice (or drink) is found by multiplying the price per ounce by the number of ounces.

Product	Price	Number of Ounces	Value of Product
Juice	5	x	$5x$
Drink	2	$8 - x$	$2(8 - x)$
Juice drink	4	8	$4(8)$

$$\text{Value of juice} + \text{value of drink} = \text{value of juice-drink mixture}$$
$$5x + 2(8 - x) = 4(8)$$
$$5x + 16 - 2x = 32$$
$$3x + 16 = 32$$
$$3x = 16$$
$$x = \frac{16}{3} = 5\frac{1}{3}$$

Therefore, the company must use $5\frac{1}{3}$ ounces of apple juice and $8 - 5\frac{1}{3} = 2\frac{2}{3}$ ounces of apple drink in each 8-ounce can. ■

EXAMPLE 4   Paul Toland has a total of 30 dimes and quarters. The total value of these coins is $4.50. How many of each coin does he have?

Solution: We are asked to find the number of each type of coin.

$$\text{Let } x = \text{number of dimes,}$$
$$\text{then } 30 - x = \text{number of quarters}$$

The total value of the coins is found by multiplying the value of the coin by the number of coins.

Coin	Value of Coin	Number of Coins	Total Value of Coins
Dime	0.10	x	$0.10x$
Quarter	0.25	$30 - x$	$0.25(30 - x)$

$$\text{Value of dimes} + \text{value of quarters} = \text{total value}$$
$$0.10x + 0.25(30 - x) = 4.50$$
$$0.10x + 7.5 - 0.25x = 4.50$$
$$-0.15x + 7.5 = 4.50$$
$$-0.15x = -3.0$$
$$x = \frac{-3.0}{-0.15} = 20$$

Thus there are 20 dimes and $30 - 20 = 10$ quarters.

Check: 20 dimes = $2.00

10 quarters = $2.50

total = $4.50   true   ■

EXAMPLE 5 How many liters of a 25% salt solution must be added to 80 liters of a 40% salt solution to get a solution that is 30% salt?

Solution: We are asked to find the number of liters of the 25% salt solution.

Let x = number of liters of 25% salt solution

The amount of pure salt in a given solution is found by multiplying the percent strength by the number of liters.

Strength of Solution	Liters of Solution	Amount of Pure Salt
25%	x	$0.25x$
40%	80	$0.40(80)$
30%	$x + 80$	$0.30(x + 80)$

$$\begin{matrix} \text{amount of pure salt} \\ \text{in 25\% solution} \end{matrix} + \begin{matrix} \text{amount of pure salt} \\ \text{in 40\% solution} \end{matrix} = \begin{matrix} \text{amount of pure salt} \\ \text{in 30\% mixture} \end{matrix}$$

$$0.25x + 0.40(80) = 0.30(x + 80)$$
$$0.25x + 32 = 0.30x + 24$$
$$0.25x + 8 = 0.30x$$
$$8 = 0.05x$$
$$\frac{8}{0.05} = x$$
$$160 = x$$

Therefore, 160 liters of 25% salt solution must be added to the 80 liters of 40% solution to get a 30% salt solution. The total number of liters that will be obtained is $160 + 80$ or 240.   ■

Geometric Problems

❷ Now let us do some geometric problems.

EXAMPLE 6 Mrs. O'Connor is planning to build a sandbox for her daughter. She has 26 feet of wood to build the perimeter. What should be the dimensions of the rectangular sandbox if the length is to be 3 feet longer than the width?

Solution: We are asked to find the dimensions of the sandbox.

$$\text{Let } x = \text{width of sandbox}$$
$$\text{then } x + 3 = \text{length of sandbox (Fig. 3.2)}$$

From Section 3.1 we know that

$$P = 2l + 2w$$
$$26 = 2(x + 3) + 2x$$
$$26 = 2x + 6 + 2x$$
$$26 = 4x + 6$$
$$20 = 4x$$
$$5 = x$$

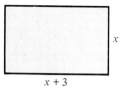

x + 3

Figure 3.2

Thus the width is 5 feet, and the length $= x + 3 = 5 + 3 = 8$ feet.

Check: $P = 2l + 2w$
$$26 = 2(8) + 2(5)$$
$$26 = 16 + 10$$
$$26 = 26 \qquad \text{true} \qquad \blacksquare$$

EXAMPLE 7   The sum of the angles of a triangle measure 180 degrees (180°). If two angles are the same and the third is 30° greater than the other two, find the three angles of the triangle.

Solution: We are asked to find the three angles.

$$\text{Let } x = \text{each smaller angle}$$
$$\text{then } x + 30 = \text{larger angle} \quad \text{(Fig. 3.3)}$$
$$\text{Sum of the 3 angles} = 180$$
$$x + x + (x + 30) = 180$$
$$3x + 30 = 180$$
$$3x = 150$$
$$x = \frac{150}{3} = 50°$$

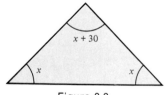

Figure 3.3

Therefore, the three angles are 50°, 50°, and 50° + 30° or 80°.

Check: $50 + 50 + 80 = 180$
$$180 = 180 \qquad \text{true} \qquad \blacksquare$$

EXAMPLE 8   A bookcase is to have four shelves, including the top, as shown in Fig. 3.4. The height of the bookcase is to be 3 feet more than the width. Find the dimensions of the bookcase if only 30 feet of wood are available.

Solution: We are asked to find the dimensions of the bookcase.

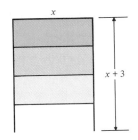

Let x = length of a shelf

then $x + 3$ = height of bookcase

4 shelves + 2 sides = total wood available

$$4x + 2(x + 3) = 30$$
$$4x + 2x + 6 = 30$$
$$6x + 6 = 30$$
$$6x = 24$$
$$x = 4$$

The length of a shelf is 4 feet and the height of the bookcase is $4 + 3$ or 7 feet.

Check: $4 + 4 + 4 + 4 + 7 + 7 = 30$

$$30 = 30 \qquad \text{true} \qquad \blacksquare$$

Exercise Set 3.5

Set up an equation that can be used to solve each problem. Solve the equation, and find the values desired.

1. Teledyne stock is selling at $140 per share. Barber Oil stock is selling at $22 a share. An investor has a total of $8000 to invest. She wishes to purchase four times as many shares of Teledyne as Barber Oil.
   (a) How many shares of each will she purchase?
   (b) How much money will be left over?

2. Mr. Temple invested $8900, part at 8% and the rest at 11% annual interest. How much does he invest at each rate if his total annual interest from both investments is $874? Use interest rate = principal · rate · time.

3. Rich invested $6000, part at 9% and the rest at 10% per year. If he received a total annual interest of $562 from both investments, how much did he invest at each rate?

4. Barbara Wilson invested $5000, part at 10% and part at 15%. How much did she invest at each rate if the same amount of interest was received from each account?

5. The Clars invested $12,500, part at 8% and part at 12%. How much was invested at each rate if the same amount of interest was received from each account?

6. Steven has a total of 28 dimes and quarters. The total value of the coins is $3.55. How many of each coin does he have?

See Exercise 1.

7. Kathleen has a total of 62 dimes and nickels. The total value of all the coins is $3.75. How many of each coin does she have?

8. Phil has a total of 12 bills in his wallet. Some are $1 bills and some are $10 bills. The total value of the 12 bills is $39. How many of each type does he have?

9. Cheryl has a total of 25 quarters and half-dollars. The total value of the coins is $9.00. How many of each coin does she have?

10. Casey holds two part-time jobs. One job pays $4.00 per hour, and the other pays $4.50 per hour. Last week Casey worked a total of 18 hours and earned $78.00. How many hours did he work on each job?

11. Almonds cost $2.00 per pound. Walnuts cost $1.60 per pound. How many pounds of each should Bridget mix to produce a 30-pound mixture that costs $1.75 per pound?

12. How many pounds of coffee costing $3.20 per pound must Jack mix with 18 pounds of coffee costing $2.60 per pound to produce a mixture that costs $2.80 per pound?

13. There were 600 people at a movie. Adult admission was $4. Children's admission was $3. How many adults and how many children attended if the total receipts were $2050 for the day?

14. In chemistry class, Ramon has 1 liter of a 20% sulfuric acid solution. How much of a 12% sulfuric acid solution must be mixed with the 1 liter of 20% solution to make a 15% sulfuric acid solution?

15. A pharmacist has a 60% solution of the drug sodium iodite. He also has a 25% solution of the same drug. He gets a prescription calling for a 40% solution of the drug. How much of each solution should he mix to make 0.5 liter of the 40% solution?

16. Six quarts of the punch for the class party contains 12% alcohol. How much pure punch must Jason add to reduce the alcohol content to 10%?

17. The label on a 12-ounce can of frozen concentrate Hawaiian Punch indicates that when the can of concentrate is mixed with 3 cans of cold water, the resulting mixture is 10% juice. Find the percentage of pure juice in the concentrate.

18. The length of a rectangle is to be 8 feet more than its width. What should the dimensions of the rectangle be if the perimeter is to be 48 feet?

19. In an isosceles triangle two sides are equal. The third side is 2 meters less than one of the other sides. Find the length of each side if the perimeter is 10 meters.

20. The perimeter of a rectangle is 120 feet. Find the length and width of the rectangle if the length is twice the width.

21. The sum of the angles of a triangle is 180°. If one angle is 20° larger than the smallest angle and the third angle is twice as large as the smallest angle, find the three angles.

22. The sum of the angles of a parallelogram is 360°. If the two smaller angles are equal and the two larger angles are each 30° larger than a smaller angle, find the measure of each angle.

23. A bookcase is to have four shelves as shown. The height of the bookcase is to be 2 feet more than the width, and only 20 feet of wood is available. What should be the dimensions of the bookcase?

24. What should be the dimensions of the bookcase in Exercise 23 if the height is to be twice its width?

---

1. The labels on bottles of liquor state the "proof" of the contents. For example, a bottle of rum may be 80 proof. The formula

$$\text{proof} = 2(\% \text{ alcohol by volume})$$

is used to determine the proof of the liquor. Thus 80-proof rum is 40% alcohol by volume.

Mr. DeLeo wishes to make a fruit punch containing only 10% alcohol by volume. If he is going to use a 1-quart bottle of whiskey that is 90 proof to make the punch, how much pure fruit punch will he have to use?

2. The radiator of an automobile has a capacity of 16 quarts. If it is presently filled with a 20% antifreeze solution, how many quarts must be drained and replaced by pure antifreeze to make the radiator contain a 50% antifreeze solution?

Summary

Glossary

Area: The total surface area within a figure's boundaries.
Circumference: The length of the curve that forms a circle.
Diameter: A line segment through the center of a circle with both endpoints on the circle.

Formula: An equation commonly used to express a specific physical concept mathematically.
Perimeter: The sum of the lengths of the sides of a figure.
Radius: A line segment from the center of a circle to any point on the circle.

Important Facts

$i = prt$ (simple interest formula)

amount = rate · time (motion formula)

To Solve a Word Problem

1. Read the question carefully.
2. If possible, draw a sketch to help visualize the problem.
3. Determine which quantity you are being asked to find. Choose a letter to represent this unknown quantity, write down exactly what this letter represents.
4. Write the word problem as an equation.
5. Solve the equation for the unknown quantity.
6. Answer the question or questions asked.
7. Check the solution.

Review Exercises

[3.1] Use the formula to find the value of the indicated variable for the values given.

1. $C = \pi d$, find C when $d = 4$ and $\pi = 3.14$.
2. $A = \frac{1}{2}bh$, find A when $b = 12$ and $h = 8$.
3. $P = 2l + 2w$, find P when $l = 6$ and $w = 4$.
4. $i = prt$, find i when $p = 1000$, $r = 15\%$ and $t = 2$.
5. $E = IR$, find E when $I = 0.12$ and $R = 2000$.
6. $A = \pi r^2$, find A when $r = 3$ and $\pi = 3.14$.
7. $d = rt$, find t when $d = 120$ and $r = 30$.
8. $Fd^2 = km$, find k when $F = 60$, $m = 12$, and $d = 2$.

9. $y = mx + b$, find b when $y = 15, m = 3$, and $x = -2$.
10. $2x + 3y = -9$, find y when $x = 12$.
11. $4x - 3y = 15 + x$, find y when $x = -3$.
12. $2x = y + 3z + 4$, find y when $x = 5$ and $z = -3$.
13. $IR = E + Rr$, find r when $I = 5.0$, $E = 100$, and $R = 200$.

Solve the given equation for y, then find the value of y for the given value of x.

14. $2x - y = 12$, $x = 10$
15. $3x - 2y = -4$, $x = 2$
16. $3x = 5 + 2y$, $x = -3$

17. $-6x - 2y = 20$, $x = 0$
18. $6 = -3x - 2y$, $x = -6$

Solve for the variable indicated.

19. $F = ma$, for m
20. $A = \frac{1}{2}bh$, for h
21. $PV = kT$, for K

22. $i = prt$, for t
23. $P = 2l + 2w$, for w
24. $2x - 3y = 6$, for y

25. $y - 4x = x^2 - 3$, for y
26. $y + 6 = 2x^2 - 3x$, for y
27. $A = \frac{1}{2}h(b + d)$, for h
28. $F = \frac{9}{5}C + 32$, for C

29. $C = \frac{5}{9}(F - 32)$, for F
30. $A = \dfrac{B - C}{2}$ for C

Solve each problem.

31. How much interest will Karen pay if she borrows $600 for 2 years at 15% simple interest? (Use $i = prt$)
32. The Fishers borrow $5000 at 15% simple interest for a period of 4 years.

(a) How much interest will they pay on their loan?
(b) What will be the total amount the Fishers must repay the bank?

[3.2–3.3] Solve each problem.

33. One number is 4 more than the other. Find the two numbers if their sum is 62.
34. The sum of two consecutive integers is 255. Find the two integers.
35. The sum of two consecutive odd integers is 208. Find the two integers.
36. The larger of two integers is 3 more than five times the smaller integers. Find the two numbers if the smaller subtracted from the larger is 31.
37. Dreyel Company plans to increase its number of employees by 25 per year. If the company presently has 427 employees, how long will it take before they reach 638 employees?

38. Mr. and Mrs. Lendel rent a car for $18 a day and 16 cents a mile. If they plan to use the car for 2 days, how many miles can they drive if they have only $100?
39. What is the cost of a car before tax if the total cost including a 5% tax is $8400?
40. Mr. McAdams sells water softeners. He receives a weekly salary of $300 plus a 5% commission on the sales he makes. If Mr. McAdams earned $900 last week, what was his dollar volume in sales?

[3.4] Solve each problem.

41. A train travels at 70 miles per hour. How long will it take the train to travel 280 miles?
42. How fast must a plane fly to travel 3500 miles in 6.5 hours?
43. Madison can fertilize 0.7 acre per hour on his farm. How long will it take him to fertilize a 40-acre farm?
44. Two trains leave from the same station on parallel tracks going in opposite directions. One train

travels at 50 miles per hour, and the other at 60 miles per hour. How long will it take for the trains to be 440 miles apart?
45. Two joggers follow the same route. Marty jogs at 8 kilometers per hour and Nick at 6 kilometers per hour. If they leave at the same time, how long will it take for them to be 4 kilometers apart?

[3.5] Solve each problem.

46. Kathy Platico wishes to place part of her $12,000 into a money-market fund earning 13% interest and part into a savings account earning $7\frac{1}{4}$% interest. How much should she invest in each if she wishes her interest for the year to be $1100?
47. A gasoline distribution company needs 1200 gallons of 89-octane gasoline. The distributor only has

gasoline rated at 87 octane and gasoline rated at 91 octane. How much of each should be mixed to obtain the desired gasoline?
48. A chemist wishes to make 2 liters of an 8% acid solution by mixing a 10% acid solution and a 5% acid solution. How many liters of each should the chemist use?

49. A butcher combined hamburger that cost $2.50 per pound with hamburger that cost $3.10 per pound. How many pounds of each were used to make 80 pounds of a mixture that sells for $2.65 per pound?

50. Joan has a total of 30 stamps. Some are 22-cent stamps and some are 15-cent stamps. How many of each type does she have if the total value of her stamps is $5.90?

51. Mrs Appleby wants a garden whose length is 4 feet more than its width. The perimeter of the garden is to be 70 feet. What will be the dimensions of the garden?

Practice Test

1. Use $P = 2l + 2w$ to find P when $l = 6$ feet and $w = 3$ feet.

2. Use $A = P + Prt$ to find A when $P = 100, r = 0.15$, and $t = 3$.

Solve for the variable indicated.

3. $P = IR$, for R

4. $3x - 2y = 6$, for y

5. $A = \dfrac{a + b}{3}$, for a

6. $D = R(c + a)$, for c

7. The sum of two integers is 158. Find the two integers if the larger is 10 less than twice the smaller.

8. A triangle has a perimeter of 75 inches. Find the three sides if one side is 15 inches larger than the smallest side, and the third side is twice the smallest side.

9. Mr. Herron has only $20. If he wishes to leave a 15% tip and must pay 7% tax, find the price of the most expensive meal that he can order.

10. The sum of three consecutive integers is 42. Find the three integers.

11. Train A travels 60 miles per hour for 4 hours. If train B is to travel the same distance in 3 hours, find the speed of train B.

12. How many liters of 20% salt solution must be added to 60 liters of 40% salt solution to get a solution that is 35% salt?

4

Exponents and Polynomials

See Section 4.2, Exercise 57.

4.1

Exponents

1 *Review exponents.*

2 *Learn the rules of exponents.*

3 *Evaluate expressions containing negative exponents.*

1   In order to understand and work with polynomials we need to expand our knowledge of exponents. Exponents were introduced in Chapter 1 (Section 1.7). Let us review the fundamental concepts. In the expression x^n, x is referred to as the **base** and n is called the **exponent.** x^n is read "x to the nth power."

$$x^2 = \underbrace{x \cdot x}_{2 \text{ factors of } x}$$

$$x^4 = \underbrace{x \cdot x \cdot x \cdot x}_{4 \text{ factors of } x}$$

$$x^m = \underbrace{x \cdot x \cdot x \cdots x}_{m \text{ factors of } x}$$

EXAMPLE 1   Write $xxxxyy$ using exponents.

Solution: $\underbrace{x \; x \; x \; x}_{\substack{4 \text{ factors} \\ \text{of } x}} \; \underbrace{y \; y}_{\substack{2 \text{ factors} \\ \text{of } y}} = x^4 y^2$ ■

Remember that when a term containing a variable is given without a numerical coefficient, the numerical coefficient of the term is assumed to be 1. For example, $x = 1x$ and $x^2 y = 1x^2 y$.

Also recall that when a variable or numerical value is given without an exponent, the exponent of that variable or numerical value is assumed to be 1. For example, $x = x^1$, $xy = x^1 y^1$, $x^2 y = x^2 y^1$, and $2xy^2 = 2^1 x^1 y^2$.

Rules of Exponents   **2**   Now we will learn the rules of exponents.

EXAMPLE 2   Multiply $x^4 \cdot x^3$.

Solution: $\underbrace{x^4 \qquad \cdot \qquad x^3}_{x \cdot x \cdot x \cdot x \cdot x \cdot x \cdot x} = x^7$ ■

Example 2 illustrates that when multiplying expressions with the same base, keep the base and add the exponents. This is the product rule of exponents.

Product Rule
$$x^m \cdot x^n = x^{m+n}$$

In Example 2 we showed that $x^4 \cdot x^3 = x^7$. This problem could also be done using the product rule: $x^4 \cdot x^3 = x^{4+3} = x^7$.

EXAMPLE 3   Multiply using the product rule.

(a) $3^2 \cdot 3$ (b) $2^4 \cdot 2^2$ (c) $x \cdot x^4$ (d) $x^2 \cdot x^5$ (e) $y^4 \cdot y^7$

Solution: (a) $3^2 \cdot 3 = 3^2 \cdot 3^1 = 3^{2+1} = 3^3$ or 27
(b) $2^4 \cdot 2^2 = 2^{4+2} = 2^6$ or 64
(c) $x \cdot x^4 = x^1 \cdot x^4 = x^{1+4} = x^5$
(d) $x^2 \cdot x^5 = x^{2+5} = x^7$
(e) $y^4 \cdot y^7 = y^{4+7} = y^{11}$ ■

COMMON STUDENT ERROR

Note in Example 3(a) that $3^2 \cdot 3^1$ is 3^3 and not 9^3. When multiplying powers of the same base, *do not multiply the bases.*

Correct	*Wrong*
$3^2 \cdot 3^1 = 3^3$	$3^2 \cdot 3^1 = 9^3$
$x^y \cdot x^z = x^{y+z}$	$x^y \cdot x^z = (x^2)^{y+z}$

Example 4 will be helpful in explaining the quotient rule of exponents.

EXAMPLE 4   Divide $x^5 \div x^3$.

Solution: $\dfrac{x^5}{x^3} = \dfrac{x \cdot x \cdot x \cdot x \cdot x}{x \cdot x \cdot x} = \dfrac{1x^2}{1} = x^2$ ■

When dividing expressions with the same base, keep the base and subtract the exponent in the denominator from the exponent in the numerator.

Quotient Rule

$$\frac{x^m}{x^n} = x^{m-n}, \qquad x \neq 0$$

In Example 4 we showed that $x^5/x^3 = x^2$. This problem could also be done using the quotient rule: $x^5/x^3 = x^{5-3} = x^2$.

EXAMPLE 5   Divide each expression.

(a) $\dfrac{3^5}{3^2}$ (b) $\dfrac{5^4}{5}$ (c) $\dfrac{x^{12}}{x^5}$ (d) $\dfrac{x^9}{x^5}$ (e) $\dfrac{x^7}{x}$

Solution: (a) $\dfrac{3^5}{3^2} = 3^{5-2} = 3^3$ or 27   (b) $\dfrac{5^4}{5} = \dfrac{5^4}{5^1} = 5^{4-1} = 5^3$ or 125

(c) $\dfrac{x^{12}}{x^5} = x^{12-5} = x^7$ (d) $\dfrac{x^9}{x^5} = x^{9-5} = x^4$

(e) $\dfrac{x^7}{x} = x^{7-1} = x^6$ ■

COMMON STUDENT ERROR

Note in Example 5(a) that $3^5/3^2$ is 3^3 and not 1^3. When dividing powers of the same base, *do not divide out the bases.*

	Correct	*Wrong*
	$\dfrac{3^3}{3^1} = 3^2$ or 9	$\dfrac{\cancel{3}^3}{\cancel{3}^1} = 1^2$
	$\dfrac{x^5}{x^3} = x^{5-3} = x^2$	$\dfrac{\cancel{x}^5}{\cancel{x}^3} = 1^2$

The power rule will be explained with the aid of Example 6.

EXAMPLE 6   Simplify $(x^3)^2$.

Solution: $(x^3)^2 = \underbrace{x^3 \cdot x^3}_{\substack{2 \text{ factors} \\ \text{of } x^3}} = x^{3+3} = x^6$ ■

Power Rule

$$(x^m)^n = x^{m \cdot n}$$

Example 6 could also be solved using the power rule.

$$(x^3)^2 = x^{3 \cdot 2} = x^6$$

Note that the answers are the same.

EXAMPLE 7   Simplify each term.
(a) $(x^3)^4$ (b) $(3^4)^2$ (c) $(y^3)^8$

Solution: (a) $(x^3)^4 = x^{3 \cdot 4} = x^{12}$
(b) $(3^4)^2 = 3^{4 \cdot 2} = 3^8$
(c) $(y^3)^8 = y^{3 \cdot 8} = y^{24}$ ■

HELPFUL HINT

Students often confuse the product and power rules. Note the difference carefully.

	Product rule	*Power rule*
	$x^m \cdot x^n = x^{m+n}$	$(x^m)^n = x^{m \cdot n}$
	$2^3 \cdot 2^5 = 2^{3+5} = 2^8$	$(2^3)^5 = 2^{3 \cdot 5} = 2^{15}$

EXAMPLE 8   Simplify $\dfrac{x^3}{x^3}$.

Solution: By the quotient rule,

$$\frac{x^3}{x^3} = x^{3-3} = x^0$$

However,

$$\frac{x^3}{x^3} = \frac{\cancel{x} \cdot \cancel{x} \cdot \cancel{x}}{\cancel{x} \cdot \cancel{x} \cdot \cancel{x}} = 1$$

Since $x^3/x^3 = x^0$ and $x^3/x^3 = 1$, then x^0 must equal 1.   ▪

Example 8 illustrates the zero exponent rule.

Zero Exponent Rule

$$x^0 = 1, \qquad x \neq 0$$

EXAMPLE 9   Simplify each expression.

(a) 3^0 (b) x^0 (c) $3x^0$ (d) $(3x)^0$

Solution: (a) $3^0 = 1$
(b) $x^0 = 1$
(c) $3x^0 = 3(x^0)$ Remember: the exponent refers only to the immediately
      $= 3 \cdot 1 = 3$ preceding symbol unless parentheses are used.
(d) $(3x)^0 = 1$ ▪

The zero exponent rule specifies that $x \neq 0$. Why can x not be 0? Consider the problem $0^2/0^2$.

$$\frac{0^2}{0^2} = \frac{0 \cdot 0}{0 \cdot 0} = \frac{0}{0} = \text{indeterminate} \qquad \text{(from Section 1.6)}$$

By the quotient rule,

$$\frac{0^2}{0^2} = 0^{2-2} = 0^0$$

Notice that $0^2/0^2 = 0^0$ and $0^2/0^2 = \text{indeterminate}$. Therefore, 0^0 must be indeterminate.

❸   The negative exponent rule will be developed with the aid of Example 10.

EXAMPLE 10   Simplify $\dfrac{x^3}{x^5}$.

Solution: By the quotient rule,

$$\frac{x^3}{x^5} = x^{3-5} = x^{-2}$$

However,

$$\frac{x^3}{x^5} = \frac{\cancel{x} \cdot \cancel{x} \cdot \cancel{x}}{\cancel{x} \cdot \cancel{x} \cdot \cancel{x} \cdot x \cdot x} = \frac{1}{x^2}$$

Since $x^3/x^5 = x^{-2}$ and $x^3/x^5 = 1/x^2$, then x^{-2} must equal $1/x^2$.   ▪

Negative Exponent Rule

$$x^{-m} = \frac{1}{x^m}, \qquad x \neq 0$$

EXAMPLE 11   Use the negative exponent rule to write each expression with positive exponents.

(a) x^{-2} (b) y^{-4} (c) 3^{-2} (d) 5^{-1} (e) $\dfrac{1}{x^{-2}}$

Solution: (a) $x^{-2} = \dfrac{1}{x^2}$ (b) $y^{-4} = \dfrac{1}{y^4}$

(c) $3^{-2} = \dfrac{1}{3^2} = \dfrac{1}{9}$ (d) $5^{-1} = \dfrac{1}{5}$

(e) $\dfrac{1}{x^{-2}} = \dfrac{1}{1/x^2} = \dfrac{1}{1} \cdot \dfrac{x^2}{1} = x^2$ ∎

HELPFUL HINT

> When a factor is moved from the denominator to the numerator or from the numerator to the denominator, the sign of the exponent changes.
>
> $$x^{-4} = \frac{1}{x^4} \qquad \frac{1}{x^{-4}} = x^4$$
>
> $$3^{-5} = \frac{1}{3^5} \qquad \frac{1}{3^{-5}} = 3^5$$

Now let's look at examples that combine two or more of the rules presented so far.

EXAMPLE 12   Simplify each term. Write each answer without negative exponents.

(a) $(y^{-3})^8$ (b) $(4^2)^{-3}$

Solution: (a) $(y^{-3})^8 = y^{(-3)(8)}$ by the power rule

$= y^{-24}$

$= \dfrac{1}{y^{24}}$ by the negative exponent rule

(b) $(4^2)^{-3} = 4^{(2)(-3)}$ by the power rule

$= 4^{-6}$

$= \dfrac{1}{4^6}$ by the negative exponent rule   ∎

EXAMPLE 13   Simplify each of the following. Write each answer without negative exponents.

(a) $x^3 \cdot x^{-5}$ (b) $3^{-4} \cdot 3^{-7}$

Solution: (a) $x^3 \cdot x^{-5} = x^{3+(-5)}$ by the product rule

$= x^{-2}$

$= \dfrac{1}{x^2}$ by the negative exponent rule

(b) $3^{-4} \cdot 3^{-7} = 3^{-4+(-7)}$ by the product rule

$= 3^{-11}$

$= \dfrac{1}{3^{11}}$ by the negative exponent rule   ∎

COMMON STUDENT ERROR

What is the sum of $3^2 + 3^{-2}$? Look carefully at the correct solution.

$$\textit{Correct} \qquad\qquad \textit{Wrong}$$

$$3^2 + 3^{-2} = 9 + \tfrac{1}{9} \qquad \cancel{3^2 + 3^{-2} = 0}$$
$$= 9\tfrac{1}{9}$$

Note that $3^2 \cdot 3^{-2} = 3^{2+(-2)} = 3^0 = 1$

EXAMPLE 14   Simplify each of the following. Write each answer without negative exponents.

(a) $\dfrac{x^7}{x^{10}}$ (b) $\dfrac{5^{-7}}{5^{-4}}$

Solution: (a) $\dfrac{x^7}{x^{10}} = x^{7-10}$ by the quotient rule

$$= x^{-3}$$

$$= \frac{1}{x^3} \qquad \text{by the negative exponent rule}$$

(b) $\dfrac{5^{-7}}{5^{-4}} = 5^{-7-(-4)}$ by the quotient rule

$$= 5^{-7+4}$$

$$= 5^{-3}$$

$$= \frac{1}{5^3} \text{ or } \frac{1}{125} \qquad \text{by the negative exponent rule} \quad \blacksquare$$

EXAMPLE 15   Simplify each expression. Write your answer without negative exponents.

(a) $4x^2(5x^{-5})$ (b) $\dfrac{8x^3y^2}{4xy^2}$ (c) $\dfrac{2x^2y^5}{8x^7y^{-3}}$

Solution: (a) $4x^2(5x^{-5}) = 4 \cdot 5 \cdot x^2 \cdot x^{-5} = 20x^{-3} = \dfrac{20}{x^3}$

(b) $\dfrac{\overset{2}{\cancel{8}}x^3y^2}{\underset{1}{\cancel{4}}xy^2} = 2 \cdot \dfrac{x^3}{x} \cdot \dfrac{y^2}{y^2} = 2x^2$

(c) $\dfrac{\overset{1}{\cancel{2}}x^2y^5}{\underset{4}{\cancel{8}}x^7y^{-3}} = \dfrac{1}{4} \cdot \dfrac{x^2}{x^7} \cdot \dfrac{y^5}{y^{-3}}$

$$= \frac{1}{4} \cdot \frac{1}{x^5} \cdot y^{5-(-3)}$$

$$= \frac{1}{4} \cdot \frac{1}{x^5} \cdot y^8$$

$$= \frac{y^8}{4x^5} \quad \blacksquare$$

Example 16 will help us in explaining the expanded power rule.

EXAMPLE 16   Simplify $(ax/y)^4$.

Solution:
$$\left(\frac{ax}{y}\right)^4 = \frac{ax}{y} \cdot \frac{ax}{y} \cdot \frac{ax}{y} \cdot \frac{ax}{y}$$

$$= \frac{a \cdot a \cdot a \cdot a \cdot x \cdot x \cdot x \cdot x}{y \cdot y \cdot y \cdot y} = \frac{a^4 \cdot x^4}{y^4} \quad \blacksquare$$

Example 16 illustrates the expanded power rule.

Expanded Power Rule

$$\left(\frac{ax}{y}\right)^m = \frac{a^m x^m}{y^m}, \quad y \neq 0$$

The expanded power rule illustrates that every factor within parentheses is affected by an exponent outside the parentheses.

EXAMPLE 17   Simplify each expression. Write each answer without negative exponents.

(a) $(2x)^2$ (b) $(-x)^3$ (c) $(2xy)^3$

(d) $\left(\dfrac{-3x}{y}\right)^2$ (e) $(2x)^{-1}$ (f) $(3x)^{-2}$

Solution:

(a) $(2x)^2 = 2^2 x^2 = 4x^2$

(b) $(-x)^3 = (-1x)^3 = (-1)^3 x^3 = -1x^3 = -x^3$

(c) $(2xy)^3 = 2^3 x^3 y^3 = 8x^3 y^3$

(d) $\left(\dfrac{-3x}{y}\right)^2 = \dfrac{(-3)^2 x^2}{y^2} = \dfrac{9x^2}{y^2}$

(e) $(2x)^{-1} = 2^{-1} x^{-1} = \dfrac{1}{2} \cdot \dfrac{1}{x} = \dfrac{1}{2x}$

(f) $(3x)^{-2} = 3^{-2} x^{-2} = \dfrac{1}{3^2} \cdot \dfrac{1}{x^2} = \dfrac{1}{9x^2}$

Parts (e) and (f) could also be found by using the negative exponent rule first.

$$(2x)^{-1} = \frac{1}{(2x)^1} = \frac{1}{2x}$$

$$(3x)^{-2} = \frac{1}{(3x)^2} = \frac{1}{3^2 x^2} = \frac{1}{9x^2} \quad \blacksquare$$

EXAMPLE 18   Simplify each expression. Write your answer without negative exponents.

(a) $(2x^2y^3)^4(xy^2)$ (b) $(2x^{-1}y^5)^{-2}$ (c) $\left(\dfrac{xy^{-2}}{z}\right)^3$

Solution: (a) $(2x^2y^3)^4(xy^2) = (16x^8y^{12})(x^1y^2)$

$$= 16 \cdot x^8 \cdot x^1 \cdot y^{12} \cdot y^2$$

$$= 16x^9y^{14}$$

(b) $(2x^{-1}y^5)^{-2} = 2^{-2}(x^{-1})^{-2}(y^5)^{-2}$

$$= \frac{1}{2^2}x^2y^{-10}$$

$$= \frac{1}{4}x^2\frac{1}{y^{10}}$$

$$= \frac{x^2}{4y^{10}}$$

(c) $\left(\dfrac{xy^{-2}}{z}\right)^3 = \dfrac{x^3(y^{-2})^3}{z^3}$

$$= \frac{x^3y^{-6}}{z^3}$$

$$= \frac{x^3}{y^6z^3}\quad\blacksquare$$

Summary of Rules of Exponents

1. $x^m \cdot x^n = x^{m+n}$ product rule

2. $\dfrac{x^m}{x^n} = x^{m-n}, \quad x \neq 0$ quotient rule

3. $(x^m)^n = x^{m \cdot n}$ power rule

4. $x^0 = 1, \quad x \neq 0$ zero exponent rule

5. $x^{-m} = \dfrac{1}{x^m}, \quad x \neq 0$ negative exponent rule

6. $\left(\dfrac{ax}{y}\right)^m = \dfrac{a^mx^m}{y^m}, \quad y \neq 0$ expanded power rule

---

Exercise Set 4.1

Simplify each of the following.

1. $x^4 \cdot x^3$ **2.** $x^5 \cdot x^2$ **3.** $y^2 \cdot y$ **4.** $4^2 \cdot 4$

5. $3^2 \cdot 3^3$ **6.** $x^4 \cdot x^2$ **7.** $y^3 \cdot y^2$ **8.** $x^3 \cdot x^4$

9. $y^4 \cdot y$ **10.** $\dfrac{x^4}{x^3}$ **11.** $\dfrac{x^{15}}{x^7}$ **12.** $\dfrac{y^3}{y}$

13. $\dfrac{5^4}{5^2}$ **14.** $\dfrac{3^5}{3^2}$ **15.** $\dfrac{x^9}{x^5}$ **16.** $\dfrac{x^3}{x^5}$

17. $\dfrac{y^2}{y}$

18. $\dfrac{x^{13}}{x^4}$

19. $\dfrac{x^2}{x^2}$

20. $\dfrac{3^4}{3^4}$

21. x^0

22. 5^0

23. $3x^0$

24. $-2x^0$

25. $(3x)^0$

26. $-(4x)^0$

27. $(-4x)^0$

28. $(x^2)^3$

29. $(x^5)^2$

30. $(x^2)^2$

31. $(x^5)^5$

32. $(x^4)^2$

33. $(x^3)^1$

34. $(x^3)^2$

35. $(x^3)^4$

36. $(x^5)^4$

37. $(x^4)^2$

38. $(2x)^2$

39. $(3x)^2$

40. $(-3x)^2$

41. $(-x)^2$

42. $(-x)^3$

43. $(4x^2)^3$

44. $(5x^3)^2$

45. $(-3x^3)^3$

46. $(xy)^4$

47. $(2x^2y)^3$

48. $(4x^3y^2)^3$

49. $(5x^2y^5)^2$

50. $(2xy^4)^3$

51. $\left(\dfrac{x}{y}\right)^2$

52. $\left(\dfrac{x}{3}\right)^2$

53. $\left(\dfrac{x}{5}\right)^3$

54. $\left(\dfrac{2}{x}\right)^3$

55. $\left(\dfrac{y}{x}\right)^5$

56. $\left(\dfrac{3}{y}\right)^4$

57. $\left(\dfrac{6}{x}\right)^3$

58. $\left(\dfrac{2x}{y}\right)^3$

59. $\left(\dfrac{3x}{y}\right)^3$

60. $\left(\dfrac{5x^2}{y}\right)^2$

61. $\left(\dfrac{2x}{5}\right)^2$

62. $\left(\dfrac{3x^4}{2}\right)^3$

63. $\left(\dfrac{4y^3}{x}\right)^3$

64. $\left(\dfrac{-4x^2}{5}\right)^2$

65. $\left(\dfrac{-3x^3}{4}\right)^3$

66. $\left(\dfrac{-x^5}{y^2}\right)^3$

Simplify each of the following. Write the answer without negative exponents.

67. x^{-4}

68. $\dfrac{1}{x^{-4}}$

69. 3^{-5}

70. $\dfrac{1}{3^{-5}}$

71. $x^4 \cdot x^{-3}$

72. $x^5 \cdot x^{-7}$

73. $x^{-3} \cdot x^{-5}$

74. $3^{-4} \cdot 3^2$

75. $4^{-3} \cdot 4^5$

76. $(x^2)^{-3}$

77. $(x^{-2})^4$

78. $(x^{-2})^{-3}$

79. $x^{-4} \cdot x^7$

80. $\dfrac{x^5}{x^7}$

81. $\dfrac{x^4}{x^9}$

82. $\dfrac{5^{-2}}{5^3}$

83. $\dfrac{2^4}{2^{-3}}$

84. $\dfrac{3^8}{3^9}$

85. $\dfrac{x^{-3}}{x^5}$

86. $(3x)^{-1}$

87. $(2y^2)^{-2}$

88. $(4y^2)^{-3}$

89. $(2y^2)^{-1}$

90. $(2y^{-1})^{-1}$

91. $(y^{-1})^{-1}$

92. $(3x^{-2})^{-2}$

93. $(3x^{-4})^{-3}$

Simplify each of the following. Write the answer without negative exponents.

94. $\dfrac{x^2y^6}{x^4y}$

95. $\dfrac{x^5y^7}{x^{12}y^3}$

96. $\dfrac{x^4y^5}{x^7y^{12}}$

97. $\dfrac{10x^3y^8}{2xy^{10}}$

98. $\dfrac{5x^{12}y^2}{10xy^9}$

99. $\dfrac{4xy}{16x^3y^2}$

100. $\dfrac{20x^4y^6}{5xy^9}$

101. $\dfrac{35x^4y^7}{10x^9y^{12}}$

102. $\dfrac{x^2y^{-2}}{x^4y^4}$

103. $\dfrac{2x^{-3}y^4}{xy}$

104. $\dfrac{5xy^5}{2x^{-3}y^8}$

105. $\dfrac{12x^{-4}y}{24x^{-3}y}$

106. $(2x^4)(5x^6)$

107. $(3x^2y)(-2x^4y^6)$

108. $(2x^3)(6xy^2)$

109. $(x^2y^3)(3x^4y^5)$

110. $(2x^{-4})(3x^6)$

111. $(5x^2y^{-4})(2x^3y^5)$

112. $(3x^4y^2)(-2x^5y^{-3})$

113. $(2x^{-4}y^{-5})(x^4y^5)$

114. $(-4x^{-3}y^2)(-2x^4y^5)$

115. $(-3x^{-2}y^{-2})(2x^4y^6)$

Simplify each of the following. Write the answer without negative exponents.

116. $(4x^2y)^{-2}$ **117.** $(xy^2)^{-1}$ **118.** $(4x^3y^{-2})^{-2}$ **119.** $(6x^{-1}y^{-1})^{-2}$

120. $(4x^{-1}y^2)^{-3}$ **121.** $(xy^{-3})^{-3}$ **122.** $(2x^4y^{-2})^4$ **123.** $(3xy^{-5})^{-2}$

124. $(3x)^2(6xy)$ **125.** $(2x)^3(3x)^2$ **126.** $(5x^3)^2(2x)^2$ **127.** $(2x^{-2})^2(4x)$

128. $(3x^{-1})^3(x^{-2})$ **129.** $(4x)^2(2x^2)^3$ **130.** $(x^4)^3(3x^2)$ **131.** $4x^2(2x^{-3})^2$

132. $\left(\dfrac{6x^2y}{z^2}\right)^3$ **133.** $\left(\dfrac{-2x^3y^2}{z}\right)^2$ **134.** $\left(\dfrac{3x^2y^4}{z^2}\right)^3$ **135.** $\left(\dfrac{2x^3y^4}{5}\right)^2$

136. $\left(\dfrac{2x^{-3}y}{5}\right)^2$ **137.** $\left(\dfrac{3x^2y^{-3}}{4}\right)^2$ **138.** $\left(\dfrac{x^{-4}y^{-4}}{z^3}\right)^2$ **139.** $\left(\dfrac{4x^{-3}y^2}{3z}\right)^3$

140. For what value of x is $x^0 \neq 1$?

141. Explain the difference between the product rule and power rule. Give an example of each.

142. Explain what is *wrong* with the following: $3x^{-2} = 1/3x^2$. Write $3x^{-2}$ without negative exponents.

143. Explain what is *wrong* with the following: $2x^{-1} = 1/2x$. Write $2x^{-1}$ without negative exponents.

JUST FOR FUN

1. Simplify $(3x^{-2})^2(4x^2)^{-2}(-x^{-5})^{-5}$

2. Simplify $\dfrac{(-6x^2y^3)^3}{(3xy^4)^3}$.

3. Often problems involving exponents can be done in more than one way. Simplify $\left(\dfrac{3x^2y^3}{z}\right)^{-2}$

   (a) by first using the expanded power rule.

   (b) by first using the negative exponent rule.

4.2

Scientific Notation (Optional)

1 *Convert a decimal number to and from scientific notation.*

2 *Do calculations with numbers in scientific notation form.*

---

1   When working with scientific problems, one often deals with very large and very ·small numbers. For example, the distance from the earth to the sun is about 93,000,000 miles. The wavelength of a yellow color of light is about 0.0000006 meters. Because it is difficult to work with many zeros, scientists often express such numbers with exponents. For example, the number 93,000,000 might be written 9.3×10^7 and the number 0.0000006 might be written 6.0×10^{-7}. Numbers such as 9.3×10^7 and 6.0×10^{-7} are in a form called **scientific notation.** Each number written in scientific notation is written as a number greater than or equal to 1 and less than 10 $(1 \leq a < 10)$ multiplied by some power of 10.

Examples of numbers in scientific notation

$$1.2 \times 10^6$$
$$3.762 \times 10^3$$
$$8.07 \times 10^{-2}$$
$$1 \times 10^{-5}$$

If we divide and multiply a given number by any nonzero number, the value of the given number does not change; only its form changes. We will use this concept to help explain how to write a number in scientific notation form.

Consider the number 68,400.

$$68,400 = \frac{68,400}{10,000} \times 10,000 \qquad \text{divide and multiply by 10,000}$$
$$= 6.84 \times 10,000$$
$$= 6.84 \times 10^4 \qquad (\text{Note that } 10,000 = 10 \cdot 10 \cdot 10 \cdot 10 = 10^4)$$

Therefore $68,400 = 6.84 \times 10^4$. Note that to go from 68,400 to 6.84 the decimal point was moved 4 places to the left. Also note that the exponent on the 10, the 4, is the same as the number of places the decimal point was moved to the left. Here is a simplified procedure for writing a number in scientific notation:

To write a number in scientific notation

1. Move the decimal in the original number to the right or left until you obtain a number greater than or equal to 1 and less than 10.
2. Count the number of places you have moved the decimal to obtain the number in step 1. If the decimal was moved to the left, the count is to be considered positive. If the decimal was moved to the right, the count is to be considered negative.
3. Multiply the number obtained in step 1 by 10 raised to the count (power) found in step 2.

EXAMPLE 1   Write the following numbers using scientific notation.

(a) 10,700
(b) 0.000386
(c) 972,000
(d) 0.0083

Solution: (a) 10,700 means 10,700.

$$10,700. = 1.07 \times 10^4$$

4 places
to left

(b) $0.000386 = 3.86 \times 10^{-4}$

4 places
to right

(c) $972,000. = 9.72 \times 10^5$

5 places
to left

(d) $0.0083 = 8.3 \times 10^{-3}$ ■

3 places
to right

> **To Convert from a Number Given in Scientific Notation**
>
> **1.** Observe the exponent of the power of 10.
>
> **2. (a)** If the exponent is positive, move the decimal in the number (greater than or equal to 1 and less than 10) to the right the same number of places as the exponent. It may be necessary to add zeros to the number.
>
>   **(b)** If the exponent is negative, move the decimal in the number to the left the same number of places as the exponent. It may be necessary to add zeros.

EXAMPLE 2   Write each number without exponents.

(a) 3.2×10^4 (b) 6.28×10^{-3} (c) 7.95×10^8

Solution: (a) Moving the decimal four places to the right gives

$$3.2 \times 10^4 = 3.2 \times 10{,}000 = 32000$$

(b) Move the decimal three places to the left.

$$6.28 \times 10^{-3} = 0.00628$$

(c) Move the decimal eight places to the right.

$$7.95 \times 10^8 = 795{,}000{,}000 \qquad \blacksquare$$

2   We can use the rules of exponents presented in Section 4.1 when working with numbers written in scientific notation.

EXAMPLE 3   Multiply $(4.2 \times 10^6)(2 \times 10^{-4})$.

Solution: $(4.2 \times 10^6)(2 \times 10^{-4}) = (4.2 \times 2)(10^6 \times 10^{-4})$

$$= 8.4 \times 10^2$$
$$= 840 \qquad \blacksquare$$

EXAMPLE 4   Divide $\dfrac{6.2 \times 10^{-5}}{2 \times 10^{-3}}$

Solution: $\dfrac{6.2 \times 10^{-5}}{2 \times 10^{-3}} = \left(\dfrac{6.2}{2}\right)\left(\dfrac{10^{-5}}{10^{-3}}\right)$

$$= 3.1 \times 10^{-5-(-3)}$$
$$= 3.1 \times 10^{-5+3}$$
$$= 3.1 \times 10^{-2}$$
$$= 0.031 \qquad \blacksquare$$

EXAMPLE 5   Multiply $(42{,}100{,}000)(0.008)$.

Solution: Change each number to scientific notation form.

$$(42{,}100{,}000)(0.008) = (4.21 \times 10^7)(8 \times 10^{-3})$$
$$= (4.21 \times 8)(10^7 \times 10^{-3})$$
$$= 33.68 \times 10^4$$
$$= 336{,}800 \qquad \blacksquare$$

☐ *Calculator
Corner*

What will your calculator show when you multiply very large or very small numbers? The answer depends on whether your calculator has the ability to display an answer in scientific notation form. On calculators without the ability to express numbers in scientific notation, you will probably get an error message because the answer will be too large or too small for the display.

Example: **On a calculator without scientific notation:**

$$\boxed{C}\;\; 8000000 \;\boxed{\times}\; 600000 \;\boxed{=}\; \text{Error}$$

If your calculator has the ability to give an answer in scientific notation form, you would probably get the following:

Example: **On a calculator that uses scientific notation form:**

$$\boxed{C}\;\; 8000000 \;\boxed{\times}\; 600000 \;\boxed{=}\; 4.8 \qquad 12$$

This 4.8   12 means 4.8×10^{12}.

Example: **On a calculator that uses scientific notation form:**

$$\boxed{C}\;\; .0000003 \;\boxed{\times}\; .004 \;\boxed{=}\; 1.2 \qquad -9$$

This 1.2 -9 means 1.2×10^{-9}.

Exercise Set 4.2

Express each number in scientific notation form.

1. 42,000	**2.** 3,610,000	**3.** 900	**4.** 0.00062
5. 0.053	**6.** 0.0000462	**7.** 19,000	**8.** 5,260,000,000
9. 0.00000186	**10.** 0.0003	**11.** 0.00000914	**12.** 37,000
13. 107	**14.** 0.02	**15.** 0.153	**16.** 416,000

Express each number without exponents.

17. 4.2×10^3	**18.** 1.63×10^{-4}	**19.** 4×10^7	**20.** 6.15×10^5
21. 2.13×10^{-5}	**22.** 9.64×10^{-7}	**23.** 3.12×10^{-1}	**24.** 4.6×10^1
25. 9×10^6	**26.** 7.3×10^4	**27.** 5.35×10^2	**28.** 1.04×10^{-2}
29. 3.5×10^4	**30.** 2.17×10^{-6}	**31.** 1×10^4	**32.** 1×10^{-3}

Perform the indicated operation and express each number without exponents.

33. $(4 \times 10^2)(3 \times 10^5)$

34. $(2 \times 10^{-3})(3 \times 10^2)$

35. $(5.1 \times 10^1)(3 \times 10^{-4})$

36. $(1.6 \times 10^{-2})(4 \times 10^{-3})$

37. $\dfrac{6.4 \times 10^5}{2 \times 10^3}$

38. $\dfrac{8 \times 10^{-3}}{2 \times 10^1}$

39. $\dfrac{8.4 \times 10^{-6}}{4 \times 10^{-3}}$

40. $\dfrac{25 \times 10^3}{5 \times 10^{-2}}$

41. $\dfrac{4 \times 10^5}{2 \times 10^4}$

42. $\dfrac{16 \times 10^3}{8 \times 10^{-3}}$

Perform the indicated operation by first converting each number to scientific notation form. Write the answer in scientific notation form.

43. (700,000)(6,000,000)

44. (0.0006)(5,000,000)

45. (0.003)(0.00015)

46. $(230,000)(3000)$

47. $\dfrac{1,400,000}{700}$

48. $\dfrac{20,000}{0.0005}$

49. $\dfrac{0.00004}{200}$

50. $\dfrac{0.0012}{0.000006}$

51. $\dfrac{150,000}{0.0005}$

52. List the numbers from smallest to largest.
4.8×10^5, 3.2×10^{-1}, 4.6, 8.3×10^{-4}

53. List the numbers from smallest to largest.
9.2×10^{-5}, 8.4×10^3, 1.3×10^{-1}, 6.2×10^4

54. The distance from the earth to the planet Jupiter is approximately 4.5×10^8 miles. If a spacecraft travels at a speed of 25,000 miles per hour, how long, in hours, would it take the spacecraft to travel from the earth to Jupiter. Use distance = rate × time.

55. If a computer can do a calculation in 0.000004 seconds, how long, in seconds, would it take the computer to do 8 trillion (8,000,000,000,000) calculations?

56. The half life of a radioactive isotope is the time required for half the quantity of the isotope to decompose. The half life of uranium-238 is 4.5×10^9 years, and the half life of uranium-234 is 2.5×10^5 years. How many times greater is the half life of uranium-238 than uranium-234?

57. A treaty between the United States and Canada requires that during the tourist season a minimum of 100,000 cubic feet of water per second flow over Niagara Falls (another 130,000–160,000 cubic feet/sec are diverted for power generation). Find the minimum amount of water that will flow over the falls in a 24-hour period during the tourist season.

JUST FOR FUN

1. (a) Light travels at a speed of 1.86×10^5 miles per second. A *light year* is the distance that light travels in one year. Determine the number of miles in a light year.

(b) The earth is approximately 93,000,000 miles from the sun. How long does it take light from the sun to reach the earth?

4.3

Addition and Subtraction of Polynomials

1 *Identify polynomials.*

2 *Add polynomials.*

3 *Subtract polynomials.*

---

1 A **polynomial in x** is an expression containing the sum of a finite number of terms of the form ax^n, for any real number a and any whole number n.

Examples of polynomials	*Not polynomials*
$2x$	$4x^{1/2}$ (fractional exponent)
$\dfrac{1}{3}x - 4$	$3x^2 + 4x^{-1} + 5$ (negative exponent)
$x^2 - 2x + 1$	$4 + \dfrac{1}{x}$ $\left(\dfrac{1}{x} = x^{-1},\ \text{negative exponent}\right)$

A polynomial is written in **descending order** (or **descending powers**) **of the variable** when the exponents on the variable decrease from left to right.

Example of polynomial in descending order

$$2x^3 + 4x^2 - 6x + 3$$

Note that in the example the constant, 3, can be written as $3x^0$ since x^0 is 1.

A polynomial of one term is called a **monomial.** Examples of monomials are $4x$, $3x^2$, and 5.

A **binomial** is a two-termed polynomial. Examples of binomials are $x + 4$, and $x^2 - 6x$.

A **trinomial** is a three-termed polynomial. An example of a trinomial is $x^2 - 2x + 1$.

Polynomials containing more than three terms are not given special names. The term "poly" is a prefix meaning "many."

The **degree of a term** of a polynomial in one variable is the exponent on the variable.

Term	*Degree of term*
$4x^2$	Second
$2y^5$	Fifth
$-5x$	First $(-5x$ can be written $-5x^1)$
3	Zero (3 can be written $3x^0$)

The **degree of a polynomial** in one variable is the same as that of its highest-degree term.

Polynomial	*Degree of polynomial*
$8x^3 + 2x^2 - 3x + 4$	Third (x^3 is highest-degree term)
$x^2 - 4$	Second (x^2 is highest-degree term)
$2x - 1$	First ($2x$ or $2x^1$ is highest-degree term)
4	Zero (4 or $4x^0$ is highest-degree term)

Addition of Polynomials

2   In Section 2.1 we stated that like terms are terms having the same variables and same exponents.

Examples of like terms

$$3, \quad -5$$
$$2x, \quad x$$
$$-2x^2, \quad 4x^2$$
$$3y^2, \quad 5y^2$$
$$3xy^2, \quad 5xy^2$$

To add polynomials, combine the like terms of the polynomials.

EXAMPLE 1   Simplify $(4x^2 + 6x + 3) + (2x^2 + 5x - 1)$.

Solution:
$$(4x^2 + 6x + 3) + (2x^2 + 5x - 1)$$
$$= 4x^2 + 6x + 3 + 2x^2 + 5x - 1 \quad \text{remove parentheses}$$
$$= \underbrace{4x^2 + 2x^2} \underbrace{+ 6x + 5x} \underbrace{+ 3 - 1} \quad \text{rearrange terms}$$
$$= \quad 6x^2 \quad + \quad 11x \quad + \quad 2 \quad \text{combine like terms} \quad \blacksquare$$

EXAMPLE 2   Simplify $(4x^2 + 3x + y) + (x^2 - 6x + 3)$.

Solution:
$$(4x^2 + 3x + y) + (x^2 - 6x + 3)$$
$$= 4x^2 + 3x + y + x^2 - 6x + 3 \quad \text{remove parentheses}$$
$$= \underbrace{4x^2 + x^2} \underbrace{+ 3x - 6x} + y + 3 \quad \text{rearrange terms}$$
$$= \quad 5x^2 \quad - \quad 3x \quad + y + 3 \quad \text{combine like terms} \quad \blacksquare$$

EXAMPLE 3   Simplify $(3x^2y - 4xy + y) + (x^2y + 2xy + 3y)$

Solution:
$$(3x^2y - 4xy + y) + (x^2y + 2xy + 3y)$$
$$= 3x^2y - 4xy + y + x^2y + 2xy + 3y \quad \text{remove parentheses}$$
$$= \underbrace{3x^2y + x^2y} \underbrace{- 4xy + 2xy} \underbrace{+ y + 3y} \quad \text{rearrange terms}$$
$$= \quad 4x^2y \quad - \quad 2xy \quad + \quad 4y \quad \text{combine like terms} \quad \blacksquare$$

Addition of Polynomials in Columns

To Add Polynomials in Columns
1. Arrange polynomials in descending order one under the other with like terms in the same columns.
2. Find the sum of the terms in each column.

EXAMPLE 4   Add $4x^2 - 2x + 2$ and $-2x^2 - x + 4$ using columns.

Solution:
$$\begin{array}{r} 4x^2 - 2x + 2 \\ -2x^2 - \ x + 4 \\ \hline 2x^2 - 3x + 6 \end{array} \quad \blacksquare$$

EXAMPLE 5   Add $(3x^3 + 2x - 4)$ and $(2x^2 - 6x - 3)$ using columns.

Solution: Since the polynomial $3x^3 + 2x - 4$ does not have an x^2 term, we will add the term $0x^2$ to the polynomial. This procedure sometimes helps in aligning like terms.

$$\begin{array}{r} 3x^3 + 0x^2 + 2x - 4 \\ 2x^2 - 6x - 3 \\ \hline 3x^3 + 2x^2 - 4x - 7 \end{array} \quad \blacksquare$$

Subtraction of Polynomials

3

To Subtract Polynomials
1. Remove parentheses. (This will have the effect of changing the sign of *every* term within the parentheses of the polynomial being subtracted.)
2. Combine like terms

EXAMPLE 6   Simplify $(3x^2 - 2x + 5) - (x^2 - 3x + 4)$.

Solution: $(3x^2 - 2x + 5) - (x^2 - 3x + 4)$

$= 3x^2 - 2x + 5 - x^2 + 3x - 4$ remove parentheses (change sign of each term being subtracted)

$= \underbrace{3x^2 - x^2} \; \underbrace{- 2x + 3x} \; \underbrace{+ 5 - 4}$ rearrange terms

$= \qquad 2x^2 \qquad + \quad x \qquad + 1$ combine like terms   ■

EXAMPLE 7   Subtract $(-x^2 - 2x + 3)$ from $(x^3 + 4x + 6)$.

Solution: $(x^3 + 4x + 6) - (-x^2 - 2x + 3)$

$= x^3 + 4x + 6 + x^2 + 2x - 3$ remove parentheses

$= x^3 + x^2 \underbrace{+ 4x + 2x} \underbrace{+ 6 - 3}$ rearrange terms

$= x^3 + x^2 + \quad 6x \qquad + \quad 3$ combine like terms   ■

Subtraction
of Polynomials
in Columns

> **To Subtract Polynomials in Columns**
>
> **1.** Write *the polynomial being subtracted* below the polynomial from which it is being subtracted. List like terms in the same column.
> **2.** Change the sign of each term in the polynomial being subtracted. (This step can be done mentally, if you like.)
> **3.** Find the sum of the terms in each column.

EXAMPLE 8   Subtract $(x^2 - 6x + 5)$ from $(3x^2 + 5x + 7)$ using columns.

Solution: Align like terms in columns (Step 1).

$$\begin{array}{l} 3x^2 + 5x + 7 \\ \underline{x^2 - 6x + 5} \end{array} \quad \text{align like terms}$$

Change all signs in the second row (Step 2), then add (Step 3).

$$\begin{array}{l} 3x^2 + 5x + 7 \\ \underline{- x^2 + 6x - 5} \quad \text{change all signs} \\ 2x^2 + 11x + 2 \quad \text{add} \quad ■ \end{array}$$

EXAMPLE 9   Using columns, subtract $(2x^2 - 6)$ from $(-3x^3 + 4x - 3)$.

Solution: To help with aligning like terms, we will write each expression with descending powers of x. If a given power of x is missing, we will write that term with a numerical coefficient of 0.

$$-3x^3 + 4x - 3 = -3x^3 + 0x^2 + 4x - 3$$
$$2x^2 - 6 = 2x^2 + 0x - 6$$

Align like terms.

$$\begin{array}{l} -3x^3 + 0x^2 + 4x - 3 \\ \underline{ 2x^2 + 0x - 6} \end{array}$$

Change all signs in the second row, then add.

$$\begin{array}{r} -3x^3 + 0x^2 + 4x - 3 \\ -\ 2x^2 - 0x + 6 \\ \hline -3x^3 - 2x^2 + 4x + 3 \end{array}\quad\blacksquare$$

Note: Many of you will find that you can change the signs mentally, and can therefore align and change the signs in one step.

Exercise Set 4.3

Indicate the expressions that are polynomials. If the polynomial has a specific name—for example, monomial or binomial—give that name.

1. $4x$

2. $3x^2 - 6x + 7$

3. -12

4. $3x^{-2}$

5. $-4x - 6x^2$

6. $4x^3 - 8$

7. $6x^2 - 2x + 8$

8. $x - 3$

9. $3x^{1/2} + 2x$

10. $-2x^2 + 5x^{-1}$

11. $2x + 5$

12. $4 - 3x$

13. 6

14. $x^{1/3} + x^{2/3}$

15. $3x^3 - 2x^2 + 4x - 7$

16. $x^2 - 3$

17. $4 - x^2 - 6x$

18. $-3x$

19. $2x^{-2}$

20. $6x^2 + 3x - 5$

Express each polynomial in descending order. If the polynomial is already in descending order, so state. Give the degree of each polynomial.

21. $3x$

22. 6

23. $2x^2 - 6 + x$

24. $-4 + x^2 - 2x$

25. $-8 - 4x - x^2$

26. $2x + 4 - x^2$

27. $x^3 - 6$

28. $-x - 1$

29. $2x^2 + 5x - 8$

30. $3x^3 - x + 4$

31. $4 - 6x^3 + x^2 - 3x$

32. $-4 + x - 3x^2 + 4x^3$

33. $-2x + 5x^2 - 4$

34. $1 - x^3 + 3x$

35. $5x + 3x^2 - 6 - 2x^3$

36. $4 - 2x - 3x^2 + 5x^4$

Add as indicated.

37. $(2x + 3) + (4x - 2)$

38. $(3x - 6) + (2x - 3)$

39. $(-4x + 8) + (2x + 3)$

40. $(-5x - 3) + (-2x + 3)$

41. $(5x + 8) + (-6x - 10)$

42. $(-8x + 4) + (3x - 12)$

43. $(9x - 12) + (12x - 9)$

44. $(3x - 8) + (-8x + 5)$

45. $(x^2 + 2x - 3) + (4x + 6)$

46. $(-2x^2 + 3x - 9) + (-2x - 3)$

47. $(5x - 7) + (2x^2 + 3x + 12)$

48. $(-3x + 8) + (-2x^2 - 3x - 5)$

49. $(3x^2 - 4x + 8) + (2x^2 + 5x + 12)$

50. $(x^2 - 6x + 7) + (-x^2 + 3x + 5)$

51. $(-3x^2 - 4x + 8) + (2x^2 + 5x + 12)$

52. $(9x^2 + 3x - 12) + (5x^2 - 8x - 3)$

53. $(8x^2 + 4) + (-2x^2 - 5x)$

54. $(8x^3 + 4x^2 + 6) + (2x^2 + 5x)$

55. $(-7x^3 - 3x^2 + 4) + (5x^3 + 4x - 7)$

56. $(9x^3 - 2x^2 + 4x - 7) + (2x^3 - 6x^2 - 4x + 3)$

57. $(x^2 + xy - y^2) + (2x^2 - 3xy + y^2)$

58. $(x^2y + 6x^2 - 3xy^2) + (-x^2y - 12x^2 + 4xy^2)$

59. $(4x^2y + 2x - 3) + (3x^2y - 5x + 5)$

60. $(x^2y + x - y) + (2x^2y + 2x - 6y + 3)$

Add in columns.

61. Add $x + 3$ and 4.

62. Add $2x + 4$ and $3x$.

63. Add $2x - 3$ and 4.

64. Add $2x - 3$ and -4.

65. Add $3x - 6$ and $4x + 5$.

66. Add $-2x + 5$ and $-3x - 5$.

67. Add $-x + 7$ and $-2x + 9$.

68. Add $3x$ and $4x - 3$.

69. Add $x^2 - 2x + 4$ and $3x + 12$.

70. Add $4x^2 - 6x + 5$ and $-2x - 8$.

71. Add $-2x^2 + 4x - 12$ and $-x^2 - 2x$.
73. Add $3x^2 + 4x - 5$ and $4x^2 + 3x - 8$.
75. Add $-x^2 - 2x - 3$ and $-x^2 - 2x - 3$.
77. Add $2x^3 + 3x^2 + 6x - 9$ and $-4x^2 + 7$.
79. Add $6x^3 - 4x^2 + x - 9$ and $-x^3 - 3x^2 - x + 7$.
81. Add $xy + 6x + 4$ and $2xy - 3x - 1$.

72. Add $5x^2 + x + 9$ and $2x^2 - 12$.
74. Add $-5x^2 - 3$ and $x^2 + 2x - 9$.
76. Add $-x^2 + 4x - 7$ and $-2x^2 + 4x + 12$.
78. Add $-3x^3 + 3x + 9$ and $2x^2 - 4$.
80. Add $4x^3 + 7$ and $-2x^3 - 4x - 1$.
82. Add $x^2y - 6x + 3$ and $-2x^2y - 4x - 8$.

Subtract as indicated.

83. $(3x - 4) - (2x + 2)$
85. $(-2x - 3) - (-5x - 7)$
87. $(-x + 4) - (-x + 9)$
89. $(9 - 2x) - (6x + 2)$
91. $(6 - 12x) - (3 - 5x)$
93. $(9x^2 + 7x - 5) - (3x^2 + 5)$
95. $(-2x^2 + 4x - 5) - (5x^2 + 3x + 7)$
97. $(x^2 - 6x) - (2x^2 + 4)$
99. $(5x^2 - x + 12) - (x + 5)$
101. $(9x - 6) - (-2x^2 + 4x - 8)$
103. $(4x^3 - 6x^2 + 5x - 7) - (2x^2 + 6x - 3)$
105. $(9x^3 - 4) - (x^2 + 5x)$
107. Subtract $(4x - 6)$ from $(3x + 5)$.
109. Subtract $(5x - 6)$ from $(2x^2 - 4x + 8)$.
111. Subtract $(4x^3 - 6x^2)$ from $(3x^3 + 5x^2 + 9x - 7)$.

84. $(6x + 3) - (4x - 2)$
86. $(12x - 3) - (-2x + 7)$
88. $(4x + 8) - (3x + 9)$
90. $(3 + 5x) - (9 - 7x)$
92. $(4x^2 - 6x + 3) - (3x + 7)$
94. $(-x^2 + 3x - 8) - (-2x + 3)$
96. $(5x^2 - x - 1) - (-3x^2 - 2x - 5)$
98. $(5x^2 - 7) - (4x - 3)$
100. $(-5x^2 - 2x) - (2x^2 - 7x + 9)$
102. $(8x^3 + 5x^2 - 4) - (6x^2 + 4x - 3)$
104. $(-3x^2 + 4x - 7) - (x^3 + 4x^2 - 8x + 5)$
106. $(3x^3 - 6x^2 + 5x) - (4x^3 - 2x^2 + 5)$
108. Subtract $(-4x + 7)$ from $(-3x - 9)$.
110. Subtract $(2x^2 - 6x + 4)$ from $(5x^2 + 6x + 8)$.
112. Subtract $(-4x^2 + 8x - 7)$ from $(-5x^3 - 6x^2 + 7)$.

Perform each subtraction using columns.

113. Subtract $(2x - 7)$ from $(5x + 10)$.
115. Subtract $(-9x - 4)$ from $(-5x + 3)$.
117. Subtract $(4x^2 - 7)$ from $(9x^2 + 7x - 9)$.
119. Subtract $(-4x^2 + 6x)$ from $(x - 6)$.
121. Subtract $(x^2 + 6x - 7)$ from $(4x^3 - 6x^2 + 7x - 9)$.

114. Subtract $(6x + 8)$ from $(2x - 5)$.
116. Subtract $(-3x + 8)$ from $(6x^2 - 5x + 3)$.
118. Subtract $(4x^2 + 7x - 9)$ from $(x^2 - 6x + 3)$.
120. Subtract $(x^2 - 6)$ from $(x^2 + 4x)$.
122. Subtract $(2x^3 + 4x^2 - 9x)$ from $(-5x^3 + 4x - 12)$.

4.4

Multiplication of Polynomials

1   *Multiply polynomials using the distributive property.*
2   *Multiply two binomials using the FOIL method.*
3   *Identify and multiply special products.*
4   *Multiply any two polynomials.*

---

1   To multiply polynomials, we make use of the distributive property and the properties of exponents:

Multiplying a Polynomial by a Monomial

Distributive Property

$$a(b + c) = ab + ac$$

The distributive property can be expanded to

$$a(b + c + d + \cdots + n) = ab + ac + ad + \cdots + an$$

EXAMPLE 1   Multiply $2x(3x^2 + 4)$.

Solution: $2x(3x^2 + 4) = (2x)(3x^2) + (2x)(4)$
$$= 6x^3 + 8x \quad \blacksquare$$

EXAMPLE 2   Multiply $-3x(4x^2 - 2x - 1)$.

Solution: $-3x(4x^2 - 2x - 1) = (-3x)(4x^2) + (-3x)(-2x) + (-3x)(-1)$
$$= -12x^3 + 6x^2 + 3x \quad \blacksquare$$

EXAMPLE 3   Multiply $3x^2(4x^3 - 2x + 7)$.

Solution: $3x^2(4x^3 - 2x + 7) = (3x^2)(4x^3) + (3x^2)(-2x) + (3x^2)(7)$
$$= 12x^5 - 6x^3 + 21x^2 \quad \blacksquare$$

EXAMPLE 4   Multiply $2x(3x^2y - 6xy + 5)$.

Solution: $2x(3x^2y - 6xy + 5) = (2x)(3x^2y) + (2x)(-6xy) + (2x)(5)$
$$= 6x^3y - 12x^2y + 10x \quad \blacksquare$$

EXAMPLE 5   Multiply $(3x^2 - 2xy + 3)4x$.

Solution: $(3x^2 - 2xy + 3)4x = (3x^2)(4x) + (-2xy)(4x) + (3)(4x)$
$$= 12x^3 - 8x^2y + 12x$$

This problem could be written as $4x(3x^2 - 2xy + 3)$ by the commutative property of multiplication, and then solved as in Examples 1 through 4.   ∎

Multiplying a Binomial by a Binomial

Consider multiplying $(a + b)(c + d)$. Treating $(a + b)$ as a single term and using the distributive property, we get

$$(a + b)(c + d) = (a + b)c + (a + b)d$$

Using the distributive property a second time gives

$$= ac + bc + ad + bd$$

When multiplying a binomial by a binomial, each term of the first binomial must be multiplied by each term of the second binomial, and all of the results added together.

EXAMPLE 6   Multiply $(3x + 2)(x - 5)$.

Solution: $(3x + 2)(x - 5) = (3x + 2)x + (3x + 2)(-5)$
$$= 3x(x) + 2(x) + 3x(-5) + 2(-5)$$
$$= 3x^2 \underbrace{+ 2x - 15x} - 10$$
$$= 3x^2 - 13x - 10$$

Note that after performing the multiplication, like terms must be combined.   ∎

EXAMPLE 7   Multiply $(x - 4)(y + 3)$.

Solution: $(x - 4)(y + 3) = (x - 4)y + (x - 4)3$
$$= xy - 4y + 3x - 12 \quad \blacksquare$$

FOIL Method

2

A convenient method for finding the product of two binomials is the **FOIL** method. Consider
$$(a + b)(c + d)$$

F stands for **first**—multiply the first terms of each binomial together:

$$\overset{\text{F}}{(a + b)(c + d)} \qquad \text{product } ac$$

O stands for **outer**—multiply the two outer terms together:

$$\overset{\text{O}}{(a + b)(c + d)} \qquad \text{product } ad$$

I stands for **inner**—multiply the two inner terms together:

$$\overset{\text{I}}{(a + b)(c + d)} \qquad \text{product } bc$$

L stands for **last**—multiply the last terms together

$$\overset{\text{L}}{(a + b)(c + d)} \qquad \text{product } bd$$

The answer will be the sum of the products.
$$(a + b)(c + d) = ac + ad + bc + bd$$

EXAMPLE 8   Using the FOIL method, multiply $(2x - 3)(x + 4)$

Solution:

$$(2x - 3)(x + 4)$$

$$\overset{\text{F}}{(2x)(x)} + \overset{\text{O}}{(2x)(4)} + \overset{\text{I}}{(-3)(x)} + \overset{\text{L}}{(-3)(4)}$$
$$= \quad 2x^2 \quad + \quad 8x \quad - \quad 3x \quad - \quad 12$$
$$= 2x^2 + 5x - 12$$

Thus $(2x - 3)(x + 4) = 2x^2 + 5x - 12$ ■

EXAMPLE 9   Multiply $(4 - 2x)(6 - 5x)$.

Solution:

$$(4 - 2x)(6 - 5x)$$

with L, F, I, O labels

$$
\begin{array}{cccc}
F & O & I & L \\
4(6) + 4(-5x) + (-2x)(6) + (-2x)(-5x)
\end{array}
$$

$$= 24 - 20x - 12x + 10x^2$$

$$= 10x^2 - 32x + 24$$

Thus $(4 - 2x)(6 - 5x) = 10x^2 - 32x + 24$ ∎

EXAMPLE 10   Multiply $(x - 5)(x + 5)$.

Solution:

$$
\begin{array}{cccc}
F & O & I & L \\
(x)(x) + (x)(5) + (-5)(x) + (-5)(+5)
\end{array}
$$

$$= x^2 + 5x - 5x - 25$$

$$= x^2 - 25$$

Thus $(x - 5)(x + 5) = x^2 - 25$ ∎

EXAMPLE 11   Multiply $(2x + 3)(2x - 3)$.

Solution:

$$
\begin{array}{cccc}
F & O & I & L \\
(2x)(2x) + (2x)(-3) + (3)(2x) + (3)(-3)
\end{array}
$$

$$= 4x^2 - 6x + 6x - 9$$

$$= 4x^2 - 9$$

Thus $(2x + 3)(2x - 3) = 4x^2 - 9$ ∎

Examples 10 and 11 are examples of the special product of the sum and difference of two quantities.

Special Products

3

Product of Sum and Difference of Two Quantities

$$(a + b)(a - b) = a^2 - b^2 \qquad \text{for any } a \text{ and } b$$

The above product is also referred to as the **difference-of-squares formula** because the expression on the right side of the equal sign is the difference of two squares.

EXAMPLE 12   Use the rule for finding the product of the sum and difference of two quantities to multiply each expression.

(a) $(x + 3)(x - 3)$ (b) $(2x + 4)(2x - 4)$ (c) $(3x + 2y)(3x - 2y)$

Solution: (a) If we let $x = a$ and $3 = b$, then

$$(x + 3)(x - 3) = (a + b)(a - b) = a^2 - b^2$$
$$= (x)^2 - (3)^2$$
$$= x^2 - 9$$

(b) $(2x + 4)(2x - 4)$

$$(a + b)(a - b) = a^2 - b^2$$
$$= (2x)^2 - (4)^2$$
$$= 4x^2 - 16$$

(c) $(3x + 2y)(3x - 2y)$

$$(a + b)(a - b) = a^2 - b^2$$
$$= (3x)^2 - (2y)^2$$
$$= 9x^2 - 4y^2$$

This problem could also be done using the FOIL method.   ■

EXAMPLE 13   Using the FOIL method, multiply $(x + 3)^2$.

Solution: $(x + 3)^2 = (x + 3)(x + 3)$

$$\begin{array}{cccc} F & O & I & L \end{array}$$
$$x(x) + x(3) + 3(x) + (3)(3)$$
$$= x^2 + 3x + 3x + 9$$
$$= x^2 + 6x + 9 \quad ■$$

Example 13 is an example of the square of a binomial, another special product.

Square of Binomial Formulas

$$(a + b)^2 = (a + b)(a + b) = a^2 + 2ab + b^2$$
$$(a - b)^2 = (a - b)(a - b) = a^2 - 2ab + b^2$$

EXAMPLE 14   Use the square of the binomial formula to multiply each expression.

(a) $(x + 5)^2$ (b) $(2x + 4)^2$
(c) $(3x + 2y)(3x + 2y)$ (d) $(x - 3)(x - 3)$

Solution: (a) If we let $x = a$ and $5 = b$, then

$$(x + 5)(x + 5) = (a + b)(a + b) = a^2 + 2ab + b^2$$
$$= (x)^2 + 2(x)(5) + (5)^2$$
$$= x^2 + 10x + 25$$

(b) $(2x + 4)(2x + 4)$

$$(a + b)(a + b) = a^2 + 2ab + b^2$$
$$= (2x)^2 + 2(2x)(4) + (4)^2$$
$$= 4x^2 + 16x + 16$$

(c) $(3x + 2y)(3x + 2y)$

$$(a + b)(a + b) = a^2 + 2ab + b^2$$
$$= (3x)^2 + 2(3x)(2y) + (2y)^2$$
$$= 9x^2 + 12xy + 4y^2$$

(d) $(x - 3)(x - 3)$

$$(a - b)(a - b) = a^2 - 2ab + b^2$$
$$= (x)^2 - 2(x)(3) + (3)^2$$
$$= x^2 - 6x + 9$$

This problem could also be done using the FOIL method. ∎

COMMON STUDENT ERROR

Correct	Wrong
$(a + b)^2 = a^2 + 2ab + b^2$	$(a + b)^2 = a^2 + b^2$
$(a - b)^2 = a^2 - 2ab + b^2$	$(a - b)^2 = a^2 - b^2$

Do not forget the middle term when you square a binomial.

$$(x + 2)^2 \neq x^2 + 4$$
$$(x + 2)^2 = (x + 2)(x + 2)$$
$$= x^2 + 4x + 4$$

Multiplying a Polynomial by a Polynomial

4   When multiplying a binomial by a binomial, we saw that every term in the first binomial was multiplied by every term in the second binomial. When multiplying the number 23 by the number 12, we multiply each digit in the number 23 by each digit in the number 12, as illustrated below.

$$
\begin{array}{r}
23 \\
12 \\
\hline
\end{array}
$$

2(2) ⟶ 46 ⟵ 2(3)
1(2) ⟶ 23 ⟵ 1(3)
$$
\begin{array}{r}
\hline
276 \qquad \text{add in columns}
\end{array}
$$

We can follow a similar procedure when multiplying a polynomial by a polynomial. We must be careful, however, to align like terms in the same columns.

EXAMPLE 15   Multiply $(3x + 4)(2x + 5)$.

Solution: First write the polynomials one beneath the other.

$$
\begin{array}{r}
3x + 4 \\
2x + 5 \\
\hline
\end{array}
$$

Next, multiply each term in $(3x + 4)$ by 5.

$$\begin{array}{r} 3x + 4 \\ 2x + 5 \\ \hline 15x + 20 \end{array}$$

$5(3x + 4) \longrightarrow 15x + 20$

Next, multiply each term in $(3x + 4)$ by $2x$ and align like terms.

$$\begin{array}{r} 3x + 4 \\ 2x + 5 \\ \hline 15x + 20 \end{array}$$

$2x(3x + 4) \longrightarrow 6x^2 + 8x$

$$\underline{6x^2 + 8x} \qquad$$
$$6x^2 + 23x + 20 \qquad \text{add like terms in columns}$$

The same answer is obtained using the FOIL method.   ∎

EXAMPLE 16   Multiply $(5x - 2)(2x^2 + 3x - 4)$.

Solution: For convenience we place the shorter expression on the bottom, as illustrated.

$$\begin{array}{r} 2x^2 + 3x - 4 \\ 5x - 2 \\ \hline \end{array}$$

$-4x^2 - 6x + 8$ multiply top expression by -2

$\underline{10x^3 + 15x^2 - 20x}$ multiply top expression by $5x$ and align like terms

$10x^3 + 11x^2 - 26x + 8$ add like terms in columns   ∎

EXAMPLE 17   Multiply $x^2 - 3x + 2$ by $2x^2 - 3$.

Solution:

$$\begin{array}{r} x^2 - 3x + 2 \\ 2x^2 - 3 \\ \hline \end{array}$$

$-3x^2 + 9x - 6$ multiply top expression by -3

$\underline{2x^4 - 6x^3 + 4x^2}$ multiply top expression by $2x^2$ and align like terms

$2x^4 - 6x^3 + x^2 + 9x - 6$. add like terms in columns   ∎

EXAMPLE 18   Multiply $(3x^3 - 2x^2 + 4x + 6)(x^2 - 5x)$.

Solution:

$$\begin{array}{r} 3x^3 - 2x^2 + 4x + 6 \\ x^2 - 5x \\ \hline \end{array}$$

$-15x^4 + 10x^3 - 20x^2 - 30x$ multiply top expression by $-5x$

$\underline{3x^5 - 2x^4 + 4x^3 + 6x^2}$ multiply top expression by x^2

$3x^5 - 17x^4 + 14x^3 - 14x^2 - 30x$ add like terms in columns   ∎

Exercise Set 4.4

Multiply.

1. $3(x + 4)$

2. $3(x - 4)$

3. $2x(x - 3)$

4. $-5x(x + 2)$

5. $-4x(-2x + 6)$

6. $-x(3x - 5)$

7. $2x(x^2 + 3x - 1)$

8. $-x(2x^2 - 6x + 5)$

9. $-2x(x^2 - 2x + 5)$

10. $-3x(-2x^2 + 5x - 6)$

11. $5x(-4x^2 + 6x - 4)$

12. $x(x^2 - x + 1)$

13. $(3x^2 + 4x - 5)8x$

14. $x(3x^2 + y)$

15. $3x(2xy + 5x - 6y)$

16. $-2x(x^2 + 4x - y^2)$

17. $(x - y - 3)y$

18. $y(-2y^2 + 2y - x)$

Multiply.

19. $(x + 3)(x + 4)$

20. $(2x - 3)(x + 5)$

21. $(2x + 5)(3x - 6)$

22. $(x + 5)(x - 5)$

23. $(2x - 4)(2x + 4)$

24. $(4 + 3x)(2 - x)$

25. $(5 - 3x)(6 + 2x)$

26. $(4 + 6x)(x - 3)$

27. $(-x + 3)(2x + 5)$

28. $(6x - 1)(-2x + 5)$

29. $(x + 4)(x + 3)$

30. $(x - 3)(x + 5)$

31. $(x + 4)(x - 2)$

32. $(2x + 3)(x + 5)$

33. $(3x + 4)(2x + 5)$

34. $(3x - 6)(4x - 2)$

35. $(3x + 4)(2x - 3)$

36. $(4x + 4)(x + 1)$

37. $(x - 1)(x + 1)$

38. $(3x - 8)(2x + 3)$

39. $(2x - 3)(2x - 3)$

40. $(6x - 1)(2x + 1)$

41. $(4 - x)(3 + 2x)$

42. $(6 - 2x)(5x - 3)$

43. $(2x + 3)(4 - 2x)$

44. $(5 - 6x)(2x - 7)$

45. $(x + y)(x - y)$

46. $(x + 2y)(2x - 3)$

47. $(2x - 3y)(3x + 2y)$

48. $(x + 3)(2y - 5)$

49. $(4x - 3y)(2y - 3)$

50. $(x - y)(y - x)$

51. $(x + 4)(3x + 5)$

52. $(2x + 1)(4x + 5)$

53. $(5x + 4)(2x + 3)$

54. $(4x + 3)(2x - 2)$

Multiply using a special product formula.

55. $(x + 4)(x - 4)$

56. $(x + 3)^2$

57. $(2x - 1)(2x + 1)$

58. $(x + 2)(x + 2)$

59. $(x + y)^2$

60. $(2x - 3)(2x + 3)$

61. $(x - 2)^2$

62. $(x + 3y)(x + 3y)$

63. $(3x + 5)(3x - 5)$

64. $(5x + 4)(5x - 4)$

Multiply.

65. $(x + 3)(2x^2 + 4x - 1)$

66. $(2x + 3)(4x^2 - 5x + 6)$

67. $(5x + 4)(x^2 - x + 4)$

68. $(2x - 5)(3x^2 - 4x + 7)$

69. $(-2x^2 - 4x + 1)(7x - 3)$

70. $(4x^2 + 9x - 2)(x - 2)$

71. $(-3x + 9)(-6x^2 + 5x - 3)$

72. $(4x^2 + 1)(2x + 1)$

73. $(3x^2 - 2x + 4)(2x^2 + 3x + 1)$

74. $(x^2 - 2x + 3)(x^2 - 4)$

75. $(x^2 - x + 3)(x^2 - 2x)$

76. $(-3x^2 - 2x + 4)(2x^2 - 4x + 3)$

77. $(3x^3 + 2x^2 - x)(x - 3)$

78. $(-x^3 + x^2 - 6x + 3)(x^2 + 2x)$

79. $(a + b)(a^2 - ab + b^2)$

80. $(a - b)(a^2 + ab + b^2)$

JUST FOR FUN

Perform each polynomial multiplication.

1. $\sqrt{5}x\left(2x^2 + \sqrt{5}x - \dfrac{1}{2}\right)$ *Hint:* $(\sqrt{5})2 = 2\sqrt{5}$ and $\sqrt{5} \cdot \sqrt{5} = \sqrt{25} = 5$

2. $\left(\dfrac{x}{2} + \dfrac{2}{3}\right)\left(\dfrac{2x}{3} - \dfrac{2}{5}\right)$

3. $(2x^3 - 6x^2 + 5x - 3)(3x^3 - 6x + 4)$

4. $[x + (y - 1)][x - (y - 1)]$ *Hint:* Use product of sum and difference formula.

5. $[x + (y - 1)][x + (y - 1)]$ *Hint:* Use square of a binomial formula.

4.5 _____

**Division of
Polynomials**

■ Divide a polynomial by a monomial.

■ Divide a polynomial by a binomial.

■ Check the division of polynomial problems.

─────────────────────────────

■

> **To divide a polynomial by a monomial,** divide each term of the polynomial by the monomial.

Dividing a Polynomial
by a Monomial

EXAMPLE 1 $\dfrac{2x + 16}{2}$

Solution: $\dfrac{2x + 16}{2} = \dfrac{2x}{2} + \dfrac{16}{2}$

$= x + 8$ ■

EXAMPLE 2 $\dfrac{4x^2 - 8x}{2x}$

Solution: $\dfrac{4x^2 - 8x}{2x} = \dfrac{4x^2}{2x} - \dfrac{8x}{2x}$

$= 2x - 4$ ■

EXAMPLE 3 $\dfrac{4x^3 - 6x^2 + 8x - 3}{2x}$

Solution: $\dfrac{4x^3 - 6x^2 + 8x - 3}{2x} = \dfrac{4x^3}{2x} - \dfrac{6x^2}{2x} + \dfrac{8x}{2x} - \dfrac{3}{2x}$

$= 2x^2 - 3x + 4 - \dfrac{3}{2x}$ ■

EXAMPLE 4 $\dfrac{3x^3 - 6x^2 + 4x - 1}{-3x}$

Solution: In this problem a negative sign appears in the denominator. One method to simplify this problem is to multiply both numerator and denominator by -1.

$$\dfrac{(-1)(3x^3 - 6x^2 + 4x - 1)}{(-1)(-3x)} = \dfrac{-3x^3 + 6x^2 - 4x + 1}{3x}$$

$$= \dfrac{-3x^3}{3x} + \dfrac{6x^2}{3x} - \dfrac{4x}{3x} + \dfrac{1}{3x}$$

$$= -x^2 + 2x - \dfrac{4}{3} + \dfrac{1}{3x}$$ ■

COMMON STUDENT ERROR

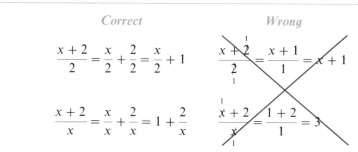

Correct

$$\frac{x+2}{2} = \frac{x}{2} + \frac{2}{2} = \frac{x}{2} + 1$$

$$\frac{x+2}{x} = \frac{x}{x} + \frac{2}{x} = 1 + \frac{2}{x}$$

Wrong

Can you explain why the procedures on the right are wrong?

Dividing a Polynomial by a Binomial

2 We divide a polynomial by a binomial in much the same way as we perform long division. This procedure will be explained in Example 5.

EXAMPLE 5

$$\frac{x^2 + 6x + 8}{x + 2} \xleftarrow{\text{dividend}}$$
$$\xleftarrow{\text{divisor}}$$

Solution: Rewrite the divison problem as

$$x + 2 \overline{)\, x^2 + 6x + 8}$$

Divide x^2 (the first term in the dividend) by x (the first term in the divisor).

$$\frac{x^2}{x} = x$$

Place the quotient, x, above the like term containing x in the dividend.

$$x + 2 \overline{)\, x^2 + 6x + 8} \qquad \overset{x}{}$$

Next, multiply the x by $x + 2$ as you would do in long division and place the terms of the product under their like terms.

$$\begin{array}{r} \overset{times}{\overset{\frown}{\quad}} \; \textcircled{x} \\ (x + 2) \overline{)\, x^2 + 6x + 8} \\ \underset{equals}{\searrow}\; x^2 + 2x \xleftarrow{} x(x + 2) \end{array}$$

Now subtract $x^2 + 2x$ from $x^2 + 6x$. When subtracting, remember to change the sign of the terms being subtracted, and then add the like terms.

$$\begin{array}{r} x \\ x + 2 \overline{)\, x^2 + 6x + 8} \\ \underline{x^2 \mp 2x } \\ 4x \end{array}$$

Next, bring down the 8.

$$\begin{array}{r} x \\ x + 2 \overline{)\, x^2 + 6x + 8} \\ \underline{x^2 + 2x } \\ 4x + 8 \end{array}$$

Determine the quotient of $4x$ divided by x.

$$\frac{4x}{x} = +4$$

Place the $+4$ above the constant in the dividend.

$$\begin{array}{r} x + 4 \\ x + 2 \overline{\smash{\big)}\ x^2 + 6x + 8} \\ \underline{x^2 + 2x} \\ 4x + 8 \end{array}$$

Multiply the $x + 2$ by 4 and place the terms of the product under their like terms.

$$\begin{array}{r} \overset{\text{times}}{x + \textcircled{4}} \\ \boxed{x + 2} \overline{\smash{\big)}\ x^2 + 2x} \\ x^2 + 2x \\ \hline 4x + 8 \\ \underset{\text{equals}}{\longrightarrow} 4x + 8 \longleftarrow 4(x + 2) \end{array}$$

Now subtract.

$$\begin{array}{r} x + 4 \longleftarrow \text{quotient} \\ x + 2 \overline{\smash{\big)}\ x^2 + 6x + 8} \\ \underline{x^2 + 2x} \\ 4x + 8 \\ \underline{\overset{-}{}4x \overset{-}{+} 8} \\ 0 \longleftarrow \text{remainder} \end{array}$$

Thus

$$\frac{x^2 + 6x + 8}{x + 2} = x + 4$$

There is no remainder. ■

EXAMPLE 6 $\dfrac{6x^2 - 5x + 5}{2x + 3}$

Solution:

$$\frac{6x^2}{2x} \quad \frac{-14x}{2x}$$

$$\downarrow \qquad$$

$$3x - 7$$

$$\begin{array}{r} 2x + 3 \overline{\smash{\big)}\ 6x^2 - 5x + 5} \\ \underline{\overset{-}{}6x^2 \overset{-}{+} 9x} \longleftarrow 3x(2x + 3) \\ -14x + 5 \\ \underline{\overset{+}{}14x \overset{+}{+} 21} \longleftarrow -7(2x + 3) \\ 26 \qquad \text{remainder} \end{array}$$

Thus

$$\frac{6x^2 - 5x + 5}{2x + 3} = 3x - 7 + \frac{26}{2x + 3} \quad ■$$

3   The answer to a division problem can be checked. Consider the division problem $\frac{13}{5}$.

$$\begin{array}{r} 2 \\ 5\overline{\smash{\big)}\,13} \\ \underline{10} \\ 3 \end{array}$$

Note that the divisor times the quotient, plus the remainder, equals the dividend:

$$(\text{divisor} \times \text{quotient}) + \text{remainder} = \text{dividend}$$
$$(5 \cdot 2) + 3 = 13$$
$$10 + 3 = 13$$
$$13 = 13 \qquad \text{true}$$

This same procedure can be used to check all division problems. Let us check the answer to Example 6. The divisor is $2x + 3$, the quotient is $3x - 7$, the remainder is 26, and the dividend is $6x^2 - 5x + 5$.

Check: $(\text{divisor} \times \text{quotient}) + \text{remainder} = \text{dividend}$
$$(2x + 3)(3x - 7) + 26 = 6x^2 - 5x + 5$$
$$(6x^2 - 5x - 21) + 26 = 6x^2 - 5x + 5$$
$$6x^2 - 5x + 5 = 6x^2 - 5x + 5 \qquad \text{true}$$

When dividing a polynomial by a binomial, both the polynomial and binomial should be listed in descending order. If a given power term is missing, it is often helpful to include that term with a numerical coefficient of 0. For example, when dividing $(6x^2 + x^3 - 4)/(x - 2)$, we rewrite the problem as $(x^3 + 6x^2 + 0x - 4)/(x - 2)$ before beginning the division.

EXAMPLE 7 $(9x^3 - x - 8) \div (3x - 4)$.

Solution: Since there is no x^2 term in the dividend, we will add $0x^2$ to help align like terms.

$$\frac{9x^3}{3x}, \quad \frac{12x^2}{3x}, \quad \frac{15x}{3x}$$

$$\begin{array}{r}
3x^2 + 4x + 5 \\
3x - 4 \overline{\smash{\big)}\, 9x^3 + 0x^2 - x - 8} \\
\underline{9x^3 - 12x^2} \longleftarrow 3x^2(3x - 4) \\
12x^2 - x \\
\underline{12x^2 - 16x} \longleftarrow 4x(3x - 4) \\
15x - 8 \\
\underline{15x - 20} \longleftarrow 5(3x - 4) \\
12 \longleftarrow \text{remainder}
\end{array}$$

$$\frac{9x^3 - x - 8}{3x - 4} = 3x^2 + 4x + 5 + \frac{12}{3x - 4} \qquad \blacksquare$$

COMMON STUDENT ERROR

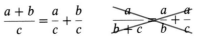

| Correct | Wrong |

$$\frac{a+b}{c} = \frac{a}{c} + \frac{b}{c}$$

The expression $a/(b+c)$ cannot be simplified.

Exercise Set 4.5

Divide as indicated.

1. $\dfrac{2x+4}{2}$ 2. $\dfrac{4x-6}{2}$ 3. $\dfrac{2x+6}{2}$ 4. $\dfrac{4x+3}{2}$

5. $\dfrac{3x+8}{2}$ 6. $\dfrac{5x-12}{6}$ 7. $\dfrac{-6x+4}{2}$ 8. $\dfrac{-4x+5}{-3}$

9. $\dfrac{-9x-3}{-3}$ 10. $\dfrac{5x-4}{-5}$ 11. $\dfrac{3x+6}{x}$ 12. $\dfrac{4x-3}{2x}$

13. $\dfrac{9-3x}{-3x}$ 14. $\dfrac{6-5x}{-3x}$ 15. $\dfrac{3x^2+6x-9}{3}$

16. $\dfrac{12x^2-6x+3}{3}$ 17. $\dfrac{-4x^2+6x+8}{2}$ 18. $\dfrac{5x^2+4x-8}{2}$

19. $\dfrac{x^2+4x-3}{x}$ 20. $\dfrac{4x^2-6x+7}{x}$ 21. $\dfrac{6x^2-4x+12}{2x}$

22. $\dfrac{8x^2-5x+10}{-2x}$ 23. $\dfrac{4x^3+6x^2-8}{-4x}$ 24. $\dfrac{-12x^3+6x^2-15x}{-3x}$

25. $\dfrac{9x^3+3x^2-12}{3x^2}$ 26. $\dfrac{-10x^3-6x^2+15}{-5x}$

Divide as indicated.

27. $\dfrac{x^2+4x+3}{x+1}$ 28. $\dfrac{x^2+7x+10}{x+5}$ 29. $\dfrac{2x^2+13x+15}{x+5}$

30. $\dfrac{2x^2+x-10}{x-2}$ 31. $\dfrac{6x^2+16x+8}{3x+2}$ 32. $\dfrac{2x^2+13x+15}{2x+3}$

33. $\dfrac{2x^2+x-10}{2x+5}$ 34. $\dfrac{8x^2-26x+15}{4x-3}$ 35. $\dfrac{2x^2+7x-18}{2x-3}$

36. $\dfrac{x^2-25}{x-5}$ 37. $\dfrac{4x^2-9}{2x-3}$ 38. $\dfrac{9x^2-4}{3x+2}$

39. $\dfrac{6x+8x^2-25}{4x+9}$ 40. $\dfrac{10x+3x^2+6}{x+2}$ 41. $\dfrac{6x+8x^2-12}{2x+3}$

42. $\dfrac{6x^2 - 13 - 11x}{2x - 5}$

43. $\dfrac{x^3 + 3x^2 + 5x + 3}{x + 1}$

44. $\dfrac{4x^3 + 12x^2 + 7x - 3}{2x + 3}$

45. $\dfrac{2x^3 - 3x^2 - 3x + 6}{x - 1}$

46. $\dfrac{9x^3 - 3x^2 - 9x + 4}{3x + 2}$

47. $\dfrac{2x^3 + 6x - 4}{x + 4}$

48. $\dfrac{x^3 - 8}{x - 3}$

49. $\dfrac{x^3 + 8}{x + 2}$

50. $\dfrac{x^3 - 27}{x - 3}$

51. $\dfrac{x^3 + 27}{x + 3}$

52. $\dfrac{4x^3 - 5x}{2x - 1}$

53. $\dfrac{9x^3 - x + 3}{3x - 2}$

54. $\dfrac{-x^3 - 6x^2 + 2x - 3}{x - 1}$

JUST FOR FUN

1. $\dfrac{4x^3 - 4x + 6}{2x + 3}$ *Hint:* The quotient will contain fractions.

2. $\dfrac{3x^3 - 5}{3x - 2}$

Summary

Glossary

Binomial: A two-termed polynomial.
Degree of a polynomial: The same as the highest-degree term in the polynomial.
Degree of a term: The exponent on the variable when the polynomial is in one variable.
Descending order, or power, of the variable: Polynomial written so that the exponents decrease from left to right.

Monomial: A one-term polynomial.
Polynomial in x: An expression containing the sum of only a finite number of terms of the form ax^n, for any real number a and any whole number n.
Scientific notation form: A number greater than or equal to one and less than 10 multiplied by some power of 10.
Trinomial: A three-termed polynomial.

Important Facts

Rules of Exponents

1. $x^m x^n = x^{m+n}$ product rule

2. $\dfrac{x^m}{x^n} = x^{m-n}$ quotient rule

3. $(x^m)^n = x^{mn}$ power rule

4. $x^0 = 1, x \neq 0$ zero exponent rule

5. $x^{-m} = \dfrac{1}{x^m}, x \neq 0$ negative exponent rule

6. $\left(\dfrac{ax}{y}\right)^m = \dfrac{a^m x^m}{y^m}, y \neq 0$ expanded power rule

FOIL   Method to Multiply Two Binomials (First, Outer, Inner, Last)

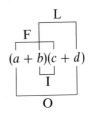

Product of sum and difference of two quantities (also called the **difference of two squares**)

$$(a + b)(a - b) = a^2 - b^2$$

The square of a binomial

$$(a + b)^2 = a^2 + 2ab + b^2$$
$$(a - b)^2 = a^2 - 2ab + b^2$$

Review Exercises

[4.1] Simplify, and express your answer without negative exponents.

1. $x^4 \cdot x^2$

2. $x^3 \cdot x^5$

3. $3^2 \cdot 3^3$

4. $2^4 \cdot 2$

5. $x^3 \cdot x^{-5}$

6. $x^{-2} \cdot x^{-3}$

7. $x^4 \cdot x^{-7}$

8. $x^{13} \cdot x^{-5}$

9. $\dfrac{x^4}{x}$

10. $\dfrac{x^6}{x^6}$

11. $\dfrac{3^5}{3^3}$

12. $\dfrac{4^5}{4^3}$

13. $\dfrac{x^6}{x^8}$

14. $\dfrac{x^2}{x^{-3}}$

15. $\dfrac{x^5}{x^{-2}}$

16. $\dfrac{x^{-3}}{x^3}$

17. $\dfrac{x^{-1}}{x^{-3}}$

18. $\dfrac{x^4}{x^{-2}}$

19. x^0

20. $3x^0$

21. $(3x)^0$

22. 4^0

23. $(2x)^2$

24. $(3x)^3$

25. $(-2x)^2$

26. $(-3x)^3$

27. $(2x^2)^4$

28. $(3x^4)^{-2}$

29. $(-x^4)^3$

30. $(-x^3)^4$

31. $\left(\dfrac{2x^3}{y}\right)^2$

32. $(4x^{-3}y)^{-3}$

33. $(-5x^{-2})^3$

34. $\left(\dfrac{3x^4}{2y}\right)^3$

35. $6x^2 \cdot 4x^3$

36. $-2x^3 \cdot 4x^5$

37. $(-x^4)(-x^2)$

38. $\dfrac{16x^2y}{4xy^2}$

39. $(2x^2y)^2 \cdot 3x$

40. $(3x^{-2}y)^3$

41. $(4x^{-2}y^3)^{-2}$

42. $\left(\dfrac{9x^2y}{3xy}\right)^2$

[4.2] Express each number in scientific notation form.

43. 364,000

44. 1,640,000

45. 0.00763

46. 0.176

47. 2080

48. 0.000314

Express each number without exponents.

49. 4.2×10^{-3}

52. 4.38×10^{-6}

50. 1.65×10^4

53. 9.14×10^{-1}

51. 9.7×10^5

54. 5.36×10^2

Perform the indicated operation and write each answer without exponents.

55. $(2.3 \times 10^2)(2 \times 10^4)$

58. $\dfrac{6.8 \times 10^3}{2 \times 10^{-2}}$

56. $(4.2 \times 10^{-3})(3 \times 10^5)$

59. $\dfrac{36 \times 10^4}{4 \times 10^6}$

57. $(6.4 \times 10^{-3})(3.1 \times 10^3)$

60. $\dfrac{15 \times 10^{-3}}{5 \times 10^2}$

Perform the indicated operation by first converting each number to scientific notation form. Write the answer in scientific notation form.

61. $(60,000)(20,000)$

64. $\dfrac{40,000}{0.0002}$

62. $(0.00004)(600,000)$

65. $\dfrac{0.000068}{0.02}$

63. $(0.00023)(40,000)$

66. $\dfrac{1,500,000}{0.003}$

[4.3] Indicate if each expression is a polynomial. If the polynomial has a specific name, give that name. If the polynomial is not written in descending order, rewrite it in descending order. State the degree of each polynomial.

67. $x + 3$

70. $-3 - x + 4x^2$

73. $x - 4x^2$

68. -2

71. $-5x^2 + 3$

74. $x^3 + x^{-2} + 3$

69. $x^2 - 4 + 3x$

72. $4x^{1/2} - 6$

75. $x^3 - 2x - 6 + 4x^2$

[4.3–4.5] Perform the operations indicated.

76. $(x + 3) + (2x + 4)$

78. $(-3x + 4) + (5x - 9)$

80. $(-x^2 + 6x - 7) + (-2x^2 + 4x - 8)$

82. $(2x - 3) - (x + 4)$

84. $(9x^2 - 6x) - (2x - 4)$

86. $(-2x^2 + 8x - 7) - (3x^2 + 12)$

88. $2(x + 5)$

90. $2x(x^2 - 3x)$

92. $-x(3x^2 - 6x - 1)$

94. $(x + 4)(x + 5)$

96. $(4x + 6)^2$

98. $(x + 4)(x - 4)$

100. $(x - 1)(3x^2 + 4x - 6)$

102. $\dfrac{2x + 4}{2}$

104. $\dfrac{8x^2 + 4x}{x}$

106. $\dfrac{8x^2 + 6x - 4}{x}$

108. $\dfrac{16x - 4}{-2}$

77. $(5x - 5) + (4x + 6)$

79. $(4x^2 + 6x + 5) + (-6x + 9)$

81. $(12x^2 + 4x - 8) + (-x^2 - 6x + 5)$

83. $(-4x + 8) - (-2x + 6)$

85. $(6x^2 - 6x + 1) - (12x + 5)$

87. $(x^2 + 7x - 3) - (x^2 + 3x - 5)$

89. $x(2x - 4)$

91. $3x(2x^2 - 4x + 7)$

93. $-4x(-6x^2 + 4x - 2)$

95. $(2x + 4)(x - 3)$

97. $(6 - 2x)(2 + 3x)$

99. $(3x + 1)(x^2 + 2x + 4)$

101. $(-5x + 2)(-2x^2 + 3x - 6)$

103. $\dfrac{4x - 8}{4}$

105. $\dfrac{6x^2 + 9x - 4}{3}$

107. $\dfrac{8x^2 - 4x}{2x}$

109. $\dfrac{12 + 6x}{-3}$

110. $\dfrac{5x^2 + 10x + 2}{2x}$

111. $\dfrac{x^2 + x - 12}{x - 3}$

112. $\dfrac{6x^2 - 11x + 3}{3x - 1}$

113. $\dfrac{5x^2 + 28x - 10}{x + 6}$

114. $\dfrac{4x^3 + 12x^2 + x - 12}{2x + 3}$

115. $\dfrac{4x^3 - 5x + 4}{2x - 1}$

Practice Test

Simplify, and express your answer without negative exponents.

1. $2x^2 \cdot 3x^4$

2. $(3x^2)^3$

3. $\dfrac{8x^4}{2x}$

4. $(2x^3y^{-2})^{-2}$

5. $\left(\dfrac{3x^2y}{6xy^3}\right)^3$

6. $\dfrac{2x^4y^{-2}}{10x^7y^4}$

In Problems 7 through 9, determine whether each expression is a polynomial. If the polynomial has a specific name, give that name.

7. $x^2 - 4 + 6x$

8. -3

9. $x^{-2} + 4$

10. Write the polynomial $-5 + 6x^3 - 2x^2 + 5x$ in descending order, and give its degree.

Perform the operations indicated.

11. $(2x + 4) + (3x^2 - 5x - 3)$

12. $(x^2 - 4x + 7) - (3x^2 - 8x + 7)$

13. $(4x^2 - 5) - (x^2 + x - 8)$

14. $3x(4x^2 - 2x + 5)$

15. $(4x + 7)(2x - 3)$

16. $(6 - 4x)(5 + 3x)$

17. $(2x - 4)(3x^2 + 4x - 6)$

18. $\dfrac{16x^2 + 8x - 4}{4}$

19. $\dfrac{3x^2 - 6x + 5}{-3x}$

20. $\dfrac{8x^2 - 2x - 15}{2x - 3}$

*Perform the indicated operation by first converting each number to scientific notation form. Write the answer in scientific notation form.

21. $(42{,}000)(30{,}000)$

22. $\dfrac{0.0008}{4{,}000}$

* From optional section

5 Factoring

See Section 5.6, Just For Fun.

Factoring A Monomial from A Polynomial

1   *Identify factors.*

2   *Find the greatest common factor of two or more numbers.*

3   *Find the greatest common factor of two or more terms.*

4   *Factor a monomial from a polynomial.*

1   In Chapter 4 we learned how to multiply polynomials. In this chapter we focus on factoring, the reverse process of multiplication. In Section 4.4 we showed that $2x(3x^2 + 4) = 6x^3 + 8x$. In this chapter we start with an expression like $6x^3 + 8x$ and determine that its factors are $2x$ and $3x^2 + 4$, and write $6x^3 + 8x = 2x(3x^2 + 4)$. To **factor an expression** means to write the expression as a product of its factors. Factoring is important because it can be used to solve equations.

    If $a \cdot b = c$, then a and b are said to be *factors* of c.

$3 \cdot 5 = 15$; thus 3 and 5 are factors of 15.

$x^3 \cdot x^4 = x^7$; thus x^3 and x^4 are factors of x^7.

$x(x + 2) = x^2 + 2x$; thus x and $(x + 2)$ are factors of $x^2 + 2x$.

$(x - 1)(x + 3) = x^2 + 2x - 3$; thus $(x - 1)$ and $(x + 3)$ are factors of $x^2 + 2x - 3$.

    A given number or expression may have many factors. Consider the number 30.

$$1 \cdot 30 = 30, \qquad 2 \cdot 15 = 30, \qquad 3 \cdot 10 = 30, \qquad 5 \cdot 6 = 30$$

Thus the positive factors of 30 are 1, 2, 3, 5, 6, 10, 15, and 30. Factors can also be negative. Since $(-1)(-30) = 30$, -1 and -30 are also factors of 30. In fact, for each factor a of an expression, $-a$ must also be a factor. Other factors of 30 are therefore $-1, -2, -3, -5, -6, -10, -15$, and -30. When asked to list the factors of an expression with a positive numerical coefficient that contains a variable, we generally list only positive factors.

EXAMPLE 1   List the factors of $6x^3$.

Solution:

factors   factors

$1 \cdot 6x^3 = 6x^3$ $x \cdot 6x^2 = 6x^3$

$2 \cdot 3x^3 = 6x^3$ $2x \cdot 3x^2 = 6x^3$

$3 \cdot 2x^3 = 6x^3$ $3x \cdot 2x^2 = 6x^3$

$6 \cdot x^3 \ \ = 6x^3$ $6x \cdot x^2 \ \ = 6x^3$

The factors of $6x^3$ are 1, 2, 3, 6, x, $2x$, $3x$, $6x$, x^2, $2x^2$, $3x^2$, $6x^2$, x^3, $2x^3$, $3x^3$, and $6x^3$. The opposite (or negative) of each of these factors is also a factor, but these opposites are generally not listed unless specifically asked for.   ■

    Here are examples of multiplying and factoring:

$$2(3x + 4) = 6x + 8 \qquad \text{multiplying}$$
$$6x + 8 = 2(3x + 4) \qquad \text{factoring}$$

$$5x(x + 4) = 5x^2 + 20x \qquad \text{multiplying}$$
$$5x^2 + 20x = 5x(x + 4) \qquad \text{factoring}$$
$$(x + 1)(x + 3) = x^2 + 4x + 3 \qquad \text{multiplying}$$
$$x^2 + 4x + 3 = (x + 1)(x + 3) \qquad \text{factoring}$$

2 To factor a monomial from a polynomial we make use of the **greatest common factor (GCF).** The greatest common factor of two or more numbers is the greatest number that divides all of the numbers. The greatest common factor of the numbers 6 and 8 is 2. Two is the greatest number that divides both 6 and 8. What is the GCF of 48 and 60? When the GCF of two or more numbers is not easily found we can determine the GCF by writing each number as a product of prime numbers. A **prime number** is an integer greater than 1 that has exactly two factors, itself and one. The first 13 prime numbers are

2, 3, 5, 7, 11, 13, 17, 19, 23, 29, 31, 37, 41.

A positive integer (other than 1) which is not prime is called **composite.** The number 1 is neither prime nor composite, it is called a **unit.**

To write a number as a product of prime numbers follow the procedure illustrated in Examples 2 and 3.

EXAMPLE 2   Write 48 as a product of prime numbers.

Solution: Select any two numbers whose product is 48. Two possibilities are $6 \cdot 8$ and $4 \cdot 12$, but there are other choices. Continue breaking down the factors until all the factors are prime, as illustrated in Fig. 5.1.

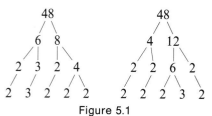

Figure 5.1

Note that no matter how you select your initial factors,

$$48 = 2 \cdot 2 \cdot 2 \cdot 2 \cdot 3 = 2^4 \cdot 3 \quad \blacksquare$$

EXAMPLE 3   Write 60 as a product of prime numbers.

Solution: One breakdown is shown in Fig. 5.2.

Figure 5.2

Therefore $60 = 2 \cdot 2 \cdot 3 \cdot 5 = 2^2 \cdot 3 \cdot 5.$ $\quad \blacksquare$

> **To Determine the GCF of Two or More numbers**
>
> 1. Write each number as a product of prime numbers.
> 2. Determine the prime factors common to all the numbers.
> 3. Multiply the common factors found in Step 2. The product of these factors will be the GCF.

EXAMPLE 4   Determine the greatest common factor of 48 and 60.

Solution: In Examples 1 and 2 we found that

$$48 = 2 \cdot 2 \cdot 2 \cdot 2 \cdot 3 = 2^4 \cdot 3$$
$$60 = 2 \cdot 2 \cdot 3 \cdot 5 = 2^2 \cdot 3 \cdot 5$$

There are two factors of 2 and one factor of 3 common to both numbers. The product of these factors is the GCF of 48 and 60

$$GCF = 2 \cdot 2 \cdot 3 = 12$$

The GCF of 48 and 60 is 12. Twelve is the greatest number that divides both 48 and 60. ■

EXAMPLE 5   Find the GCF of 18 and 24.

Solution: $18 = 2 \cdot 3 \cdot 3 = 2 \cdot 3^2$
$24 = 2 \cdot 2 \cdot 2 \cdot 3 = 2^3 \cdot 3$

One factor of 2 and one factor of 3 are common to both numbers.

$$GCF = 2 \cdot 3 = 6 \quad ■$$

3   The GCF of a collection of terms containing variables is easily found. Consider the terms x^3, x^4, x^5, and x^6. The GCF of these terms is x^3, since x^3 is the highest power of x that divides all four terms. We can illustrate this by writing the terms in factored form.

$$x^3 = x^3 \cdot 1$$
$$x^4 = x^3 \cdot x$$
$$x^5 = x^3 \cdot x^2$$
$$x^6 = x^3 \cdot x^3$$

$$\underbrace{\qquad\qquad\qquad}$$

GCF of all four terms

EXAMPLE 6   Find the GCF of the terms p^6, p^2, p^7, and p^4.

Solution: $GCF = p^2$ since p^2 is the highest power of p that divides each term. ■

EXAMPLE 7   Find the GCF of the terms x^2y^3, x^3y^2, and xy^4.

Solution: The highest power of x common to all three terms is x^1 or x. The highest power of y common to all three terms is y^2. So the GCF of the three terms is xy^2. ■

> **Greatest Common Factor of Two or More Terms**
>
> To find the GCF of two or more terms, take each factor the *fewest* number of times that it appears in any of the terms.

EXAMPLE 8   Find the GCF of the terms xy, x^2y^2, and x^3.

Solution: The GCF is x. The smallest power of x that appears in any of the terms is x. Since the x^3 term does not contain a power of y, the GCF does not contain a y. ▪

EXAMPLE 9   Find the GCF of each pair of terms.

(a) $x(x + 3)$ and $2(x + 3)$
(b) $x(x - 2)$ and $x - 2$
(c) $2(x + y)$ and $3x(x + y)$

Solution: (a) The GCF is $(x + 3)$.
(b) $x - 2$ can be written as $1(x - 2)$. The GCF of $x(x - 2)$ and $x - 2$ is therefore $x - 2$.
(c) The GCF is $(x + y)$. ▪

EXAMPLE 10   Find the GCF of each set of terms.

(a) $18y^2$, $15y^3$, $21y^5$ (b) $-20x^2$, $8x$, $40x^3$ (c) $4x^2$, x^2, x^3

Solution: (a) The GCF of 18, 15, and 21 is 3. The GCF of the three terms is $3y^2$.
(b) The GCF of -20, 8, and 40 is 4. The GCF of the three terms is $4x$.
(c) The GCF is x^2. ▪

4

> **To Factor a Monomial from a Polynomial**
>
> **1.** Determine the greatest common factor of all terms in the polynomial.
> **2.** Write each term as the product of the GCF and its other factor.
> **3.** Use the distributive property to factor out the GCF.

EXAMPLE 11   Factor $6x + 12$.

Solution: The GCF is 6.

$$6x + 12 = 6 \cdot x + 6 \cdot 2 \qquad \text{write each term as a product of the GCF and some other factor}$$

$$= 6(x + 2) \qquad \text{distributive property}$$

$6x + 12$ in factored form is $6(x + 2)$. To check the factoring process, multiply the factors using the distributive property.

Check: $6(x + 2) = 6x + 12$ ▪

EXAMPLE 12   Factor $8x - 10$.

Solution: 2 is the GCF of $8x$ and -10.

$$8x - 10 = 2 \cdot 4x - 2 \cdot 5$$
$$= 2(4x - 5) \quad ▪$$

EXAMPLE 13   Factor $6y^2 + 9y^5$.

Solution: The GCF is $3y^2$.
$$6y^2 + 9y^5 = 3y^2 \cdot 2 + 3y^2 \cdot 3y^3$$
$$= 3y^2(2 + 3y^3) \quad \blacksquare$$

EXAMPLE 14   Factor $8x^3 + 12x^2 - 16x$.

Solution: The GCF is $4x$.
$$8x^3 + 12x^2 - 16x = 4x \cdot 2x^2 + 4x \cdot 3x - 4x \cdot 4$$
$$= 4x(2x^2 + 3x - 4)$$

Check: $4x(2x^2 + 3x - 4) = 8x^3 + 12x^2 - 16x \quad \blacksquare$

EXAMPLE 15   Factor $45x^2 - 30x + 5$.

Solution: The GCF is 5.
$$45x^2 - 30x + 5 = 5 \cdot 9x^2 - 5 \cdot 6x + 5 \cdot 1$$
$$= 5(9x^2 - 6x + 1) \quad \blacksquare$$

EXAMPLE 16   Factor $4x^3 + x^2 + 8x^2y$.

Solution: The GCF is x^2.
$$4x^3 + x^2 + 8x^2y = x^2 \cdot 4x + x^2 \cdot 1 + x^2 \cdot 8y$$
$$= x^2(4x + 1 + 8y) \quad \blacksquare$$

Notice in Examples 15 and 16 that when one of the terms is itself the GCF, we express it in factored form as the product of the term itself and 1.

EXAMPLE 17   Factor $3x(2x - 1) + 4(2x - 1)$.

Solution: The GCF is $(2x - 1)$. Factoring out the GCF gives
$$3x(2x - 1) + 4(2x - 1) = (3x + 4)(2x - 1)$$

The answer may also be expressed as $(2x - 1)(3x + 4)$ by the commutative property of multiplication. $\blacksquare$

EXAMPLE 18   Factor $x(x + 3) - 5(x + 3)$.

Solution: The GCF is $(x + 3)$. Factoring out the GCF gives
$$x(x + 3) - 5(x + 3) = (x - 5)(x + 3) \quad \blacksquare$$

EXAMPLE 19   Factor $3x^2 + 5xy + 7y^2$.

Solution: The only factor common to all three terms is 1. Whenever the only factor common to all the terms in a polynomial is 1, the polynomial cannot be factored by the method presented in this section. $\blacksquare$

Whenever you are factoring a polynomial by any of the methods presented in this chapter, the first step will always be to see if there is a common factor (other than 1) to all the terms in the polynomial. If so, factor the greatest common factor from each term, using the distributive property.

HELPFUL HINT

> A factoring problem may be checked by multiplying the factors together. The product of the factors should be identical to the expression that was originally factored.

Exercise Set 5.1

Write each number as a product of prime numbers.

1. 40

2. 70

3. 90

4. 180

5. 200

6. 96

Find the greatest common factor for the two given numbers.

7. 36, 20

8. 45, 27

9. 60, 84

10. 120, 96

11. 72, 90

12. 76, 68

Determine the greatest common factor for each set of terms.

13. x^2, x, x^3

14. x^2, x^5, x^7

15. $3x, 6x^2, 9x^3$

16. $6p, 4p^2, 8p^3$

17. x, y, z

18. xy, x, x^2

19. xy, xy^2, xy^3

20. $x^2y, x^3y, 4x$

21. $x^3y^7, x^7y^{12}, x^5y^5$

22. $4x^3, 12x, 8y$

23. $-8, 24x, 48x^2$

24. $24p^3, 18p^2q, 9q^2$

25. $12x^4y^7, 6x^3y^9, 9x^{12}y^7$

26. $14x^8y^2, 16x^7y^7, 12xy^5$

27. $36x^2y, 15x^3, 14x^2y$

28. $-8x^2, -9x^3, 12xy^3$

29. $2(x + 3), 3(x + 3)$

30. $4(x - 5), 3x(x - 5)$

31. $x^2(2x - 3), 5(2x - 3)$

32. $x(2x + 5), 2x + 5$

33. $3x - 4, y(3x - 4)$

34. $x(x + 5), x + 5$

Factor the GCF from each term in the expression. If an expression cannot be factored, so state.

35. $2x + 4$

36. $4x + 2$

37. $15x - 5$

38. $12x + 15$

39. $13x + 5$

40. $6x^2 + 3x$

41. $16x^2 - 12x$

42. $27y - 9y^2$

43. $20p - 18p^2$

44. $9x + 18x^2$

45. $6x^3 - 8x$

46. $7x^5 - 9x^4$

47. $36x^{12} - 24x^8$

48. $45y^{12} + 30y^{10}$

49. $24y^{15} - 9y^3$

50. $38x^4 - 16x^5$

51. $x + 3xy^2$

52. $2x^2y - 6x$

53. $6x + 5y$

54. $3x^2y + 6x^2y^2$

55. $16xy^2z + 4x^3y$

56. $80x^5y^3z^4 - 36x^2yz^3$

57. $34x^2y^2 + 16xy^4$

58. $42xy^6z^{12} - 18y^4z^2$

59. $36xy^2z^3 + 36x^3y^2z$

60. $19x^4y^{12}z^{13} - 8x^5y^3z^9$

61. $14y^3z^5 - 9xy^3z^5$

62. $7x^4y^9 - 21x^3y^7z^5$

63. $3x^2 + 6x + 9$

64. $x^3 + 6x^2 - 4x$

65. $9x^2 + 18x + 3$

66. $4x^2 - 16x + 24$

67. $3x^3 - 6x^2 + 12x$

68. $15x^2 + 9x - 9$

69. $45x^2 - 16x + 10$

70. $5x^3 - 6x^2 + x$

71. $15p^2 - 6p + 9$

72. $35y^3 - 7y^2 + 14y$

73. $24x^6 + 8x^4 - 4x^3$

74. $44x^5y + 11x^3y + 22x^2$

75. $48x^2y + 16xy^2 + 33xy$

76. $52x^2y^2 + 16xy^3 + 26z$

77. $x(x + 2) + 3(x + 2)$

78. $5x(2x - 5) + 3(2x - 5)$

79. $7x(4x - 3) - 4(4x - 3)$

80. $3x(7x + 1) - 2(7x + 1)$

81. $4x(2x + 1) + 1(2x + 1)$

82. $3x(4x - 5) + 1(4x - 5)$

83. $4x(2x + 1) + 2x + 1$

84. $3x(4x - 5) + 4x - 5$

JUST FOR FUN

1. Factor $4x^2(x-3)^3 - 6x(x-3)^2 + 4(x-3)$.

2. Factor $6x^5(2x+7) + 4x^3(2x+7) - 2x^2(2x+7)$.

3. Consider the expression

$$1 + 2 - 3 + 4 + 5 - 6 + 7 + 8 - 9 + 10 + 11 - 12 + 13 + 14 - 15$$

   (a) Construct groups of three terms (for example, the first group is $1 + 2 - 3$), and write the sum of each group as a product of 3 and another factor [for example, the first group would be $3(0)$].

   (b) Factor out the common factor of 3.

   (c) Find the sum of the numbers.

   (d) Use the procedure above to find the sum of the numbers if the process above was continued until $\cdots + 31 + 32 - 33$.

5.2

Factoring by Grouping

1 *Factor a polynomial containing four terms by grouping.*

---

1   It may be possible to factor a polynomial containing four or more terms by removing common factors from groups of terms. This process is called factoring by grouping. We will use factoring by grouping to factor trinomials in Section 5.4. It is therefore important that you understand this section. Example 1 illustrates the procedure for factoring by grouping. The procedure is summarized after Example 1.

EXAMPLE 1   Factor $ax + ay + bx + by$.

Solution: There is no factor (other than 1) common to all four terms. However, a is common to the first two terms and b is common to the last two terms. Factor a from the first two terms and b from the last two terms.

$$ax + ay + bx + by = a(x + y) + b(x + y)$$

Now $(x + y)$ is common to both terms. Factor out $(x + y)$.

$$a(x + y) + b(x + y) = (a + b)(x + y)$$

Thus $ax + ay + bx + by = (a + b)(x + y)$.   ∎

To Factor a Four-Term Polynomial Using Grouping

1. Determine if there are any factors common to all four terms. If so, factor the greatest common factor from each of the four terms.

2. If necessary, arrange the four terms so that the first two terms have a common factor and the last two have a common factor.

3. Use the distributive property to factor each group of two terms.

4. Factor the greatest common factor from the results of Step 3.

EXAMPLE 2   Factor $x^2 + 3x + 4x + 12$ by grouping.

Solution: Factor an x from the first two terms and a 4 from the last two terms.

$$x^2 + 3x + 4x + 12 = x(x + 3) + 4(x + 3)$$

Now factor the common $(x + 3)$.

$$x(x + 3) + 4(x + 3) = (x + 4)(x + 3)$$

Thus $x^2 + 3x + 4x + 12 = (x + 4)(x + 3)$.   ▪

EXAMPLE 3   Factor $6x^2 + 9x + 8x + 12$ by grouping.

Solution: $6x^2 + 9x + 8x + 12 = 3x(2x + 3) + 4(2x + 3)$
$$= (3x + 4)(2x + 3)   ▪$$

A factoring by grouping problem can be checked by multiplying the factors using the FOIL method. If you have not made a mistake your result will be the polynomial you began with. Following is a check of Example 3.

$$(3x + 4)(2x + 3) = (3x)(2x) + (3x)(3) + 4(2x) + 4(3)$$
$$= 6x^2 + 9x + 8x + 12$$

Note that this is the polynomial we started with. Therefore, the factoring is correct.

EXAMPLE 4   Factor $6x^2 + 8x + 9x + 12$ by grouping.

Solution: $6x^2 + 8x + 9x + 12 = 2x(3x + 4) + 3(3x + 4)$
$$= (2x + 3)(3x + 4)   ▪$$

Notice that Example 4 is the same as Example 3 with the two middle terms switched. The answers to Examples 3 and 4 are the same. When factoring by grouping the two like terms may be switched and the answer will remain the same.

EXAMPLE 5   Factor $x^2 + 3x + x + 3$ by grouping.

Solution: x is the common factor of the first two terms. Is there a common factor of the last two terms? Yes; remember that 1 is a factor of every term. Factor a 1 from the last two terms.

$$x^2 + 3x + x + 3 = x^2 + 3x + 1 \cdot x + 1 \cdot 3$$
$$= x(x + 3) + 1(x + 3)$$
$$= (x + 1)(x + 3)$$

Note that $x + 3$ was expressed as $1(x + 3)$.   ▪

EXAMPLE 6   Factor $4x^2 - 2x - 2x + 1$ by grouping.

Solution: When $2x$ is factored from the first two terms, we get

$$4x^2 - 2x - 2x + 1 = 2x(2x - 1) - 2x + 1$$

What should we factor from the last two terms? We wish to factor $-2x + 1$ in such a manner that we end up with an expression that is a multiple of $(2x - 1)$. **Whenever we wish to change the sign of *each* term of an expression, we can factor out a negative number from each term.** In this case we factor out a negative 1.

$$-2x + 1 = -1(2x - 1)$$

Now rewrite $-2x + 1$ as $-1(2x - 1)$

$$2x(2x - 1) \; -2x + 1 = 2x(2x - 1) \; -1(2x - 1)$$

Now factor out the common term $(2x - 1)$.

$$2x(2x - 1) - 1(2x - 1) = (2x - 1)(2x - 1) \text{ or } (2x - 1)^2 \quad \blacksquare$$

EXAMPLE 7   Factor $x^2 + 3x - x - 3$ by grouping.

Solution: $x^2 + 3x - x - 3 = x(x + 3) - x - 3$
$$= x(x + 3) - 1(x + 3)$$
$$= (x - 1)(x + 3)$$

Note: $-x - 3 = -1(x + 3)$. $\blacksquare$

EXAMPLE 8   Factor $3x^2 - 6x - 4x + 8$ by grouping.

Solution: $3x^2 - 6x - 4x + 8 = 3x(x - 2) - 4(x - 2)$
$$= (3x - 4)(x - 2)$$

Note: $-4x + 8 = -4(x - 2)$. $\blacksquare$

EXAMPLE 9   Factor $2x^2 + 4xy + 3xy + 6y^2$.

Solution: This problem contains two variables, x and y. The procedure to factor is basically the same. We will factor out a $2x$ from the first two terms and a $3y$ from the last two terms.

$$2x^2 + 4xy + 3xy + 6y^2 = 2x(x + 2y) + 3y(x + 2y)$$

Now factor out the common terms $(x + 2y)$.

$$2x(x + 2y) + 3y(x + 2y) = (2x + 3y)(x + 2y) \quad \blacksquare$$

If Example 9 were given as $2x^2 + 3xy + 4xy + 6y^2$, would the results be the same? Try it and see.

EXAMPLE 10   Factor $6r^2 - 9rs + 8rs - 12s^2$.

Solution: $6r^2 - 9rs + 8rs - 12s^2 = 3r(2r - 3s) + 4s(2r - 3s)$
$$= (3r + 4s)(2r - 3s) \quad \blacksquare$$

EXAMPLE 11   Factor $3x^2 - 15x + 6x - 30$.

Solution: The first step in any factoring problem is to determine if all the terms have a common factor. If so we factor out that common factor. In the polynomial above, a 3 is common to every term. We therefore begin by factoring out the 3.

$$3x^2 - 15x + 6x - 30 = 3(x^2 - 5x + 2x - 10)$$

Now we factor the remaining expression by grouping.

$$3(x^2 - 5x + 2x - 10) = 3[x(x - 5) + 2(x - 5)]$$
$$= 3[(x + 2)(x - 5)]$$
$$= 3(x + 2)(x - 5)$$

Thus $3x^2 - 15x + 6x - 30 = 3(x + 2)(x - 5)$. ■

Exercise Set 5.2

Factor by grouping.

1. $x^2 + 4x + 3x + 12$
2. $x^2 + 5x + 2x + 10$
3. $x^2 + 2x + 5x + 10$
4. $x^2 - 2x + 3x - 6$
5. $x^2 + 3x + 2x + 6$
6. $x^2 + 2x + 3x + 6$
7. $x^2 + 3x - 5x - 15$
8. $x^2 + 3x - 2x - 6$
9. $4x^2 + 6x - 6x - 9$
10. $4x^2 - 6x + 6x - 9$
11. $2x^2 + 3x + 8x + 12$
12. $5x^2 - 10x + 3x - 6$
13. $3x^2 + 9x + x + 3$
14. $x^2 + 4x + x + 4$
15. $4x^2 - 2x - 2x + 1$
16. $2x^2 + 6x - x - 3$
17. $8x^2 + 32x + x + 4$
18. $6x^2 + 9x + 8x + 12$
19. $8x^2 - 20x - 4x + 10$
20. $8x^2 - 4x - 2x + 1$
21. $3x^2 - 2x + 3x - 2$
22. $35x^2 + 21x - 40x - 24$
23. $3x^2 - 2x - 3x + 2$
24. $35x^2 - 40x + 21x - 24$
25. $15x^2 - 18x - 20x + 24$
26. $15x^2 - 20x - 18x + 24$
27. $x^2 + 2xy - 3xy - 6y^2$
28. $x^2 - 3xy + 2xy - 6y^2$
29. $6x^2 - 9xy + 2xy - 3y^2$
30. $3x^2 - 18xy + 4xy - 24y^2$
31. $10x^2 - 12xy - 25xy + 30y^2$
32. $12x^2 - 9xy + 4xy - 3y^2$
33. $2x^2 - 12x + 8x - 48$
34. $3x^2 - 3x - 3x + 3$
35. $4x^2 + 8x + 8x + 16$
36. $2x^3 - 5x^2 - 6x^2 + 15x$
37. $6x^3 + 9x^2 - 2x^2 - 3x$
38. $9x^3 + 6x^2 - 45x^2 - 30x$
39. $2x^2 - 4xy + 8xy - 16y^2$
40. $18x^2 + 27xy + 12xy + 18y^2$

41. What is the first step in any factoring by grouping problem?

42. How can you check the solution to a factoring by grouping problem?

JUST FOR FUN

Factor by grouping.

1. $3x^5 - 15x^3 + 2x^3 - 10x$
2. $x^3 + xy - x^2y - y^2$
3. $18a^2 + 3ax^2 - 6ax - x^3$
4. $2x^3y^3 + 6xyz^2 - 3x^2y^4 - 9y^2z^2$
5. $2a^4b - 2ac^2 - 3a^3bc + 3c^3$

5.3

Factoring Trinomials with $a = 1$

1.   *Factor trinomials of the form $ax^2 + bx + c$ where $a = 1$.*

2.   *Remove a common factor from the trinomial before factoring a trinomial.*

---

1.   In this section we learn how to factor trinomials of the form $ax^2 + bx + c$, where a, the numerical coefficient of the squared term, is equal to 1. Examples of such trinomials are

$$x^2 + 7x + 12 \qquad\qquad x^2 - 2x - 24$$
$$a = 1, b = 7, c = 12 \qquad a = 1, b = -2, c = -24$$

In Section 5.4 we learn how to factor trinomials when a is different from 1.

In Section 4.4 we illustrated how the FOIL method is used to multiply two binomials. Let us multiply $(x + 3)(x + 4)$ using the FOIL method.

$$
\begin{aligned}
& \quad\quad\quad\quad\quad\quad\ \text{F} \quad \text{O} \quad \text{I} \quad \text{L} \\
(x + 3)(x + 4) &= x^2 + \underline{4x + 3x} + 12 \\
&= x^2 + \quad 7x \quad + 12
\end{aligned}
$$

We see that $(x + 3)(x + 4) = x^2 + 7x + 12$.

Note that the **sum** of the outer and inner terms **is $7x$** and the **product** of the last terms **is 12.**

To factor $x^2 + 7x + 12$ we look for two numbers whose product is 12 and whose sum is 7,

Factors of 12	*Sum of factors*
$(1)(12) = 12$	$1 + 12 = 13$
$(2)(6) = 12$	$2 + 6 = 8$
$(3)(4) = 12$	$3 + 4 = 7$
$(-1)(-12) = 12$	$-1 + (-12) = -13$
$(-2)(-6) = 12$	$-2 + (-6) = -8$
$(-3)(-4) = 12$	$-3 + (-4) = -7$

The only factors of 12 whose sum is a positive 7 are 3 and 4. The factors of $x^2 + 7x + 12$ will therefore be $(x + 3)$ and $(x + 4)$.

$$x^2 + 7x + 12 = (x + 3)(x + 4)$$

To Factor Trinomials of the Form $ax^2 + bx + c$, where $a = 1$

1. Determine two numbers whose product equals the constant, c, and whose sum equals b.
2. The factors of the trinomial will be $(x + \text{first number})$, and $(x + \text{second number})$.

EXAMPLE 1   Factor $x^2 + x - 6$.

Solution: We must find two numbers whose product is -6 and whose sum is 1. Remember that x means $1x$.

Factors of -6	Sum of factors
$1(-6) = -6$	$1 + (-6) = -5$
$2(-3) = -6$	$2 + (-3) = -1$
$3(-2) = -6$	$3 + (-2) = 1$
$6(-1) = -6$	$6 + (-1) = 5$

Note that the factors 1 and -6 are different from the factors -1 and 6.

The numbers 3 and -2 have a product of -6 and a sum of 1. Thus the factors are $(x + 3)$ and $[x + (-2)]$. Since $[x + (-2)]$ is the same as $(x - 2)$,

$$x^2 + x - 6 = (x + 3)(x - 2)$$

Note that the order of the factors is not crucial. Therefore, $x^2 + x - 6 = (x - 2)(x + 3)$ is also an acceptable answer.   ■

Trinomial factoring problems of this type can be checked by multiplying the factors using the FOIL method. If the factoring is correct, the product obtained using the FOIL method will be identical to the original trinomial. Let us check the factors obtained in Example 1.

Check: Does $x^2 + x - 6 = (x + 3)(x - 2)$?
$$= x^2 - 2x + 3x - 6$$
$$= x^2 + x - 6$$

Since the product of the factors is the original trinomial, the factoring was correct.

EXAMPLE 2   Factor $x^2 - x - 6$.

Solution: The factors of -6 are illustrated in Example 1. The factors whose product is -6 and whose sum is -1 are 2 and -3.

Factors	Sum of factors
$(2)(-3) = -6$	$2 + (-3) = -1$

$$x^2 - x - 6 = (x + 2)(x - 3)$$ ■

EXAMPLE 3   Factor $x^2 - 5x + 6$.

Solution: We must find the factors of 6 whose sum is -5.

Factors of 6	Sum of factors
$(1)(6)$	$1 + 6 = 7$
$(2)(3)$	$2 + 3 = 5$
$(-1)(-6)$	$-1 + (-6) = -7$
$(-2)(-3)$	$-2 + (-3) = -5$

The factors of 6 whose sum is -5 are -2 and -3.
$$x^2 - 5x + 6 = (x - 2)(x - 3)$$ ■

EXAMPLE 4   Factor $x^2 + 2x - 8$.

Solution: We must find the factors of -8 whose sum is 2.

Factors of -8	Sum of factors
$(1)(-8)$	$1 + (-8) = -7$
$(2)(-4)$	$2 + (-4) = -2$
$(4)(-2)$	$4 + (-2) = 2$
$(8)(-1)$	

Since we have found the two numbers, 4 and -2, whose product is -8 and whose sum is 2, we need go no further.

$$x^2 + 2x - 8 = (x + 4)(x - 2) \quad \blacksquare$$

EXAMPLE 5   Factor $x^2 - 6x + 9$.

Solution: The two factors whose product is 9 and whose sum is -6 are -3 and -3.

$$x^2 - 6x + 9 = (x - 3)(x - 3)$$
$$= (x - 3)^2 \quad \blacksquare$$

HELPFUL HINT

When factoring a trinomial of the form $x^2 + bx + c$, the sign of the constant, c, is very helpful in finding the solution.

1. When the constant is positive, and the numerical coefficient of the x term is positive, both numerical factors will be positive.

   Example: $x^2 + 7x + 12 = (x + 3)(x + 4)$

   positive   positive   positive   positive

2. When the constant is positive and the numerical coefficient of the x term is negative, both numerical factors will be negative.

   Example: $x^2 - 5x + 6 = (x - 2)(x - 3)$

   negative   positive   negative   negative

3. When the constant is negative, one of the numerical factors will be positive and the other will be negative.

   Example: $x^2 + x - 6 = (x + 3)(x - 2)$

   positive   negative   positive   negative

   Example: $x^2 - x - 6 = (x + 2)(x - 3)$

   negative   negative   positive   negative

EXAMPLE 6   Factor $x^2 - 2x - 24$.

Solution: Since the constant is negative, one factor must be positive and the other negative. The desired factors are -6 and 4 because $(-6)(4) = -24$ and $-6 + 4 = -2$.

$$x^2 - 2x - 24 = (x - 6)(x + 4) \quad \blacksquare$$

EXAMPLE 7   Factor $x^2 + 4x + 12$.

Solution: Let us first find the two numbers whose product is 12 and whose sum is 4. Since both the constant and x term are positive, the two numbers must also be positive.

Factors of 12	Sum of factors
(1)(12)	$1 + 12 = 13$
(2)(6)	$2 + 6 = 8$
(3)(4)	$3 + 4 = 7$

Note that there are no two numbers whose product is 12 and whose sum is 4. When two numbers cannot be found to satisfy the given conditions, the trinomial cannot be factored by the method presented in this section. ■

There is at most one pair of numbers that satisfy the two specific conditions of the problem. For example, when factoring $x^2 - 2x - 24$ the two numbers whose product is -24 and whose sum is -2 are -6 and 4. No other pair of numbers will satisfy these specific conditions. Thus the only factors of $x^2 - 2x - 24$ are $(x - 6)(x + 4)$.

A slightly different type of problem is illustrated in Example 8.

EXAMPLE 8   Factor $x^2 + 2xy + y^2$ completely.

Solution: In this problem the second term contains two variables, x and y, and the last term is not a constant. The procedure used to solve this problem is similar to that outlined previously. You should realize, however, that the product of the first terms of the factors we are looking for must be x^2, and the product of the last terms of the factors must be y^2.

We must find two numbers whose product is 1 (from $1y^2$) and whose sum is 2 (from $2xy$). The two numbers are 1 and 1. Thus $x^2 + 2xy + y^2 = (x + 1y)(x + 1y)$ or $(x + y)(x + y)$. ■

EXAMPLE 9   Factor $x^2 - xy - 6y^2$.

Solution: Determine two numbers whose product is -6 and whose sum is -1. The numbers are -3 and $+2$. The last terms must be $-3y$ and $2y$ to obtain the $-6y^2$.

$$x^2 - xy - 6y^2 = (x - 3y)(x + 2y) \quad ■$$

2   There will be times when each term of a trinomial has a common factor. When this occurs, factor out the common factor as explained in Section 5.2 before factoring the trinomial. **Whenever the numerical coefficient of the highest-powered term is not 1, you should check for a common factor.**

EXAMPLE 10   Factor $2x^2 + 2x - 12$.

Solution: Since the numerical coefficient of the squared term is not 1, we check for a common factor. Two is common to each term of the polynomial, so we factor it out.

$$2x^2 + 2x - 12 = 2(x^2 + x - 6) \qquad \text{factor out common factor}$$

Now factor the remaining trinomial $x^2 + x - 6$ into $(x + 3)(x - 2)$.

$$2x^2 + 2x - 12 = 2(x + 3)(x - 2)$$

Note that the 2 that is factored out *is a part of the answer*. After the 2 has been factored out, it plays no part in the factoring of the remaining trinomial. ■

EXAMPLE 11   Factor $3x^3 + 24x^2 - 60x$.

Solution: $3x$ divides each term of the polynomial and therefore is a common factor.

$$3x^3 + 24x^2 - 60x = 3x(x^2 + 8x - 20) \qquad \text{factor out common factor}$$
$$= 3x(x + 10)(x - 2) \qquad \text{factor remaining trinomial} \quad ■$$

Exercise Set 5.3

Factor each expression. If an expression cannot be factored by the method presented in this section, so state.

1. $x^2 + 7x + 10$

2. $x^2 - 8x + 12$

3. $x^2 + 5x + 6$

4. $x^2 + 7x + 6$

5. $x^2 + 7x + 12$

6. $x^2 - x - 6$

7. $x^2 - 7x + 9$

8. $y^2 - 6y + 8$

9. $y^2 - 16y + 15$

10. $x^2 + 8x - 20$

11. $x^2 + x - 6$

12. $p^2 - 3p - 10$

13. $k^2 - 2k - 15$

14. $x^2 - 6x + 10$

15. $b^2 - 11b + 18$

16. $x^2 + 11x - 30$

17. $x^2 - 8x - 15$

18. $x^2 - 10x + 21$

19. $a^2 + 12a + 11$

20. $x^2 + 16x + 64$

21. $x^2 + 13x - 30$

22. $x^2 - 30x - 64$

23. $x^2 + 4x + 4$

24. $x^2 - 4x + 4$

25. $x^2 + 6x + 9$

26. $x^2 - 6x + 9$

27. $x^2 + 10x + 25$

28. $x^2 - 10x - 25$

29. $w^2 - 18w + 45$

30. $x^2 - 11x + 10$

31. $x^2 + 22x - 48$

32. $x^2 - 2x + 8$

33. $x^2 - x - 20$

34. $x^2 - 17x - 60$

35. $y^2 - 9y + 14$

36. $x^2 - 2xy + y^2$

37. $x^2 + 4xy + 3y^2$

38. $x^2 - 6xy + 8y^2$

39. $x^2 + 8xy + 15y^2$

40. $x^2 - 5xy - 14y^2$

Factor completely.

41. $2x^2 - 14x + 12$

42. $3x^2 - 9x - 30$

43. $5x^2 + 20x + 15$

44. $4x^2 + 12x - 16$

45. $2x^2 - 14x + 24$

46. $3y^2 - 33y + 54$

47. $x^3 - 3x^2 - 18x$

48. $x^3 + 11x^2 - 42x$

49. $2x^3 + 6x^2 - 56x$

50. $3x^3 - 36x^2 + 33x$

51. $x^3 + 4x^2 + 4x$

52. $2x^3 - 12x^2 + 10x$

53. How can a trinomial factoring problem be checked?

54. Explain how to determine the factors when factoring a trinomial of the form $x^2 + bx + c$.

5.4

Factoring Trinomials with $a \neq 1$

1 *Factor trinomials of the form $ax^2 + bx + c$, $a \neq 1$.*

1   In this section we determine how to factor quadratic trinomials whose highest-powered term, after removing any common factors, has a numerical coefficient not equal to 1.

Examples of trinomials with $a \neq 1$ are

$$2x^2 + 11x + 12 \quad \text{and} \quad 4x^2 - 4x + 1$$

To Factor Trinomials of the Form $ax^2 + bx + c$, $a \neq 1$

1. Find two numbers whose product is equal to the product of a times c, and whose sum is equal to b.
2. Rewrite the bx term using the factors found in Step 1.
3. Factor by grouping as explained in Section 5.2.

This process will be made clear in Example 1.

EXAMPLE 1   Factor $2x^2 + 11x + 12$.

Solution: First determine if there is a common factor to all the terms of the polynomial. There are no common factors (other than 1) to the three terms.

$$a = 2, \quad b = 11, \quad c = 12$$

1. We must find two numbers whose product is $a \cdot c$ and whose sum is b. We must therefore find two numbers whose product equals $2 \cdot 12$ or 24 and whose sum equals 11. Only the positive factors of 24 need be considered since all signs of the trinomial are positive.

Factors of 24	*Sum of factors*
(1)(24)	$1 + 24 = 25$
(2)(12)	$2 + 12 = 14$
(3)(8)	$3 + 8 = 11$
(4)(6)	$4 + 6 = 10$

The desired factors are 3 and 8.

2. Rewrite the $11x$ term using the values found in step 1. Therefore we rewrite $11x$ as $3x + 8x$.

$$2x^2 + \underbrace{11x}_{} + 12$$
$$= 2x^2 + \overbrace{3x + 8x} + 12$$

3. Factor by grouping.

x is common 4 is common
factor   factor
$$\overbrace{2x^2 + 3x} + \overbrace{8x + 12}$$
$$= x(2x + 3) + 4(2x + 3) = (x + 4)(2x + 3)$$

Note that we rewrote $11x$ as $3x + 8x$ in Step 2. Would it have made a difference if we had written $11x$ as $8x + 3x$? Let us work it out and see.

$$2x^2 + \underbrace{11x}_{} + 12$$
$$= 2x^2 + \overbrace{8x + 3x} + 12$$

$2x$ is common 3 is common
factor   factor
$$\overbrace{2x^2 + 8x} + \overbrace{3x + 12}$$
$$= 2x(x + 4) + 3(x + 4) = (2x + 3)(x + 4)$$

Since $(2x + 3)(x + 4) = (x + 4)(2x + 3)$, the $11x$ may be expressed either way. ■

EXAMPLE 2   Factor $5x^2 - 7x - 6$.

Solution: There are no common factors other than 1.

$$a = 5, \qquad b = -7, \qquad c = -6$$

The product of a times c is $5(-6) = -30$. We must find two numbers whose product is -30 and whose sum is -7.

Factors of -30	*Sum of factors*
$(-1)(30)$	$-1 + 30 = 29$
$(-2)(15)$	$-2 + 15 = 13$
$(-3)(10)$	$-3 + 10 = 7$
$(-5)(6)$	$-5 + 6 = 1$
$(-6)(5)$	$-6 + 5 = -1$
$(-10)(3)$	$-10 + 3 = -7$
$(-15)(2)$	$-15 + 2 = -13$
$(-30)(1)$	$-30 + 1 = -29$

Rewrite the $-7x$ as $-10x + 3x$.

$$5x^2 - 7x - 6$$
$$= 5x^2 \overbrace{-10x + 3x} - 6 \qquad \text{now factor by grouping}$$
$$= 5x(x - 2) + 3(x - 2)$$
$$= (5x + 3)(x - 2)$$

Note that the $-7x$ could have also been expressed as $3x - 10x$. ■

EXAMPLE 3   Factor $8x^2 + 33x + 4$.

Solution: There are no common factors other than 1. We must find two numbers whose product is $8 \cdot 4$ or 32 and whose sum is 33. The numbers are 1 and 32.

Factors of 32	*Sum of factors*
$(1)(32)$	$1 + 32 = 33$

Rewrite $33x$ as $32x + x$.

$$8x^2 + 33x + 4$$
$$= 8x^2 + 32x + x + 4$$
$$= 8x(x + 4) + 1(x + 4)$$
$$= (8x + 1)(x + 4) \quad \blacksquare$$

EXAMPLE 4   Factor $4x^2 - 4x + 1$.

Solution: There are no common factors other than 1. We must find two numbers whose product is $4 \cdot 1$ or 4 and whose sum is -4. Since the product of a times c is positive and the b term is negative, both numerical factors must be negative.

Factors of 4	*Sum of factors*
$(-1)(-4)$	$-1 + (-4) = -5$
$(-2)(-2)$	$-2 + (-2) = -4$

The desired factors are -2 and -2.

$$4x^2 - 4x + 1$$
$$= 4x^2 - 2x - 2x + 1 \qquad \text{rewrite } -4x \text{ as } -2x - 2x$$
$$= 2x(2x - 1) - 2x + 1$$
$$= 2x(2x - 1) - 1(2x - 1) \qquad \text{rewrite } -2x + 1 \text{ as } -1(2x - 1)$$
$$= (2x - 1)(2x - 1) \text{ or } (2x - 1)^2 \quad \blacksquare$$

When attempting to factor a trinomial, if there are no two integers whose product equals $a \cdot c$ and whose sum equals b, the trinomial cannot be factored by this method.

EXAMPLE 5   Factor $2x^2 + 3x + 5$.

Solution: There are no common factors other than 1. We must find two numbers whose product is 10 and whose sum is 3.

Factors of 10	*Sum of factors*
$(1)(10)$	$1 + 10 = 11$
$(2)(5)$	$2 + 5 = 7$

Since there are no two factors of 10 whose sum is 3, we conclude that this trinomial cannot be factored by this method. $\blacksquare$

EXAMPLE 6   Factor $4x^2 + 7xy + 3y^2$.

Solution: There are no common factors other than 1. This trinomial contains two variables. It is factored in basically the same manner as the previous examples. Determine two numbers whose product is $4 \cdot 3$ or 12 and whose sum is 7. The two numbers are 4 and 3.

$$4x^2 + 7xy + 3y^2$$
$$= 4x^2 + 4xy + 3xy + 3y^2$$
$$= 4x(x + y) + 3y(x + y)$$
$$= (4x + 3y)(x + y) \quad \blacksquare$$

EXAMPLE 7   Factor $6x^2 - 13xy - 8y^2$.

Solution: There are no common factors other than 1. Determine two numbers whose product is $6(-8)$ or -48 and whose sum is -13. Since the product is negative, one factor must be positive and the other negative. Some factors are:

Product of factors	Sum of factors
$(1)(-48)$	$1 + (-48) = -47$
$(2)(-24)$	$2 + (-24) = -22$
$(3)(-16)$	$3 + (-16) = -13$

There are many other factors. We have, however, found the ones we are looking for. The two numbers whose product is -48 and whose sum is -13 are -16 and $+3$.

$$6x^2 - 13xy - 8y^2$$
$$= 6x^2 - 16xy + 3xy - 8y^2$$
$$= 2x(3x - 8y) + y(3x - 8y)$$
$$= (2x + y)(3x - 8y)$$

Check: $(2x + y)(3x - 8y)$

$$\quad\;\; \text{F} \qquad\quad \text{O} \qquad\quad \text{I} \qquad\quad \text{L}$$
$$(2x)(3x) + (2x)(-8y) + (y)(3x) + (y)(-8y)$$
$$6x^2 \;\; - \;\; 16xy \;\; + \;\; 3xy \;\; - \;\; 8y^2$$
$$6x^2 - 13xy - 8y^2 \quad \blacksquare$$

Remember that in any factoring problem our first step is to determine if all terms in the polynomial have a common factor other than 1. If so, we use the distributive property to factor the GCF from each of the terms. We then continue to factor by one of the methods mentioned in this chapter, if possible.

EXAMPLE 8   Factor $4x^3 + 10x^2 + 6x$.

Solution: The factor $2x$ is common to all three terms. Factor the $2x$ from each term of the polynomial.

$$4x^3 + 10x^2 + 6x = 2x(2x^2 + 5x + 3)$$

Now continue by factoring $2x^2 + 5x + 3$. The two numbers whose product is $2 \cdot 3$ or 6 and whose sum is 5 are 2 and 3.

$$2x[2x^2 + 5x + 3]$$
$$= 2x[2x^2 + 2x + 3x + 3]$$
$$= 2x[2x(x + 1) + 3(x + 1)]$$
$$= 2x(2x + 3)(x + 1) \quad \blacksquare$$

Exercise Set 5.4 _____

Factor completely. If an expression cannot be factored, so state.

1. $3x^2 + 5x + 2$
2. $3x^2 + 4x + 1$
3. $6x^2 + 13x + 6$
4. $5x^2 + 13x + 6$
5. $2x^2 + 5x + 3$
6. $4x^2 + 4x - 3$
7. $2x^2 + 11x + 15$
8. $3x^2 - 2x - 8$
9. $3x^2 - 10x - 8$
10. $4x^2 - 11x + 7$
11. $5y^2 - 8y + 3$
12. $5m^2 - 16m + 3$
13. $5a^2 - 12a + 6$
14. $2x^2 - x - 1$
15. $4x^2 + 13x + 3$
16. $6y^2 - 19y + 15$
17. $5x^2 + 11x + 4$
18. $3x^2 - 2x - 5$
19. $5y^2 - 16y + 3$
20. $5x^2 + 2x + 7$
21. $3x^2 + 14x - 5$
22. $7x^2 + 43x + 6$
23. $7x^2 - 16x + 4$
24. $15x^2 - 19x + 6$
25. $3x^2 - 10x + 7$
26. $3y^2 - 22y + 7$
27. $5z^2 - 33z - 14$
28. $3z^2 - 11z - 6$
29. $6x^2 + 33x + 15$
30. $18x^2 - 3x - 10$
31. $6x^2 + 4x - 10$
32. $12z^2 + 32z + 20$
33. $6x^3 + 5x^2 - 4x$
34. $8x^2 + 2x - 20$
35. $4x^3 + 2x^2 - 6x$
36. $18x^3 - 21x^2 - 9x$
37. $6x^3 + 4x^2 - 10x$
38. $300x^2 - 400x - 400$
39. $60x^2 + 40x + 5$
40. $36x^2 - 36x + 9$
41. $2x^2 + 5xy + 2y^2$
42. $8x^2 - 8xy - 6y^2$
43. $2x^2 - 7xy + 3y^2$
44. $15x^2 - xy - 6y^2$
45. $18x^2 + 18xy - 8y^2$
46. $12a^2 - 34ab + 24b^2$

47. What is the first step in factoring any trinomial?
48. How may any trinomial factoring problem be checked?
49. Explain in your own words the procedure used to factor a trinomial of the form $ax^2 + bx + c, a \neq 1$.

5.5 _____

Special Factoring Formulas and A General Review of Factoring

1   *Factor the difference of two squares.*

2   *Factor the sum and difference of two cubes.*

3   *Learn the general procedure for factoring a polynomial.*

There are special formulas for certain types of factoring problems that are used very often. The special formulas we focus on in this section are the difference of two squares and the sum and difference of two cubes. There is no special formula for the sum of two squares; this is because the sum of two squares cannot be factored over the set of real numbers. You will need to memorize the three formulas in this section so that you can use them automatically.

Difference of Two Squares

1 Let us begin with the difference of two squares. Consider the binomial $4x^2 - 9$. Note that each term of the binomial can be expressed as the square of some expression

$$4x^2 - 9 = (2x)^2 - (3)^2$$

This is an example of a difference of two squares problem. We will first show in Example 1, how the expression $4x^2 - 9$ can be factored using the procedure presented in Section 5.4. We will then give a quicker technique for factoring the difference of two squares, and factor $4x^2 - 9$ using the difference of squares formula.

EXAMPLE 1   Factor $4x^2 - 9$.

Solution: There is no common factor other than 1. This expression can be rewritten as $4x^2 + 0x - 9$. We must find two numbers whose product is $4(-9)$ or -36 and whose sum is 0. The numbers are 6 and -6.

Factors	*Sum of factors*
$(6)(-6) = -36$	$6 + (-6) = 0$

$$4x^2 + 0x - 9$$
$$= 4x^2 + 6x - 6x - 9$$
$$= 2x(2x + 3) - 3(2x + 3)$$
$$= (2x - 3)(2x + 3)$$

Thus $4x^2 - 9 = (2x + 3)(2x - 3)$.   ■

To factor the difference of two squares, it is convenient to use the difference of two squares formula first introduced in Section 4.4.

Difference of Two Squares

$$a^2 - b^2 = (a + b)(a - b)$$

Let us now factor the binomial in Example 1 using the difference of two squares formula. First write $4x^2 - 9$ as a difference of two squares

$$4x^2 - 9 = (2x)^2 - (3)^2$$

If we let $a = 2x$ and $b = 3$, then

$$4x^2 - 9 = (2x)^2 - (3)^2 \quad \text{square of } 2x \text{ minus square of } 3$$
$$= a^2 - b^2 = (a + b)(a - b)$$
$$= (2x + 3)(2x - 3)$$

EXAMPLE 2   Factor each of the following using the difference of two squares formula.
(a) $x^2 - 16$ (b) $16x^2 - 9y^2$

Solution: (a) $x^2 - 16 = (x)^2 - (4)^2$
$$= (x + 4)(x - 4)$$

(b) $16x^2 - 9y^2 = (4x)^2 - (3y)^2$
$$= (4x + 3y)(4x - 3y)   \blacksquare$$

EXAMPLE 3   Factor each of the following differences of squares.
(a) $16x^4 - 9y^4$ (b) $x^6 - y^4$

Solution: (a) Let $a = 4x^2$ and $b = 3y^2$. Then
$$16x^4 - 9y^4 = (4x^2)^2 - (3y^2)^2$$
$$= (4x^2 + 3y^2)(4x^2 - 3y^2)$$

(b) Let $a = x^3$ and $b = y^2$. Then
$$x^6 - y^4 = (x^3)^2 - (y^2)^2$$
$$= (x^3 + y^2)(x^3 - y^2)   \blacksquare$$

EXAMPLE 4   Factor $4x^2 - 16y^2$ using the difference of two squares formula.

Solution: First remove the common factor, 4.
$$4x^2 - 16y^2 = 4(x^2 - 4y^2)$$

Now use the formula for the difference of two squares.
$$4(x^2 - 4y^2) = 4[(x)^2 - (2y)^2]$$
$$= 4(x + 2y)(x - 2y)   \blacksquare$$

Sum and Difference of Two Cubes

2   We begin our discussion of the sum and difference of two cubes with a multiplication of polynomials problem. Consider the product of $(a + b)(a^2 - ab + b^2)$.

$$\begin{array}{r} a^2 - ab + b^2 \\ a + b \\ \hline a^2b - ab^2 + b^3 \\ a^3 - a^2b + ab^2 \\ \hline a^3 \qquad\qquad + b^3 \end{array}$$

Thus $(a + b)(a^2 - ab + b^2) = a^3 + b^3$. Since factoring is the opposite of multiplying, we may factor $a^3 + b^3$ as follows

$$a^3 + b^3 = (a + b)(a^2 - ab + b^2).$$

It can be shown using the same procedure that $a^3 - b^3 = (a - b)(a^2 + ab + b^2)$. The expression $a^3 + b^3$ is a sum of two cubes and the expression $a^3 - b^3$ is a difference of two cubes.

The formulas for factoring the sum and difference of two cubes follow.

Sum of Two Cubes

$$a^3 + b^3 = (a + b)(a^2 - ab + b^2)$$

Difference of Two Cubes

$$a^3 - b^3 = (a - b)(a^2 + ab + b^2)$$

Now let us do some factoring problems using the sum and difference of two cubes.

EXAMPLE 5   Factor $x^3 + 8$.

Solution: Rewrite $x^3 + 8$ as a sum of two cubes.

$$x^3 + 8 = (x)^3 + (2)^3$$

If we let x be a and 2 be b, then using the sum of cubes formula we get

$$
\begin{aligned}
x^3 + 8 &= (x)^3 + (2)^3 \\
&= (x + 2)\left[x^2 - x(2) + 2^2\right] \\
&= (x + 2)(x^2 - 2x + 4) \quad \blacksquare
\end{aligned}
$$

You can check this problem by multiplying $(x + 2)(x^2 - 2x + 4)$. If factored correctly, the product of the factors will equal the original expression, $x^3 + 8$. Try it and see.

EXAMPLE 6   Factor $y^3 - 27$.

Solution: Rewrite $y^3 - 27$ as a difference of two cubes.

$$
\begin{aligned}
y^3 - 27 &= (y)^3 - (3)^3 \\
&= (y - 3)\left[y^2 + y(3) + 3^2\right] \\
&= (y - 3)(y^2 + 3y + 9) \quad \blacksquare
\end{aligned}
$$

EXAMPLE 7   Factor $8a^3 - b^3$.

Solution: Rewrite $8a^3 - b^3$ as a difference of two cubes. Note that $2^3 = 8$, thus we can write

$$
\begin{aligned}
8a^3 - b^3 &= (2a)^3 - (b)^3 \\
&= (2a - b)\left[(2a)^2 + (2a)(b) + b^2\right] \\
&= (2a - b)(4a^2 + 2ab + b^2) \quad \blacksquare
\end{aligned}
$$

EXAMPLE 8   Factor $8r^3 + 27s^3$.

Solution: Rewrite $8r^3 + 27s^3$ as a sum of two cubes. Since $8 = 2^3$ and $27 = 3^3$ we write

$$
\begin{aligned}
8r^3 + 27s^3 &= (2r)^3 + (3s)^3 \\
&= (2r + 3s)\left[(2r)^2 - (2r)(3s) + (3s)^2\right] \\
&= (2r + 3s)(4r^2 - 6rs + 9s^2) \quad \blacksquare
\end{aligned}
$$

A General Review
of Factoring

3   In this chapter we presented a number of different methods of factoring. We will now combine problems and techniques from this and previous sections.

Here is a general procedure for factoring any polynomial:

> **To Factor a Polynomial**
>
> **1.** Determine if the polynomial has a greatest common factor other than 1. If so factor out the GCF from every term in the polynomial.
> **2.** If the polynomial has two terms (or is a binomial), determine if it is a difference of two squares or a sum or difference of two cubes. If so, factor using the appropriate formula.
> **3.** If the polynomial has 3 terms, factor the trinomial using the methods discussed in Sections 5.3 and 5.4.
> **4.** If the polynomial has more than 3 terms, then try factoring by grouping.
> **5.** As a final step, examine your factored polynomial to see if any factors listed have a common factor and can be factored further. If you find a common factor, factor it out at this point.

EXAMPLE 9   Factor $3x^4 - 27x^2$.

Solution: First determine if there is a greatest common factor other than 1. Since $3x^2$ is common to both terms factor it out.

$$3x^4 - 27x^2 = 3x^2(x^2 - 9)$$
$$= 3x^2(x + 3)(x - 3)$$

Note that $x^2 - 9$ is a difference of two squares.   ■

EXAMPLE 10   Factor $3x^2y^2 - 6xy^2 - 24y^2$.

Solution: Begin by factoring the GCF, $3y^2$, from each term.

$$3x^2y^2 - 6xy^2 - 24y^2 = 3y^2(x^2 - 2x - 8)$$
$$= 3y^2(x - 4)(x + 2)$$ ■

EXAMPLE 11   Factor $18x^2 + 24x - 36x - 48$.

Solution: Always begin by determining if the polynomial has a common factor. In this example 6 is the GCF. Factor a 6 from each term.

$$18x^2 + 24x - 36x - 48 = 6(3x^2 + 4x - 6x - 8)$$

Now factor by grouping.

$$= 6[x(3x + 4) - 2(3x + 4)]$$
$$= 6(x - 2)(3x + 4)$$ ■

In Example 11 what would happen if we thought the GCF was 3 instead of 6? Let's work it out like this and see what happens.

$$18x^2 + 24x - 36x - 48 = 3(6x^2 + 8x - 12x - 16)$$
$$= 3[2x(3x + 4) - 4(3x + 4)]$$
$$= 3(2x - 4)(3x + 4)$$

In Step 5 we are told to examine the factored polynomial to see if any factor listed has a common factor. If we study the factors we see that the factor $2x - 4$ has a common factor of 2. If we factor out the 2 from $2x - 4$ we will obtain the same answer obtained in Example 11.

$$3(2x - 4)(3x + 4) = 3[2(x - 2)(3x + 4)]$$
$$= 6(x - 2)(3x + 4)$$

EXAMPLE 12   Factor $10a^2b - 15ab + 20b$.

Solution: $10a^2b - 15ab + 20b = 5b(2a^2 - 3a + 4)$

Since $2a^2 - 3a + 4$ cannot be factored we stop here.   ■

EXAMPLE 13   Factor $2x^4y + 54xy$.

Solution: First factor out the common factor $2xy$.

$$2x^4y + 54xy = 2xy(x^3 + 27)$$
$$= 2xy(x + 3)(x^2 - 3x + 9)$$

Note that $x^3 + 27$ is a sum of two cubes.   ■

Exercise Set 5.5

Factor the difference of two squares.

1. $x^2 - 4$

2. $x^2 - 9$

3. $y^2 - 25$

4. $z^2 - 64$

5. $x^2 - 49$

6. $x^2 - a^2$

7. $x^2 - y^2$

8. $4x^2 - 9$

9. $9y^2 - 16$

10. $16x^2 - 9y^2$

11. $64a^2 - 36b^2$

12. $100x^2 - 81y^2$

13. $25x^2 - 16$

14. $y^4 - 4x^2$

15. $z^4 - 81x^2$

16. $9x^4 - 16y^4$

17. $9x^4 - 81y^2$

18. $4x^4 - 25y^4$

19. $49m^4 - 16n^2$

20. $2x^4 - 50y^2$

21. $20x^2 - 180$

22. $4x^3 - xy^2$

Factor the sum or difference of two cubes.

23. $x^3 + y^3$

24. $x^3 - y^3$

25. $a^3 - b^3$

26. $a^3 + b^3$

27. $x^3 + 8$

28. $x^3 - 8$

29. $x^3 - 27$

30. $a^3 + 27$

31. $a^3 + 1$

32. $a^3 - 1$

33. $8x^3 + 27$

34. $27y^3 - 8$

35. $27a^3 - 64$

36. $64 - x^3$

37. $27 - 8y^3$

38. $1 + 27y^3$

39. $8x^3 - 27y^3$

40. $64x^3 - 27y^3$

Factor each of the following completely.

41. $2x^2 - 2x - 12$

42. $3x^2 - 9x - 12$

43. $x^2y - 16y$

44. $2x^2 - 8$

45. $3x^2 + 6x + 3$

46. $3x^2 - 9x - 12x + 36$

47. $5x^2 + 10x - 15$

48. $4x^2 - 100$

49. $3x^2 - 18x + 12x - 72$

50. $x^2y + 2xy - 6xy - 12y$

51. $2x^2 - 72$

52. $4ya^2 - 36y$

53. $3x^2y - 27y$

54. $2x^3 - 50x$

55. $3x^3y^2 + 3y^2$

56. $x^4 - 8x$

57. $2x^3 - 16$

58. $x^3 - 64x$

59. $6x^2 - 4x + 24x - 16$

60. $4x^2y - 6xy - 20xy + 30y$

61. $3x^3 - 10x^2 - 8x$

62. $4x^3 - 22x^2 + 30x$

65. $25b^2 - 100$

68. $12x^2 + 8x - 18x - 12$

71. $x^3 + 25x$

73. $y^4 - 16$

75. $10a^2 + 25ab - 60b^2$

77. $9x^2 + 12x - 5$

79. $x^3 - 25x$

63. $4x^2 + 5x - 6$

66. $3b^2 - 75c^2$

69. $3x^4 - 18x^3 + 27x^2$

72. $8y^2 + 23y - 3$

74. $16m^3 + 250$

76. $4 - 2x - 6y + 3xy$

78. $2w^3 - 6w^2 - 18w + 54$

80. $9 - 9y^4$

64. $12a^2 - 36a + 27$

67. $a^5b^2 - 4a^3b^4$

70. $a^6 + 4a^4b^2$

JUST FOR FUN

1. Factor $x^6 + 1$.

2. Factor $x^6 - 27y^9$.

3. Have you ever seen the proof that 1 is equal to 2? Here it is.

Let $a = b$, then square both sides of the equation:

$$a^2 = b^2$$

$$a^2 = b \cdot b$$

$$a^2 = ab \qquad \text{substitute } a = b$$

$$a^2 - b^2 = ab - b^2 \qquad \text{subtract } b^2 \text{ from both sides of equation}$$

$$(a + b)(a - b) = b(a - b) \qquad \text{factor both sides of equation}$$

$$\frac{(a + b)\cancel{(a - b)}}{\cancel{(a - b)}} = \frac{b\cancel{(a - b)}}{\cancel{(a - b)}} \qquad \begin{array}{l}\text{divide both sides of equation by } (a - b) \text{ and}\\ \text{divide out common factors}\end{array}$$

$$a + b = b$$

$$b + b = b \qquad \text{substitute } a = b$$

$$2b = b$$

$$\frac{\overset{1}{\cancel{2b}}}{\underset{1}{\cancel{b}}} = \frac{\overset{1}{\cancel{b}}}{\underset{1}{\cancel{b}}} \qquad \text{divide both sides of equation by } b$$

$$2 = 1$$

Obviously, $2 \neq 1$. Therefore, we must have made an error somewhere. Can you find it?

5.6

Solving Quadratic Equations Using Factoring

1 *Recognize quadratic equations.*

2 *Solve quadratic equations using factoring.*

---

1 In this section we introduce **quadratic equations,** which are equations that contain a squared term and no term of a higher degree.

Quadratic Equation

Quadratic equations have the form

$$ax^2 + bx + c = 0$$

where a, b, and c are real numbeers, $a \neq 0$.

Examples of Quadratic Equations

$$x^2 + 2x - 3 = 0$$
$$3x^2 - 4x = 0$$
$$2x^2 - 3 = 0$$

Quadratic equations like those given above, where one side of the equation is written in descending order of the variable and the other side of the equation is equal to zero are said to be in **standard form.**

Some quadratic equations can be solved by factoring. Two methods that can be used to solve quadratic equations that cannot be solved by factoring are given in Chapter 10. To solve a quadratic equation by factoring we use the zero-factor property.

Zero-factor Property

If $ab = 0$ then $a = 0$ or $b = 0$.

In other words, if the product of two factors is 0, then at least one of the factors must be 0.

EXAMPLE 1   Solve the equation $(x + 3)(x + 4) = 0$.

Solution: Since the product of the factors equals 0, according to the rule above, one or both factors must equal zero. Set each factor equal to 0, and solve each resulting equation.

$$
\begin{array}{ccc}
(x + 3) = 0 & \text{or} & (x + 4) = 0 \\
x + 3 = 0 & & x + 4 = 0 \\
x + 3 - 3 = 0 - 3 & & x + 4 - 4 = 0 - 4 \\
x = -3 & \text{or} & x = -4
\end{array}
$$

Thus if x is either -3 or -4, the product of the factors is 0. The solutions to the equation are -3 and -4.

Check:

$$
\begin{array}{cc}
x = -3 & x = -4 \\
(x + 3)(x + 4) = 0 & (x + 3)(x + 4) = 0 \\
(-3 + 3)(-3 + 4) = 0 & (-4 + 3)(-4 + 4) = 0 \\
0(1) = 0 & -1(0) = 0 \\
0 = 0 & 0 = 0 \quad \blacksquare
\end{array}
$$

EXAMPLE 2   Solve the equation $(4x - 3)(2x + 4) = 0$.

Solution: Set each factor equal to 0 and solve for x.

$$(4x - 3) = 0 \quad \text{or} \quad (2x + 4) = 0$$
$$4x - 3 = 0 \qquad\qquad 2x + 4 = 0$$
$$4x = 3 \qquad\qquad 2x = -4$$
$$x = \frac{3}{4} \quad \text{or} \quad x = -2$$

The solutions to the equation are $\frac{3}{4}$ and -2.   ■

To Solve A Quadratic Equation Using Factoring

1. Write the equation in standard form with the squared term positive. This will result in the one side of the equation being equal to 0.

2. Set each factor containing a variable equal to zero and find the solution.

EXAMPLE 3   Solve the equation $2x^2 - 12x = 0$.

Solution: Since all terms are already on the left of the equal sign and the right side equals 0, we factor the left side:

$$2x^2 - 12x = 0$$
$$2x(x - 6) = 0$$

Now set each factor equal to zero.

$$2x = 0 \quad \text{or} \quad x - 6 = 0$$
$$x = \frac{0}{2} \qquad\qquad x = 6$$
$$x = 0$$

The solutions to the equation are 0 and 6.   ■

EXAMPLE 4   Solve the equation $x^2 + 10x + 28 = 4$.

Solution: To make the right side of the equation equal to 0, we subtract 4 from both sides.

$$x^2 + 10x + 24 = 0$$

Now factor:

$$(x + 4)(x + 6) = 0$$

and solve.

$$x + 4 = 0 \quad \text{or} \quad x + 6 = 0$$
$$x = -4 \qquad\qquad x = -6$$

The solutions are -4 and -6.   ■

EXAMPLE 5   Solve the equation $3x^2 + 2x - 12 = -7x$.

Solution: Since all terms are not on the same side of the equation, add $7x$ to both sides of the equation.

$$3x^2 + 9x - 12 = 0$$

Factor out common factor.

$$3(x^2 + 3x - 4) = 0$$

Factor the remaining trinomial.

$$3(x + 4)(x - 1) = 0$$

Now solve for x.

$$x + 4 = 0 \qquad \text{or} \qquad x - 1 = 0$$
$$x = -4 \qquad\qquad\qquad x = 1$$

Since the 3 that was factored out is an expression not containing a variable, we do not have to set it equal to zero. The solutions to the equation are -4 and 1.   ■

EXAMPLE 6   Solve the equation $-x^2 + 5x + 6 = 0$.

Solution: When the squared term is negative, we generally make it positive by multiplying both sides of the equation by -1.

$$-1(-x^2 + 5x + 6) = -1 \cdot 0$$
$$x^2 - 5x - 6 = 0$$

Note that the sign of each term on the left side of the equation changed and that the right side of the equation remained zero. Now proceed as before.

$$x^2 - 5x - 6 = 0$$
$$(x - 6)(x + 1) = 0$$

$$x - 6 = 0 \qquad \text{or} \qquad x + 1 = 0$$
$$x = 6 \qquad\qquad\qquad x = -1$$

Check in original equation:

$x = 6$	$x = -1$
$-x^2 + 5x + 6 = 0$	$-x^2 + 5x + 6 = 0$
$-(6)^2 + 5(6) + 6 = 0$	$-(-1)^2 + 5(-1) + 6 = 0$
$-36 + 30 + 6 = 0$	$-1 - 5 + 6 = 0$
$0 = 0$ true	$0 = 0$ true ■

EXAMPLE 7   Solve the equation $x^2 = 9$.

Solution: Subtract 9 from both sides of the equation, then factor using the difference of two squares formula.

$$x^2 - 9 = 0$$
$$(x + 3)(x - 3) = 0$$

$$x + 3 = 0 \qquad \text{or} \qquad x - 3 = 0$$
$$x = -3 \qquad\qquad\qquad x = 3 \quad ■$$

EXAMPLE 8 The product of two numbers is 66. Find the two numbers if one number is 5 more than the other.

Solution: Let x = smaller number

$x + 5$ = larger number

$$x(x + 5) = 66$$
$$x^2 + 5x = 66$$
$$x^2 + 5x - 66 = 0$$
$$(x - 6)(x + 11) = 0$$
$$x = 6, \ x = -11$$

Remember that x represents the smaller of the two numbers. This problem has two possible solutions.

Solution 1		*Solution 2*
$x = 6$	smaller number	$x = -11$
$x + 5 = 6 + 5 = 11$	larger number	$x + 5 = -11 + 5 = -6$

One solution is: smaller number = 6, larger number = 11. A second solution is: smaller number = -11, larger number = -6.

Check:

Solution 1		*Solution 2*
$6 \cdot 11 = 66$	product is 66	$(-11) \cdot (-6) = 66$
$6 + 5 = 11$	one number is 5 more than the other	$-11 + 5 = -6$

If the question had stated: "The product of two *positive* numbers is 66," the only solution would be 6 and 11. ■

EXAMPLE 9 Find the length and width of a rectangle if its length is 3 more than the width and its area is 54.

Solution: Let x = width

$x + 3$ = length, see Fig. 5.3.

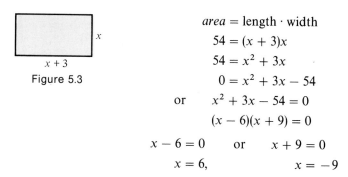

Figure 5.3

$$area = length \cdot width$$
$$54 = (x + 3)x$$
$$54 = x^2 + 3x$$
$$0 = x^2 + 3x - 54$$
or $$x^2 + 3x - 54 = 0$$
$$(x - 6)(x + 9) = 0$$
$$x - 6 = 0 \quad \text{or} \quad x + 9 = 0$$
$$x = 6, \qquad\qquad x = -9$$

Since the length of a geometric figure cannot be a negative number, the only solution is

$$\text{width} = 6, \quad \text{length} = x + 3 = 6 + 3 = 9$$

Check: $A = l \cdot w$

$A = 9 \cdot 6$

$54 = 54 \quad$ true   ■

Exercise Set 5.6

Solve each equation.

1. $x(x + 3) = 0$

2. $4x(x - 7) = 0$

3. $5x(x - 9) = 0$

4. $(x + 3)(x - 5) = 0$

5. $(2x + 5)(x - 3) = 0$

6. $(2x + 3)(x - 5) = 0$

7. $x^2 - 16 = 0$

8. $x^2 - 25 = 0$

9. $x^2 - 12x = 0$

10. $x^2 + 4x = 0$

11. $9x^2 + 18x = 0$

12. $x^2 + 6x + 5 = 0$

13. $x^2 + x - 12 = 0$

14. $x^2 + 6x + 9 = 0$

15. $x^2 - 12x = -20$

16. $3y^2 - 2 = -y$

17. $z^2 + 3z = 18$

18. $3x^2 = -21x - 18$

19. $3x^2 - 6x - 72 = 0$

20. $x^2 = 3x + 18$

21. $x^2 + 19x = 42$

22. $3x^2 - 9x - 30 = 0$

23. $2y^2 + 22y + 60 = 0$

24. $w^2 + 45 + 18w = 0$

25. $-2x - 8 = -x^2$

26. $-9x + 20 = -x^2$

27. $-x^2 + 30x + 64 = 0$

28. $-y^2 + 12y - 11 = 0$

29. $x^2 - 3x - 18 = 0$

30. $z^2 + 16z = -64$

31. $3p^2 = 22p - 7$

32. $5w^2 - 16w = -3$

33. $3r^2 + r = 2$

34. $3x^2 = 7x + 20$

35. $4x^2 + 4x - 48 = 0$

36. $6x^2 + 13x + 6 = 0$

37. $6x^2 - 5x = 4$

38. $2x^2 - 4x - 6 = 0$

39. $2x^2 - 10x = -12$

40. $x^2 - 25 = 0$

41. $x^2 - 16x = 0$

42. $4x^2 - 9 = 0$

43. $x^2 = 36$

44. $2x^2 - 32 = 0$

45. $x^2 = 9$

46. $x^2 = 4$

Express each problem as an equation, and solve.

47. The product of two consecutive positive even integers is 80. Find the two integers.

48. The product of two consecutive positive even integers is 120. Find the two integers.

49. The product of two consecutive positive odd integers is 63. Find the two integers.

50. The product of two positive integers is 108. Find the two numbers if one is 3 more than the other.

51. The product of two positive integers is 35. Find the two numbers if the larger number is 3 less than twice the smaller number.

52. The product of two positive integers is 64. Find the two integers if one number is four times the other.

53. The area of a rectangle is 36 square feet. Find the length and width if the length is four times the width.

54. The area of a rectangle is 54 square inches. Find the length and width if the length is 3 inches less than twice the width.

55. If the sides of a square are increased by 6 meters, the area becomes 64 square meters. Find the length of a side of the original square.

56. If the sides of a square are increased by 4 meters, the area becomes 121 square meters. Find the length of a side of the original square.

The sums of the first n *even numbers is given by the formula* $s = n^2 + n$. *Find* n *for the sum of the first* n *even integers given below.*

57. $s = 12$

58. $s = 30$

JUST FOR FUN

1. When a cannon is fired, under certain specific conditions, the height of the cannonball from the ground, in feet, at any time, t, can be found by the formula, $h = -16t^2 + 128t$.

(a) Find the height of the cannonball at 2 seconds.

(b) Find the time it takes for the cannonball to hit the ground.

Hint: What is the value of h at impact?

Summary

Glossary

Factor an expression: To factor an expression means to write the expression as a product of its factors.

Factors: If $a \cdot b = c$, then a and b are factors of c.

Greatest common factor (GCF): The greatest factor that divides each of the terms in an expression.

Quadratic equation: An equation of the form $ax^2 + bx + c = 0,\ a \neq 0$.

Important Facts

Difference of two squares: $a^2 - b^2 = (a + b)(a - b)$

Sum of two cubes: $a^3 + b^3 = (a + b)(a^2 - ab + b^2)$

Difference of two cubes: $a^3 - b^3 = (a - b)(a^2 + ab + b^2)$

Note: The sum of two squares, $a^2 + b^2$, cannot be factored over the set of real numbers.

Zero-factor property: If $a \cdot b = 0$, then $a = 0$ or $b = 0$.

General Procedure to Factor a Polynomial

1. Determine if the polynomial has a greatest common factor other than 1. If so factor out the GCF from every term in the polynomial.

2. If the polynomial has two terms (or is a binomial), determine if it is a difference of two squares or a sum or difference of two cubes. If so factor using the appropriate formula.

3. If the polynomial has 3 terms factor the trinomial using the methods discussed in Sections 5.3 and 5.4.

4. If the polynomial has more than 3 terms, then try factoring by grouping.

5. As a final step examine your factored polynomial to see if any factors listed have a common factor and can be factored further. If you find a common factor, factor it out at this point.

Review Exercises

[5.1] Find the greatest common factor for each set of terms.

1. $x^3, x^5, 2x^2$
4. $40x^2y^3, 36x^3y^4, 16x^5y^2z$
7. $x(2x - 5), 3(2x - 5)$

2. $3p, 6p^2, 9p^3$
5. $9xyz, 12xz, 36, x^2y$
8. $x(x + 5), x + 5$

3. $18x, 24, 36y^2$
6. $-32x^5, 16x^2, 24x^2y$

Factor each expression. If an expression cannot be factored, so state.

9. $5x - 20$

10. $9x + 33$

11. $16y^2 - 12y$

12. $55p^3 - 20p^2$

13. $24x^2y + 18x^3y^2$

14. $18x^2y - 9xy$

15. $2x^2 + 4x - 8$

16. $60x^4y^4 + 6x^9y^3 - 18x^5y^2$

17. $24x^2 - 13y^2 + 6xy$

18. $x(5x + 3) - 2(5x + 3)$

19. $3x(x - 1) - 2(x - 1)$

20. $2x(4x - 3) + 4x - 3$

[5.2] Factor by grouping.

21. $x^2 + 3x + 2x + 6$

22. $x^2 - 5x + 3x - 15$

23. $x^2 - 7x + 7x - 49$

24. $x^2 - 3x + 3x - 9$

25. $3x^2 + 9x + x + 3$

26. $3x^2 + x + 9x + 3$

27. $5x^2 + 20x - x - 4$

28. $5x^2 - xy + 20xy - 4y^2$

29. $12x^2 - 8xy + 15xy - 10y^2$

30. $12x^2 + 15xy - 8xy - 10y^2$

31. $4x^2 + 24x - x - 6$

32. $12x^2 - 9x - 4x + 3$

33. $20x^2 - 12x + 15x - 9$

34. $6x^2 + 9x - 2x - 3$

[5.3] Factor completely.

35. $x^2 + 6x + 8$

36. $x^2 - 8x + 15$

37. $x^2 - x - 20$

38. $x^2 + x - 20$

39. $x^2 - 3x - 18$

40. $x^2 - 9x + 14$

41. $x^2 - 10x + 24$

42. $x^2 - 6x - 27$

43. $x^2 - 12x - 45$

44. $x^2 + 11x + 24$

45. $x^3 + 5x^2 + 4x$

46. $x^3 - 3x^2 - 40x$

47. $x^2 + 5xy + 6y^2$

48. $x^2 - 2xy - 15y^2$

49. $2x^3 + 12x^2y + 16xy^2$

50. $4x^3 + 32x^2y + 60xy^2$

[5.4] Factor completely.

51. $2x^2 + 7x - 4$

52. $3x^2 + 13x + 4$

53. $4x^2 - 4x - 15$

54. $4x^2 - 9x + 5$

55. $3x^2 - 5x - 12$

56. $4x^2 + 4x - 15$

57. $5x^2 - 32x + 12$

58. $3x^2 + 13x + 12$

59. $6x^2 + 31x + 5$

60. $6x^2 - 33x + 36$

61. $2x^2 + 9x - 35$

62. $6x^2 + 11x - 10$

63. $8x^2 - 18x - 35$

64. $4x^2 + 20x + 25$

65. $9x^3 - 12x^2 + 4x$

66. $18x^3 - 24x^2 - 10x$

67. $2x^2 - xy - 10y^2$

68. $4x^2 - 16xy + 15y^2$

69. $6x^2 + 5xy - 21y^2$

70. $16x^2 - 22xy - 3y^2$

[5.5] Factor the difference of two squares.

71. $x^2 - 25$

72. $x^2 - 64$

73. $4x^2 - 16$

74. $81x^2 - 9y^2$

75. $64x^4 - 81y^4$

76. $16 - 25y^2$

77. $4x^4 - 9y^4$

78. $100x^4 - 121y^4$

Factor the sum or difference of two cubes.

79. $x^3 - y^3$

80. $x^3 + y^3$

81. $a^3 + 8$

82. $a^3 - 1$

83. $a^3 + 27$

84. $x^3 - 8$

85. $8x^3 - y^3$

86. $27 - 8y^3$

Factor completely.

87. $8x^2 + 16x - 24$

88. $2x^2 - 16x + 32$

89. $4x^2 - 36$

90. $3y^2 - 27$

91. $8x^3 - 8$

92. $x^3y - 27y$

93. $x^2y - xy + 4xy - 4y$

94. $6x^3 + 30x^2 + 9x^2 + 45x$

95. $4x^2 - 20xy + 25y^2$

96. $16y^2 - 49z^2$

97. $ab + 7a + 6b + 42$

98. $16y^5 - 25y^7$

99. $32x^3 + 32x^2 + 6x$

100. $y^4 - 1$

[5.6] Solve each equation.

101. $x(x - 4) = 0$

102. $(x + 3)(x + 4) = 0$

103. $(x - 5)(3x + 2) = 0$

104. $x^2 - 3x = 0$

105. $5x^2 + 20x = 0$

106. $x^2 - 2x - 24 = 0$

107. $x^2 + 8x + 15 = 0$

108. $x^2 = -2x + 8$

109. $x^2 - 12 = -x$

110. $3x^2 + 21x + 30 = 0$

111. $x^2 - 6x + 8 = 0$

112. $6x^2 + 6x - 12 = 0$

113. $8x^2 - 3 = -10x$

114. $2x^2 + 15x = 8$

115. $4x^2 - 16 = 0$

116. $36x^2 - 49 = 0$

[5.6] Express each problem as an equation, and solve.

117. The product of two consecutive positive integers is 110. Find the two integers.

118. The product of two consecutive positive even integers is 48. Find the two integers.

119. The product of two positive integers is 40. Find the integers if the larger is 2 less than twice the smaller.

120. The area of a rectangle is 63 square feet. Find the length and width of the rectangle if the length is 2 feet greater than the width.

121. One square has a side 4 inches longer than the side of a second square. If the area of the larger square is 81 square inches, find the length of a side of each square.

Practice Test _____

1. Find the greatest common factor of $4x^4$, $12x^5$, and $10x^2$.

2. Find the greatest common factor of $6x^2y^3$, $9xy^2$, and $12xy^5$.

Factor completely.

3. $4x^2y - 8xy$

4. $24x^2y - 6xy + 9x$

5. $x^2 - 3x + 2x - 6$

6. $3x^2 - 12x + x - 4$

7. $5x^2 - 15xy - 3xy + 9y^2$

8. $x^2 + 12x + 32$

9. $x^2 + 5x - 24$

10. $x^2 - 9xy + 20y^2$

11. $2x^2 - 22x + 60$

12. $2x^3 - 3x^2 + x$

13. $12x^2 - xy - 6y^2$

14. $x^2 - 9y^2$

15. $x^3 + 27$

Solve each equation.

16. $(x - 2)(2x - 5) = 0$

17. $x^2 + 6 = -5x$

18. $x^2 + 4x - 5 = 0$

Solve each problem.

19. The product of two positive integers is 36. Find the two integers if the larger is 1 more than twice the smaller.

20. The area of a rectangle is 24 square meters. Find the length and width of the rectangle if its length is 2 meters greater than its width.

6

Rational Expressions and Equations

See Section 6.8, Exercise 12

Reducing Rational Expressions

1   *Identify and define rational expressions.*

2   *Reduce rational expressions.*

3   *Factor a negative 1 from a polynomial.*

---

1   In this chapter we focus on rational expressions. To be successful with this material, you need a thorough understanding of the factoring techniques discussed in Chapter 5.

A **rational expression** (also called an **algebraic fraction**) is an algebraic expression of the form p/q, where p and q are polynomials and $q \neq 0$. Examples of rational expressions are

$$\frac{4}{5}, \quad \frac{x - 6}{x}, \quad \frac{x^2 + 2x}{x - 3}, \quad \frac{x}{x^2 - 4}$$

Note that the denominator of a rational expression cannot equal 0 since division by 0 is not permitted. In the expression $(x + 3)/x$, x cannot have a value of 0 since the denominator would then equal 0. In $(x^2 + 4x)/(x - 3)$, x cannot have a value of 3 for that would result in the denominator having a value of 0. What values of x cannot be used in the expression $x/(x^2 - 4)$? If you answered 2 and -2, you answered correctly. **Whenever we list a rational expression containing a variable in the denominator, we always assume that the value or values of the variable that make the denominator 0 are excluded.**

Three signs are associated with any fraction: the sign of the numerator, the sign of the denominator, and the sign of the fraction.

$$\text{sign of fraction} \longrightarrow + \frac{-a}{+b} \begin{array}{l} \text{sign of numerator} \\ \\ \text{sign of denominator} \end{array}$$

Whenever any of the three signs is omitted, we assume it to be positive. For example:

$$\frac{a}{b} \quad \text{means} \quad +\frac{+a}{+b}$$

$$\frac{-a}{b} \quad \text{means} \quad +\frac{-a}{+b}$$

$$-\frac{a}{b} \quad \text{means} \quad -\frac{+a}{+b}$$

> Changing any two of the three signs of a fraction does not change the value of a fraction. Thus
>
> $$\frac{-a}{b} = -\frac{a}{b} = \frac{a}{-b}$$

Generally, we do not write a fraction with a negative denominator. For example, the expression $\dfrac{2}{-5}$ would be written as either $\dfrac{-2}{5}$ or $-\dfrac{2}{5}$. The expression $\dfrac{x}{-(4-x)}$ can be written $\dfrac{x}{x-4}$ since $-(4-x) = -4 + x$ or $x - 4$.

2 A rational expression is **reduced to its lowest terms** when the numerator and denominator have no common factors other than 1. The fraction $\frac{9}{12}$ is not in reduced form because the 9 and 12 both contain the common factor of 3. When the 3 is factored out, the reduced fraction is $\frac{3}{4}$.

$$\frac{9}{12} = \frac{\overset{1}{\cancel{3}} \cdot 3}{\underset{1}{\cancel{3}} \cdot 4} = \frac{3}{4}$$

The rational expression $\dfrac{ab - b^2}{2b}$ is not in reduced form because both the numerator and denominator have a common factor of b. To reduce this expression, factor the b from each term in the numerator, then divide out the common factor b.

$$\frac{ab - b^2}{2b} = \frac{\cancel{b}(a - b)}{2\cancel{b}} = \frac{a - b}{2}$$

$\dfrac{ab - b^2}{2b}$ becomes $\dfrac{a - b}{2}$ when reduced to its lowest terms.

To Reduce Rational Expressions

1. Factor both numerator and denominator as completely as possible.

2. Divide both the numerator and denominator by any common factors.

EXAMPLE 1   Reduce $\dfrac{5x^3 + 10x^2 - 25x}{5x^2}$ to its lowest terms.

Solution: Factor the numerator. The greatest common factor of each term in the numerator is $5x$.

$$\frac{5x^3 + 10x^2 - 25x}{5x^2} = \frac{\cancel{5x}(x^2 + 2x - 5)}{\cancel{5}x^{\cancel{2}}}$$
$$= \frac{x^2 + 2x - 5}{x} \qquad \blacksquare$$

EXAMPLE 2   Reduce $\dfrac{x^2 + 2x - 3}{x + 3}$ to its lowest terms.

Solution: Factor the numerator, then divide out the common factor.

$$\frac{x^2 + 2x - 3}{x + 3} = \frac{\cancel{(x + 3)}(x - 1)}{\cancel{x + 3}} = x - 1$$

The expression reduces to $x - 1$. $\blacksquare$

EXAMPLE 3   Reduce $\dfrac{x^2 - 16}{x - 4}$ to its lowest terms.

Solution: Factor the numerator, then divide out common factors.

$$\frac{x^2 - 16}{x - 4} = \frac{(x + 4)(x - 4)}{x - 4} = x + 4 \qquad \blacksquare$$

COMMON STUDENT ERROR

Remember: Only common *factors* can be divided out when *multiplying* expressions.

Correct   *Wrong*

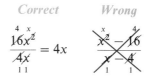

In the denominator $4x$, the 4 and x are factors since they are *multiplied* together. The 4 and the x are also both factors of the numerator $16x^2$ since $16x^2$ can be written $4 \cdot x \cdot 4x$.

Many students divide out *terms* incorrectly. In the expression $\dfrac{x^2 - 16}{x - 4}$ neither x nor 4 is a factor of either the numerator or denominator. The x and -4 are *terms* of the denominator.

EXAMPLE 4   Reduce $\dfrac{2x^2 + 7x + 6}{x^2 - x - 6}$ to lowest terms.

Solution: $\dfrac{2x^2 + 7x + 6}{x^2 - x - 6} = \dfrac{(2x + 3)(x + 2)}{(x - 3)(x + 2)} = \dfrac{2x + 3}{x - 3}.$

Note that $\dfrac{2x + 3}{x - 3}$ cannot be simplified any further. $\blacksquare$

3   **When a -1 is factored from a polynomial, the sign of each term in the polynomial changes.**

Examples

$$-3x + 5 = -1(3x - 5) = -(3x - 5)$$
$$6 - 2x = -1(-6 + 2x) = -(2x - 6)$$
$$-2x^2 + 3x - 4 = -1(2x^2 - 3x + 4) = -(2x^2 - 3x + 4)$$

Whenever the terms in a numerator and denominator differ only in their signs (one is the opposite or additive inverse of the other), we can factor out -1 from either the numerator or denominator, and then divide out the common factor. This procedure is illustrated in Examples 5 and 6.

EXAMPLE 5   Reduce $\dfrac{3x - 7}{7 - 3x}$ to lowest terms.

Solution: Since each term in the numerator differs only in sign from its like term in the denominator, we will factor -1 from each term in the denominator.

$$\frac{3x - 7}{7 - 3x} = \frac{3x - 7}{-1(-7 + 3x)}$$

$$= \frac{3x \overset{1}{\cancel{-7}}}{-(3x \cancel{-7})_1}$$

$$= -1 \quad \blacksquare$$

EXAMPLE 6   Reduce $\dfrac{x^2 - 2x - 24}{6 - x}$ to lowest terms.

Solution: $\dfrac{x^2 - 2x - 24}{6 - x} = \dfrac{(x + 4)(x - 6)}{6 - x}$

$$= \frac{(x + 4)(x \cancel{-6})}{-1(x \cancel{-6})}$$

$$= \frac{x + 4}{-1}$$

$$= -(x + 4) \quad \blacksquare$$

Exercise Set 6.1

Reduce each expression to its lowest terms.

1. $\dfrac{x}{x + xy}$

2. $\dfrac{3x}{6x + 9}$

3. $\dfrac{4x + 12}{x + 3}$

4. $\dfrac{3x^2 + 6x}{3x^2 + 9x}$

5. $\dfrac{x^3 + 6x^2 + 3x}{2x}$

6. $\dfrac{x^2y^2 - xy + 3}{y}$

7. $\dfrac{x^2 + 2x + 1}{x + 1}$

8. $\dfrac{x - 1}{x^2 + 2x - 3}$

9. $\dfrac{x^2 - 2x}{x^2 - 4x + 4}$

10. $\dfrac{x^2 + 3x - 18}{2x - 6}$

11. $\dfrac{x^2 - x - 6}{x^2 - 4}$

12. $\dfrac{x^2 + 6x + 9}{x^2 - 9}$

13. $\dfrac{2x^2 - 4x - 6}{x - 3}$

14. $\dfrac{4x^2 - 12x - 40}{2x^2 - 16x + 30}$

15. $\dfrac{2x - 3}{3 - 2x}$

16. $\dfrac{4x - 8}{4 - 2x}$

17. $\dfrac{x^2 - 2x - 8}{4 - x}$

18. $\dfrac{5 - x}{x^2 - 2x - 15}$

19. $\dfrac{x^2 + 3x - 18}{-2x^2 + 6x}$

20. $\dfrac{2x^2 + 5x - 12}{2x - 3}$

21. $\dfrac{2x^2 + 5x - 3}{1 - 2x}$

22. $\dfrac{x^2 - 9}{x^2 - 2x - 15}$

23. $\dfrac{x^2 - 8x + 16}{4 - x}$

24. $\dfrac{x^2 - 6x + 9}{(x - 3)^2}$

25. $\dfrac{x^2 - 25}{(x - 5)^2}$

26. $\dfrac{x^2 - 3x + 4x - 12}{x - 3}$

27. $\dfrac{x^2 - 2x + 4x - 8}{2x^2 + 3x + 8x + 12}$

28. $\dfrac{2x^2 - 8x + 3x - 12}{2x^2 + 8x + 3x + 12}$

29. $\dfrac{x^3 + 1}{x^2 - x + 1}$

30. $\dfrac{x^3 - 8}{x - 2}$

31. In any rational expression where there is a variable in the denominator, what do we always assume about the variable?

32. What value can x not represent in the expression $(x + 4)/(x - 6)$?

33. What value can x not represent in the expression $2/(x + 5)$?

34. What value can x not represent in the expression $(x + 3)/(x^2 - 4)$?

35. What value can x not represent in the expression $(x + 5)/(x^2 - 25)$?

36. Explain why x can represent any real number in the expression $(x + 3)/(x^2 + 4)$.

6.2

Multiplication and Division of Rational Expressions

1 Multiply rational expressions.

2 Divide rational expressions.

1 In Chapter 1 (Section 1.1) we reviewed multiplication of numerical fractions. Recall that to multiply two fractions we multiply their numerators together and multiply their denominators together.

Multiplication

$$\frac{a}{b} \cdot \frac{c}{d} = \frac{a \cdot c}{b \cdot d}, \qquad b \neq 0 \quad \text{and} \quad d \neq 0$$

EXAMPLE 1   Multiply $\left(\dfrac{3}{5}\right)\left(\dfrac{-2}{9}\right)$.

Solution: First divide out common factors, then multiply.

$$\overset{1}{\cancel{3}} \cdot \frac{-2}{\underset{3}{\cancel{9}}} = \frac{1 \cdot (-2)}{5 \cdot 3} = -\frac{2}{15} \quad \blacksquare$$

The same principles apply when multiplying rational expressions containing variables. Before multiplying, you should first divide out any factors common to both a numerator and a denominator.

> **To Multiply Rational Expressions**
> 1. Factor all numerators and denominators as completely as possible.
> 2. Divide out common factors.
> 3. Multiply numerators together and multiply denominators together.

EXAMPLE 2   Multiply $\dfrac{3x^2}{2y} \cdot \dfrac{4y^3}{3x}$.

Solution: This problem can be represented as

$$\frac{3xx}{2y} \cdot \frac{4yyy}{3x}$$

$$\frac{3\cancel{x}x}{2y} \cdot \frac{4yyy}{3\cancel{x}} \qquad \text{divide out the 3's and } x\text{'s}$$

$$\frac{3\cancel{x}x}{2y} \cdot \frac{\overset{2}{\cancel{4}}yyy}{3\cancel{x}} \qquad \text{divide both the 4 and the 2 by 2, and divide out the } y\text{'s}$$

Now multiply the remaining numerators together and the remaining denominators together.

$$\frac{2xy^2}{1} \quad \text{or} \quad 2xy^2$$

Rather than illustrating this entire process when multiplying rational expressions, we will often proceed as follows:

$$\frac{3x^2}{2y} \cdot \frac{4y^3}{3x}$$

$$= \frac{\overset{1}{\cancel{3}}\overset{x}{\cancel{x^2}}}{\underset{1}{\cancel{2}}\,\underset{1}{y}} \cdot \frac{\overset{2}{\cancel{4}}\overset{y^2}{\cancel{y^3}}}{\underset{1}{\cancel{3}}\,\underset{1}{\cancel{x}}} = 2xy^2 \qquad \blacksquare$$

EXAMPLE 3   Multiply $\dfrac{-2a^3b^2}{3x^3y} \cdot \dfrac{4a^2x}{5b^2y^3}$.

Solution: $\dfrac{-2a^3b^2}{3x^3y} \cdot \dfrac{4a^2x}{5b^2y^3} = \dfrac{-8a^5}{15x^2y^4} \qquad \blacksquare$

EXAMPLE 4   Multiply $(x - 2) \cdot \dfrac{3}{x^2 - 2x}$.

Solution: $\dfrac{\cancel{(x-2)}}{1} \cdot \dfrac{3}{x\cancel{(x-2)}} = \dfrac{3}{x} \qquad \blacksquare$

EXAMPLE 5   Multiply $\dfrac{(x + 2)^2}{6x^2} \cdot \dfrac{3x}{x^2 - 4}$.

Solution: $\dfrac{(x + 2)(x + 2)}{6x^2} \cdot \dfrac{3x}{(x + 2)(x - 2)}$

$= \dfrac{\cancel{(x + 2)}(x + 2)}{\underset{2x}{\cancel{6x^2}}} \cdot \dfrac{\overset{1}{\cancel{3x}}}{\cancel{(x + 2)}(x - 2)}$

$= \dfrac{x + 2}{2x(x - 2)}$

This answer cannot be simplified further.   ■

EXAMPLE 6   Multiply $\dfrac{x - 3}{2x} \cdot \dfrac{4x}{3 - x}$.

Solution: $\dfrac{x - 3}{\underset{1}{\cancel{2x}}} \cdot \dfrac{\overset{2}{\cancel{4x}}}{3 - x} = \dfrac{2(x - 3)}{3 - x}$

This problem is still not complete. In Section 6.1 we showed that $3 - x$ is $-1(-3 + x)$ or $-1(x - 3)$. Thus

$$\frac{2(x - 3)}{3 - x} = \frac{2\cancel{(x - 3)}}{-1\cancel{(x - 3)}} = -2 \quad ■$$

HELPFUL HINT

When only the signs differ in a numerator and denominator in a multiplication problem, factor out -1 *from either the numerator or denominator*, then divide out the common factor.

$$\frac{a - b}{x} \cdot \frac{y}{b - a} = \frac{a - b}{x} \cdot \frac{y}{-1\cancel{(a - b)}} = \frac{-y}{x}$$

EXAMPLE 7   Multiply $\dfrac{3x + 2}{2x - 1} \cdot \dfrac{4 - 8x}{3x + 2}$.

Solution: $\dfrac{3x + 2}{2x - 1} \cdot \dfrac{4 - 8x}{3x + 2} = \dfrac{3x + 2}{2x - 1} \cdot \dfrac{4(1 - 2x)}{3x + 2} = \dfrac{\cancel{(3x + 2)}}{2x - 1} \cdot \dfrac{4(1 - 2x)}{\cancel{(3x + 2)}}$

Note that the factor $(1 - 2x)$ in the numerator of the second fraction differs only in sign from the denominator of the first fraction $(2x - 1)$. We will therefore factor a -1 from the numerator.

$$= \frac{(3x + 2)}{(2x - 1)} \cdot \frac{4(-1)(2x - 1)}{\cancel{(3x + 2)}}$$

$$= \frac{\cancel{(3x + 2)}}{\cancel{(2x - 1)}} \cdot \frac{-4\cancel{(2x - 1)}}{\cancel{(3x + 2)}} = \frac{-4}{1} = -4 \quad ■$$

EXAMPLE 8   Multiply $\dfrac{2x^2 + 7x - 15}{4x^2 - 8x + 3} \cdot \dfrac{2x^2 + x - 1}{x^2 + 6x + 5}$.

Solution: $\dfrac{(2x - 3)(x + 5)}{(2x - 3)(2x - 1)} \cdot \dfrac{(2x - 1)(x + 1)}{(x + 1)(x + 5)}$

$= \dfrac{\cancel{(2x-3)}\cancel{(x+5)}}{\cancel{(2x-3)}\cancel{(2x-1)}} \cdot \dfrac{\cancel{(2x-1)}\cancel{(x+1)}}{\cancel{(x+1)}\cancel{(x+5)}} = 1$ ∎

EXAMPLE 9   Multiply $\dfrac{2x^3 - 14x^2 + 12x}{6y^2} \cdot \dfrac{-2y}{3x^2 - 3x}$.

Solution: $\dfrac{2x(x^2 - 7x + 6)}{6y^2} \cdot \dfrac{-2y}{3x(x - 1)}$

$= \dfrac{2x(x - 6)(x - 1)}{6y^2} \cdot \dfrac{-2y}{3x(x - 1)}$

$= \dfrac{2\cancel{x}(x - 6)\cancel{(x-1)}}{\underset{3\,y}{\cancel{6}y^{\cancel{2}}}} \cdot \dfrac{-2\cancel{y}}{3\cancel{x}\cancel{(x-1)}}$

$= \dfrac{-2(x - 6)}{9y}$

Note that $\dfrac{-2(x - 6)}{9y}$ and $\dfrac{-2x + 12}{9y}$ are both acceptable answers.   ▤

EXAMPLE 10   Multiply $\dfrac{x^2 - y^2}{x + y} \cdot \dfrac{x + 2y}{2x^2 - xy - y^2}$.

Solution: $\dfrac{(x + y)(x - y)}{x + y} \cdot \dfrac{x + 2y}{(2x + y)(x - y)}$

$= \dfrac{\cancel{(x+y)}\cancel{(x-y)}}{\cancel{(x+y)}} \cdot \dfrac{x + 2y}{(2x + y)\cancel{(x-y)}}$

$= \dfrac{x + 2y}{2x + y}$ ∎

Division of Rational Expressions

In Chapter 1 we learned that to divide one fraction by a second, we invert the divisior and proceed as in multiplication.

Division

$$\dfrac{a}{b} \div \dfrac{c}{d} = \dfrac{a}{b} \cdot \dfrac{d}{c} = \dfrac{ad}{bc}, \, b \neq 0, \, d \neq 0, \text{ and } c \neq 0$$

EXAMPLE 11   Divide as indicated.

$$\text{(a) } \frac{3}{5} \div \frac{4}{5} \qquad \text{(b) } \frac{2}{3} \div \frac{5}{6}$$

Solution: (a) $\dfrac{3}{\overset{}{\underset{1}{5}}} \cdot \dfrac{\overset{1}{5}}{4} = \dfrac{3 \cdot 1}{1 \cdot 4} = \dfrac{3}{4}$ (b) $\dfrac{2}{\overset{}{\underset{1}{3}}} \cdot \dfrac{\overset{2}{6}}{5} = \dfrac{2 \cdot 2}{1 \cdot 5} = \dfrac{4}{5}$ ■

The same principles are used when dividing rational expressions.

> **To Divide Rational Expressions**
>
> Invert the divisor (the second or bottom fraction) and multiply.

EXAMPLE 12 $\dfrac{5x^2}{z} \div \dfrac{4z^3}{3}.$

Solution: $\dfrac{5x^2}{z} \cdot \dfrac{3}{4z^3} = \dfrac{15x^2}{4z^4}$ ■

EXAMPLE 13 $\dfrac{x^2 - 9}{x + 4} \div \dfrac{x - 3}{x + 4}.$

Solution: $\dfrac{x^2 - 9}{x + 4} \cdot \dfrac{x + 4}{x - 3}$

$= \dfrac{(x + 3)(x - 3)}{(x + 4)} \cdot \dfrac{(x + 4)}{(x - 3)}$

$= x + 3$ ■

EXAMPLE 14 $\dfrac{-1}{2x - 3} \div \dfrac{3}{3 - 2x}.$

Solution: $\dfrac{-1}{2x - 3} \cdot \dfrac{3 - 2x}{3} = \dfrac{-1}{(2x - 3)} \cdot \dfrac{-1(2x - 3)}{3}$

$= \dfrac{(-1)(-1)}{(1)(3)} = \dfrac{1}{3}$ ■

EXAMPLE 15 $\dfrac{x^2 + 8x + 15}{x^2} \div (x + 3)^2.$

Solution: $\dfrac{x^2 + 8x + 15}{x^2} \cdot \dfrac{1}{(x + 3)^2}$

$= \dfrac{(x + 5)(x + 3)}{x^2} \cdot \dfrac{1}{(x + 3)(x + 3)}$

$= \dfrac{x + 5}{x^2(x + 3)}$ ■

EXAMPLE 16 $\dfrac{12x^2 - 22x + 8}{3x} \div \dfrac{3x^2 + 2x - 8}{2x^2 + 4x}.$

Solution: $\dfrac{12x^2 - 22x + 8}{3x} \cdot \dfrac{2x^2 + 4x}{3x^2 + 2x - 8} = \dfrac{2(6x^2 - 11x + 4)}{3x} \cdot \dfrac{2x(x + 2)}{(3x - 4)(x + 2)}$

$$= \dfrac{2(3x - 4)(2x - 1)}{3x} \cdot \dfrac{2x(x + 2)}{(3x - 4)(x + 2)}$$

$$= \dfrac{4(2x - 1)}{3} \quad \blacksquare$$

Exercise Set 6.2

Multiply as indicated.

1. $\dfrac{3x}{2y} \cdot \dfrac{y^2}{6}$

2. $\dfrac{15x^3y^2}{z} \cdot \dfrac{z}{5xy^3}$

3. $\dfrac{16x^2}{y^4} \cdot \dfrac{5x^2}{y^2}$

4. $\dfrac{12x^2}{6y^2} \cdot \dfrac{36xy^5}{12}$

5. $\dfrac{y^3}{8} \cdot \dfrac{9x^2}{y^3}$

6. $\dfrac{45a^2b^3}{12c^3} \cdot \dfrac{4c}{9a^3b^5}$

7. $\dfrac{80m^4}{49x^5y^7} \cdot \dfrac{14x^{12}y^5}{25m^5}$

8. $\dfrac{32m}{5n^3} \cdot \dfrac{-15m^2n^3}{4}$

9. $\dfrac{6x^5y^3}{5z^3} \cdot \dfrac{6x^4}{5yz^4}$

10. $\dfrac{-18x^2y}{11z^2} \cdot \dfrac{22z^3}{x^2y^5}$

11. $(2x + 5) \cdot \dfrac{1}{4x + 10}$

12. $\dfrac{1}{4x - 3} \cdot (20x - 15)$

13. $\dfrac{x - 3}{x + 5} \cdot \dfrac{2x^2 + 10x}{2x - 6}$

14. $\dfrac{x^2 - 4}{x^2 - 9} \cdot \dfrac{x + 3}{x - 2}$

15. $\dfrac{3x - 2}{3x + 2} \cdot \dfrac{4x - 1}{1 - 4x}$

16. $\dfrac{x - 6}{2x + 5} \cdot \dfrac{2x}{-x + 6}$

17. $\dfrac{4 - x}{x - 4} \cdot \dfrac{x - 3}{3 - x}$

18. $\dfrac{5 - 2x}{x + 8} \cdot \dfrac{-x - 8}{2x - 5}$

19. $\dfrac{x^2 + 7x + 12}{x + 4} \cdot \dfrac{1}{x + 3}$

20. $\dfrac{x^2 + 3x - 10}{2x} \cdot \dfrac{x^2 - 3x}{x^2 - 5x + 6}$

21. $\dfrac{x^2 - 5x - 24}{x^2 - x - 12} \cdot \dfrac{x^2 + x - 6}{x^2 - 10x + 16}$

22. $\dfrac{4x + 4y}{xy^2} \cdot \dfrac{x^2y}{3x + 3y}$

23. $\dfrac{a^2 - b^2}{a} \cdot \dfrac{a^2 + ab}{a + b}$

24. $\dfrac{x^2 - 25}{x^2 - 3x - 10} \cdot \dfrac{x + 2}{x}$

25. $\dfrac{a^2 + 6a + 9}{a^2 - 4} \cdot \dfrac{a - 2}{a + 3}$

26. $\dfrac{x^2}{x^2 - 4} \cdot \dfrac{x^2 - 5x + 6}{x^2 - 3x}$

27. $\dfrac{x^2 + x - 42}{x - 3} \cdot \dfrac{(x - 3)^2}{x + 7}$

28. $\dfrac{5x^2 + 17x + 6}{x + 3} \cdot \dfrac{x - 1}{5x^2 + 7x + 2}$

29. $\dfrac{6x^2 - 14x - 12}{6x + 4} \cdot \dfrac{x + 3}{2x^2 - 2x - 12}$

30. $\dfrac{2x^2 - 9x + 9}{8x - 12} \cdot \dfrac{2x}{x^2 - 3x}$

31. $\dfrac{2x + 4y}{x^2 + 4xy + 4y^2} \cdot \dfrac{x + 2y}{2}$

32. $\dfrac{x^2 - y^2}{x^2 + xy} \cdot \dfrac{3x^2 + 6x}{3x^2 - 2xy - y^2}$

33. $\dfrac{x^2 - y^2}{8x^2 - 16xy + 8y^2} \cdot \dfrac{4x - 4y}{x + y}$

34. $\dfrac{x^2 - 4y^2}{x^2 + 3xy + 2y^2} \cdot \dfrac{x + y}{x^2 - 4xy + 4y^2}$

Divide as indicated.

35. $\dfrac{6x^3}{y} \div \dfrac{2x}{y^2}$

36. $\dfrac{9x^3}{4} \div \dfrac{1}{16y^2}$

37. $\dfrac{25xy^2}{7z} \div \dfrac{5x^2y^2}{14z^2}$

38. $\dfrac{36y}{7z^2} \div \dfrac{3xy}{2z}$

39. $\dfrac{x^2y^5}{3z} \div \dfrac{3z}{2x}$

40. $\dfrac{12a^2}{4bc} \div \dfrac{3a^2}{bc}$

41. $\dfrac{-2xw}{y^5} \div \dfrac{6x^2}{y^6}$

42. $\dfrac{-xy}{a} \div \dfrac{-2ax}{6y}$

43. $\dfrac{7a^2b}{xy} \div \dfrac{7}{6xy}$

44. $2xz \div \dfrac{4xy}{z}$

45. $\dfrac{27x}{5y^2} \div 3x^2y^2$

46. $\dfrac{1}{7x^2y} \div \dfrac{1}{21x^3y}$

47. $\dfrac{6x + 6y}{a} \div \dfrac{12x + 12y}{a^2}$

48. $\dfrac{2a + 2b}{3} \div \dfrac{a^2 - b^2}{a - b}$

49. $\dfrac{3x^2 + 6x}{x} \div \dfrac{2x + 4}{x^2}$

50. $\dfrac{a^2b^2}{6x + 6y} \div \dfrac{ab}{x^2 - y^2}$

51. $\dfrac{1}{-x - 4} \div \dfrac{x^2 - 7x}{x^2 - 3x - 28}$

52. $\dfrac{x - 3}{4y^2} \div \dfrac{x^2 - 9}{2xy}$

53. $\dfrac{x^2 + 10x + 21}{x + 7} \div (x + 3)$

54. $\dfrac{x^2 - 9x + 14}{x^2 - 5x + 6} \div \dfrac{x^2 - 5x - 14}{x + 2}$

55. $(x - 3) \div \dfrac{x^2 + 3x - 18}{x}$

56. $\dfrac{1}{x^2 - 17x + 30} \div \dfrac{1}{x^2 + 7x - 18}$

57. $\dfrac{x^2 - 12x + 32}{x^2 - 6x - 16} \div \dfrac{x^2 - x - 12}{x^2 - 5x - 24}$

58. $\dfrac{a - b}{9a + 9b} \div \dfrac{a^2 - b^2}{a^2 + 2a + 1}$

59. $\dfrac{x^2 - 9x + 8}{x + 7} \div \dfrac{x - 1}{x^2 + 11x + 28}$

60. $\dfrac{(x + 2)^2}{x - 2} \div \dfrac{x^2 - 4}{2x - 4}$

61. $\dfrac{x^2 - 4}{2y} \div \dfrac{2 - x}{6xy}$

62. $\dfrac{x^2 + 7x + 10}{1 - x} \div \dfrac{x^2 + 2x - 15}{x - 1}$

63. $\dfrac{2x^2 + 9x + 4}{x^2 + 7x + 12} \div \dfrac{2x^2 - x - 1}{(x + 3)^2}$

64. $\dfrac{a^2 - b^2}{9} \div \dfrac{3a - 3b}{27x^2}$

65. $\dfrac{x^2 - y^2}{x^2 - 2xy + y^2} \div \dfrac{x + y}{x - y}$

66. $\dfrac{9x^2 - 9y^2}{6x^2y^2} \div \dfrac{3x + 3y}{12x^2y^5}$

6.3

Addition and Subtraction of Rational Expressions with A Common Denominator

1 Add and subtract rational expressions with a common denominator.

---

1 Recall that when adding (or subtracting) two arithmetic fractions with a common denominator, we add (or subtract) the numerators while keeping the common denominator.

Addition and Subtraction

$$\frac{a}{c} + \frac{b}{c} = \frac{a + b}{c}, \qquad c \neq 0 \qquad \frac{a}{c} - \frac{b}{c} = \frac{a - b}{c}, \qquad c \neq 0$$

EXAMPLE 1   Add $\dfrac{3}{8} + \dfrac{2}{8}$.

Solution: $\dfrac{3}{8} + \dfrac{2}{8} = \dfrac{3 + 2}{8} = \dfrac{5}{8}$

Note we did not reduce $\frac{2}{8}$ to $\frac{1}{4}$ because the common denominator is 8. Also note that we did not add the denominators; *only the numerators are added.*   ■

EXAMPLE 2 $\dfrac{5}{7} - \dfrac{1}{7}$.

Solution: $\dfrac{5}{7} - \dfrac{1}{7} = \dfrac{5 - 1}{7} = \dfrac{4}{7}$ ■

To add or subtract rational expressions, we use the same principle.

To Add or Subtract Expressions with a Common Denominator

1. Add or subtract the numerators.
2. Place the sum or difference of the numerators found in Step 1 over the common denominator.
3. Reduce the fraction if possible.

EXAMPLE 3 $\dfrac{3}{x + 2} + \dfrac{x - 4}{x + 2}$.

Solution: $\dfrac{3}{x + 2} + \dfrac{x - 4}{x + 2} = \dfrac{3 + (x - 4)}{x + 2}$

$= \dfrac{x - 1}{x + 2}$ ■

EXAMPLE 4 $\dfrac{3x + 5}{x - 3} - \dfrac{2x}{x - 3}$.

Solution: $\dfrac{3x + 5}{x - 3} - \dfrac{2x}{x - 3} = \dfrac{3x + 5 - 2x}{x - 3}$

$= \dfrac{x + 5}{x - 3}$ ■

EXAMPLE 5 $\dfrac{2x^2 + 5}{x + 3} + \dfrac{6x - 5}{x + 3}$.

Solution: $\dfrac{2x^2 + 5}{x + 3} + \dfrac{6x - 5}{x + 3} = \dfrac{2x^2 + 5 + (6x - 5)}{x + 3}$

$= \dfrac{2x^2 + 6x}{x + 3}$

Now factor $2x$ from each term in the numerator and reduce.

$$= \frac{2x(x+3)}{(x+3)}$$

$$= 2x \quad \blacksquare$$

EXAMPLE 6 $\dfrac{x^2 + 3x - 2}{(x + 5)(x - 2)} + \dfrac{4x + 12}{(x + 5)(x - 2)}.$

Solution: $\dfrac{x^2 + 3x - 2}{(x + 5)(x - 2)} + \dfrac{4x + 12}{(x + 5)(x - 2)} = \dfrac{x^2 + 3x - 2 + (4x + 12)}{(x + 5)(x - 2)}$

$$= \frac{x^2 + 7x + 10}{(x + 5)(x - 2)}$$

$$= \frac{(x+5)(x + 2)}{(x+5)(x - 2)}$$

$$= \frac{x + 2}{x - 2} \quad \blacksquare$$

When subtracting rational expressions, be sure to subtract the entire numerator of the fraction being subtracted. Study the common student error below very carefully.

COMMON STUDENT ERROR _____

Consider the problem

$$\frac{4x}{x - 2} - \frac{2x + 1}{x - 2}$$

Many students begin problems of this type incorrectly. Below are the correct and incorrect ways of working this subtraction problem.

<table>
<tr><td align="center">Correct</td><td align="center">Wrong</td></tr>
</table>

$$\frac{4x}{x - 2} - \frac{2x + 1}{x - 2} = \frac{4x - (2x + 1)}{x - 2} \qquad \frac{4x}{x-2} - \frac{2x+1}{x-2} = \frac{4x - 2x + 1}{x - 2}$$

$$= \frac{4x - 2x - 1}{x - 2}$$

$$= \frac{2x - 1}{x - 2}$$

Note that the entire numerator of the second fraction (not just the first term) **must be subtracted.** Also note that the sign of *each* term of the numerator being subtracted will change when the parentheses are removed.

EXAMPLE 7 $\dfrac{x^2 - 2x + 3}{x^2 + 7x + 12} - \dfrac{x^2 - 4x - 5}{x^2 + 7x + 12}.$

$$Solution: \quad \frac{x^2 - 2x + 3}{x^2 + 7x + 12} - \frac{x^2 - 4x - 5}{x^2 + 7x + 12} = \frac{x^2 - 2x + 3 - (x^2 - 4x - 5)}{x^2 + 7x + 12}$$

$$= \frac{x^2 - 2x + 3 - x^2 + 4x + 5}{x^2 + 7x + 12}$$

$$= \frac{2x + 8}{x^2 + 7x + 12}$$

$$= \frac{2(x + 4)}{(x + 3)(x + 4)}$$

$$= \frac{2}{x + 3} \quad \blacksquare$$

EXAMPLE 8 $\quad \dfrac{3x}{x - 6} - \dfrac{x^2 - 4x + 6}{x - 6}.$

$$Solution: \quad \frac{3x}{x - 6} - \frac{x^2 - 4x + 6}{x - 6} = \frac{3x - (x^2 - 4x + 6)}{x - 6}$$

$$= \frac{3x - x^2 + 4x - 6}{x - 6}$$

$$= \frac{-x^2 + 7x - 6}{x - 6}$$

$$= \frac{-(x^2 - 7x + 6)}{x - 6}$$

$$= \frac{-(x - 6)(x - 1)}{(x - 6)}$$

$$= -(x - 1) \quad \blacksquare$$

Exercise Set 6.3

Add or subtract as indicated.

1. $\dfrac{x - 1}{6} + \dfrac{x}{6}$

2. $\dfrac{x + 11}{8} + \dfrac{2x + 5}{8}$

3. $\dfrac{x - 7}{3} - \dfrac{4}{3}$

4. $\dfrac{2x + 3}{5} - \dfrac{x}{5}$

5. $\dfrac{x + 2}{x} - \dfrac{5}{x}$

6. $\dfrac{3x + 6}{2} - \dfrac{x}{2}$

7. $\dfrac{1}{x} + \dfrac{x + 2}{x}$

8. $\dfrac{3x + 4}{x + 1} + \dfrac{6x + 5}{x + 1}$

9. $\dfrac{4}{x + 2} + \dfrac{x + 3}{x + 2}$

10. $\dfrac{x - 3}{x} + \dfrac{x + 3}{x}$

11. $\dfrac{x - 4}{x} - \dfrac{x + 4}{x}$

12. $\dfrac{x}{x - 2} + \dfrac{2x + 3}{x - 2}$

13. $\dfrac{4x - 3}{x - 7} - \dfrac{2x + 8}{x - 7}$

14. $\dfrac{4x - 5}{3x^2} + \dfrac{2x + 5}{3x^2}$

15. $\dfrac{9x + 7}{6x^2} - \dfrac{3x + 4}{6x^2}$

16. $\dfrac{4x-6}{x^2+2x}+\dfrac{7x+5}{x^2+2x}$

17. $\dfrac{-2x-4}{x^2+2x+1}+\dfrac{3x+5}{x^2+2x+1}$

18. $\dfrac{-2x+6}{x^2+x-6}+\dfrac{3x-3}{x^2+x-6}$

19. $\dfrac{4}{x^2-2x-3}+\dfrac{x-3}{x^2-2x-3}$

20. $\dfrac{-x-4}{x^2-16}+\dfrac{2(x+4)}{x^2-16}$

21. $\dfrac{x+4}{3x+2}-\dfrac{x+4}{3x+2}$

22. $\dfrac{2x+4}{(x+2)(x-3)}-\dfrac{x+7}{(x+2)(x-3)}$

23. $\dfrac{2x+4}{x-7}-\dfrac{6x+5}{x-7}$

24. $\dfrac{x^2+2x}{3x}-\dfrac{x^2+5x+6}{3x}$

25. $\dfrac{x^2+4x+3}{x+2}-\dfrac{5x+9}{x+2}$

26. $\dfrac{3x^2}{x^2+2x}-\dfrac{4x}{x^2+2x}$

27. $\dfrac{4}{2x+3}+\dfrac{6x+5}{2x+3}$

28. $\dfrac{-2x+5}{5x-10}+\dfrac{2(x-5)}{5x-10}$

29. $\dfrac{x^2}{x+3}+\dfrac{9}{x+3}$

30. $\dfrac{x^2-2x-3}{x^2-x-6}+\dfrac{x-3}{x^2-x-6}$

31. $\dfrac{4x+12}{3-x}-\dfrac{3x+15}{3-x}$

32. $\dfrac{-x-7}{2x-9}-\dfrac{-3x-16}{2x-9}$

33. $\dfrac{x^2-2}{x^2+6x-7}-\dfrac{-4x+19}{x^2+6x-7}$

34. $\dfrac{x^2+6x}{(x+9)(x+5)}-\dfrac{27}{(x+9)(x+5)}$

35. $\dfrac{x^2-13}{x+4}-\dfrac{3}{x+4}$

36. $\dfrac{x^2-6}{2x+3}-\dfrac{-3x^2+3}{2x+3}$

37. $\dfrac{x^2+3x+5}{x^2-64}+\dfrac{7x+11}{x^2-64}$

38. $\dfrac{-x^2}{x^2+5x-14}+\dfrac{x^2+x+7}{x^2+5x-14}$

39. $\dfrac{3x^2-7x}{4x^2-8x}+\dfrac{x}{4x^2-8x}$

40. $\dfrac{3x^2+15x}{x^3+2x^2-8x}+\dfrac{2x^2+5x}{x^3+2x^2-8x}$

41. $\dfrac{2x^2-6x+5}{2x^2+18x+16}-\dfrac{8x+21}{2x^2+18x+16}$

42. $\dfrac{x^3-10x^2+35x}{x(x-6)}-\dfrac{x^2+5x}{x(x-6)}$

43. $\dfrac{x^2+3x-6}{x^2-5x+4}-\dfrac{-2x^2+4x-4}{x^2-5x+4}$

44. $\dfrac{4x^2+5}{9x^2-64}-\dfrac{x^2-x+29}{9x^2-64}$

45. When subtracting rational expressions, what must happen to the sign of each term of the numerator being subtracted?

46. State what is *wrong* with the following step. Show what the correct step should be.

$$\dfrac{4x-3}{5x+4}-\dfrac{2x-7}{5x+4}=\dfrac{4x-3-2x-7}{5x+4}$$

47. State what is *wrong* with the following step. Show what the correct step should be.

$$\dfrac{6x-2}{x^2-4x+3}-\dfrac{3x^2-4x+5}{x^2-4x+3}=\dfrac{6x-2-3x^2-4x+5}{x^2-4x+3}$$

48. State what is *wrong* with the following step. Show what the correct step should be.

$$\dfrac{4x+5}{x^2-6x}-\dfrac{-x^2+3x+6}{x^2-6x}=\dfrac{4x+5+x^2+3x+6}{x^2-6x}$$

6.4

Finding the Least Common Denominator

1 *Find the least common denominator for rational expressions.*

1 To add two numerical fractions with unlike denominators, we must first obtain a common denominator.

EXAMPLE 1   Add $\dfrac{3}{5} + \dfrac{4}{7}$.

Solution: The least common denominator (LCD) [or least common multiple (LCM)] of 5 and 7 is 35. Thirty-five is the smallest number that is divisible by both 5 and 7. Rewrite each fraction so that it has a denominator equal to the LCD, 35.

$$\frac{3}{5} + \frac{4}{7} = \frac{3}{5} \cdot \frac{7}{7} + \frac{4}{7} \cdot \frac{5}{5}$$

$$= \frac{21}{35} + \frac{20}{35} = \frac{41}{35} \quad \text{or} \quad 1\frac{6}{35} \quad \blacksquare$$

To add or subtract rational expressions, we must also write each expression with a common denominator.

To Find the Least Common Denominator of Rational Expressions

1. Factor each denominator completely. Factors in any given denominator that occur more than once should be expressed as powers [therefore, $(x + 5)(x + 5)$ should be expressed as $(x + 5)^2$].
2. List all different factors (other than 1) that appear in any of the denominators. When the same factor appears in more than one denominator, write the factor with the highest power that appears.
3. The least common denominator is the product of all the factors in Step 2.

EXAMPLE 2   Find the least common denominator.

$$\frac{1}{3} + \frac{1}{x}$$

Solution: The only factor (other than 1) of the first denominator is 3. The only factor (other than 1) of the second denominator is x. The LCD is therefore $3 \cdot x = 3x$. $\blacksquare$

EXAMPLE 3   Find the LCD.

$$\frac{3}{5x} - \frac{2}{x^2}$$

Solution: The factors that appear in the denominators are 5 and x. List each factor with its highest power. The LCD is the product of these factors.

$$\overset{\displaystyle \text{highest power of } x}{\text{LCD} = 5 \cdot x^2 = 5x^2} \quad \blacksquare$$

EXAMPLE 4   Find the LCD.

$$\frac{1}{18x^3y} + \frac{5}{27x^2y^3}$$

Solution: Write both 18 and 27 as products of prime numbers. $18 = 2 \cdot 3^2$ and $27 = 3^3$.

$$\frac{1}{18x^3y} + \frac{5}{27x^2y^3} = \frac{1}{2 \cdot 3^2x^3y} + \frac{5}{3^3x^2y^3}$$

The factors that appear are 2, 3, x, and y. List the highest powers of each of these factors.

$$\text{LCD} = 2 \cdot 3^3 \cdot x^3 \cdot y^3 = 54x^3y^3 \quad \blacksquare$$

EXAMPLE 5   Find the LCD.

$$\frac{3}{x} - \frac{2y}{x + 5}$$

Solution: The factors that appear are x and $(x + 5)$. *Note that the x in the second denominator, $x + 5$, is not a factor of that denominator since the operation is addition rather than multiplication.*

$$\text{LCD} = x(x + 5) \quad \blacksquare$$

EXAMPLE 6   Find the LCD.

$$\frac{3}{2x^2 - 4x} + \frac{x^2}{x^2 - 4x + 4}$$

Solution: Factor both denominators.

$$\frac{3}{2x(x - 2)} + \frac{x^2}{(x - 2)(x - 2)} = \frac{3}{2x(x - 2)} + \frac{x^2}{(x - 2)^2}$$

The factors that appear are 2, x, and $x - 2$. List the highest powers of each of these factors that appear.

$$\text{LCD} = 2 \cdot x \cdot (x - 2)^2 = 2x(x - 2)^2 \quad \blacksquare$$

EXAMPLE 7   Find the LCD.

$$\frac{5x}{x^2 - x - 12} - \frac{6x^2}{x^2 - 7x + 12}$$

Solution: Factor both denominators.

$$\frac{5x}{(x + 3)(x - 4)} - \frac{6x^2}{(x - 3)(x - 4)}$$

$$\text{LCD} = (x + 3)(x - 4)(x - 3)$$

Note that although $(x - 4)$ is a common factor of each denominator, the highest power of that factor that appears in either denominator is 1. $\quad \blacksquare$

EXAMPLE 8   Find the LCD.

$$\frac{3x}{x^2 - 14x + 48} + x + 9$$

Solution: Factor the denominator of the first term.

$$\frac{3x}{(x - 6)(x - 8)} + x + 9$$

The denominator of $x + 9$ is 1. The expression can be rewritten as

$$\frac{3x}{(x - 6)(x - 8)} + \frac{x + 9}{1}$$

The LCD is therefore $1(x - 6)(x - 8)$ or simply $(x - 6)(x - 8)$.   ■

Exercise Set 6.4

Find the least common denominator for each expression.

1. $\dfrac{x}{3} + \dfrac{x - 1}{3}$

2. $\dfrac{4 - x}{5} - \dfrac{12}{5}$

3. $\dfrac{1}{2x} + \dfrac{1}{3}$

4. $\dfrac{1}{x + 2} - \dfrac{3}{5}$

5. $\dfrac{3}{5x} + \dfrac{7}{2}$

6. $\dfrac{2x}{3x} + 1$

7. $\dfrac{2}{x^2} + \dfrac{3}{x}$

8. $\dfrac{5x}{x + 1} + \dfrac{6}{x + 2}$

9. $\dfrac{x + 4}{2x + 3} + 7$

10. $\dfrac{x + 4}{2x} + \dfrac{3}{7x}$

11. $\dfrac{x}{x + 1} + \dfrac{4}{x^2}$

12. $\dfrac{x}{3x^2} + \dfrac{9}{15x^4}$

13. $\dfrac{x + 3}{16x^2y} - \dfrac{5}{9x^3}$

14. $\dfrac{-4}{8x^2y^2} + \dfrac{7}{5x^4y^5}$

15. $\dfrac{x^2 + 3}{18x} - \dfrac{x - 7}{12(x + 5)}$

16. $\dfrac{x - 7}{3x + 5} - \dfrac{6}{x + 5}$

17. $\dfrac{2x - 7}{x^2 + x} - \dfrac{x^2}{x + 1}$

18. $\dfrac{9}{(x - 4)(x + 2)} - \dfrac{x + 8}{x + 2}$

19. $\dfrac{15}{36x^2y} + \dfrac{x + 3}{15xy^3}$

20. $\dfrac{x^2 - 4}{x^2 - 16} + \dfrac{3}{x + 4}$

21. $\dfrac{6}{2x + 8} + \dfrac{6x + 3}{3x - 9}$

22. $6x^2 + \dfrac{9x}{x - 3}$

23. $\dfrac{9x + 4}{x + 6} - \dfrac{3x - 6}{x + 5}$

24. $\dfrac{x + 2}{x^2 + 11x + 18} - \dfrac{x^2 - 4}{x^2 - 3x - 10}$

25. $\dfrac{x - 2}{x^2 - 5x - 24} + \dfrac{3}{x^2 + 11x + 24}$

26. $\dfrac{6x + 5}{x^2 - 4} - \dfrac{3x}{x^2 - 5x - 14}$

27. $\dfrac{6}{x + 3} - \dfrac{x + 5}{x^2 - 4x + 3}$

28. $\dfrac{3x - 8}{x^2 - 1} + \dfrac{x^2 + 5}{x + 1}$

29. $\dfrac{2x}{x^2 - x - 2} - \dfrac{3}{x^2 + 4x + 3}$

30. $\dfrac{6x + 5}{x + 2} + \dfrac{4x}{(x + 2)^2}$

31. $\dfrac{3x - 5}{x^2 + 4x + 4} + \dfrac{3}{x + 2}$

32. $\dfrac{9x + 7}{(x + 3)(x + 2)} - \dfrac{4x}{(x - 3)(x + 2)}$

33. $\dfrac{x}{x^2 + 3x} + \dfrac{6}{2x + 6}$

6.5

Addition and Subtraction of Rational Expressions

❶ *Add and subtract rational expressions.*

❶ The method used to add or subtract rational expressions with unlike denominators is outlined in Example 1.

EXAMPLE 1 Add $\dfrac{3}{x} + \dfrac{5}{y}$.

Solution: First, determine the LCD as outlined in Section 6.4.

$$LCD = xy$$

Now write each fraction with the LCD. We do this by multiplying **both** numerator and denominator of each fraction by any factors needed to obtain the LCD.

In this problem the fraction on the left must be multiplied by y/y and the fraction on the right must be multiplied by x/x.

$$\frac{y}{y} \cdot \frac{3}{x} + \frac{5}{y} \cdot \frac{x}{x} = \frac{3y}{xy} + \frac{5x}{xy}$$

By multiplying both the numerator and denominator by the same factor we are in effect multiplying by 1, which does not change the value of the fraction, only its appearance. Thus the new fraction is equivalent to the original fraction.

Now add the numerators, while leaving the LCD alone.

$$\frac{3y}{xy} + \frac{5x}{xy} = \frac{3y + 5x}{xy} \quad \text{or} \quad \frac{5x + 3y}{xy} \quad ■$$

To Add or Subtract Two Rational Expressions with Unlike Denominators

1. Determine the LCD.
2. Rewrite each fraction as an equivalent fraction with the LCD. This is done by multiplying both the numerator and denominator of each fraction by any factors needed to obtain the LCD.
3. Add or subtract the numerators while maintaining the LCD.
4. When possible factor the remaining numerator and reduce the fraction.

EXAMPLE 2 Add $\dfrac{5}{4x^2y} + \dfrac{3}{14xy^3}$.

Solution: The LCD is $28x^2y^3$. We must write each fraction with the denominator $28x^2y^3$. To do this, multiply the fraction on the left by $7y^2/7y^2$, and the fraction on the right by $2x/2x$.

$$\frac{7y^2}{7y^2} \cdot \frac{5}{4x^2y} + \frac{3}{14xy^3} \cdot \frac{2x}{2x} = \frac{35y^2}{28x^2y^3} + \frac{6x}{28x^2y^3}$$

$$= \frac{35y^2 + 6x}{28x^2y^3} \quad \text{or} \quad \frac{6x + 35y^2}{28x^2y^3} \quad ■$$

EXAMPLE 3 Add $\dfrac{3}{x + 2} + \dfrac{4}{x}$.

Solution: We must write each fraction with the LCD $x(x + 2)$. To do this, multiply the fraction on the left by x/x and the fraction on the right by $(x + 2)/(x + 2)$.

$$\frac{x}{x} \cdot \frac{3}{(x + 2)} + \frac{4}{x} \cdot \frac{(x + 2)}{(x + 2)} = \frac{3x}{x(x + 2)} + \frac{4(x + 2)}{x(x + 2)}$$

$$= \frac{3x + 4(x + 2)}{x(x + 2)}$$

$$= \frac{3x + 4x + 8}{x(x + 2)} = \frac{7x + 8}{x(x + 2)}$$ ∎

EXAMPLE 4   Subtract $\dfrac{x}{x + 5} - \dfrac{2}{x - 3}$.

Solution: The LCD is $(x + 5)(x - 3)$. The fraction on the left must be multiplied by $(x - 3)/(x - 3)$ to obtain the LCD. The fraction on the right must be multiplied by $(x + 5)/(x + 5)$ to obtain the LCD.

$$\frac{(x - 3)}{(x - 3)} \cdot \frac{x}{(x + 5)} - \frac{2}{(x - 3)} \cdot \frac{(x + 5)}{(x + 5)} = \frac{x(x - 3)}{(x - 3)(x + 5)} - \frac{2(x + 5)}{(x - 3)(x + 5)}$$

$$= \frac{x^2 - 3x}{(x - 3)(x + 5)} - \frac{2x + 10}{(x - 3)(x + 5)}$$

$$= \frac{x^2 - 3x - (2x + 10)}{(x - 3)(x + 5)}$$

$$= \frac{x^2 - 3x - 2x - 10}{(x - 3)(x + 5)}$$

$$= \frac{x^2 - 5x - 10}{(x - 3)(x + 5)}$$

The numerator and denominator have no common factors, so we can leave the answer in this form. If you wish, you can multiply the factors in the denominator and give the answer as $(x^2 - 5x - 10)/(x^2 + 2x - 15)$. Generally, we will leave the answers in factored form.   ∎

EXAMPLE 5   Subtract $\dfrac{x + 2}{x - 4} - \dfrac{x + 3}{x + 4}$.

Solution: The LCD is $(x - 4)(x + 4)$.

$$\frac{(x + 4)}{(x + 4)} \cdot \frac{(x + 2)}{(x - 4)} - \frac{(x + 3)}{(x + 4)} \cdot \frac{(x - 4)}{(x - 4)} = \frac{(x + 4)(x + 2)}{(x + 4)(x - 4)} - \frac{(x + 3)(x - 4)}{(x + 4)(x - 4)}$$

Use the FOIL method to multiply each numerator.

$$= \frac{x^2 + 6x + 8}{(x + 4)(x - 4)} - \frac{x^2 - x - 12}{(x + 4)(x - 4)}$$

$$= \frac{x^2 + 6x + 8 - (x^2 - x - 12)}{(x + 4)(x - 4)}$$

$$= \frac{x^2 + 6x + 8 - x^2 + x + 12}{(x + 4)(x - 4)}$$

$$= \frac{7x + 20}{(x + 4)(x - 4)}$$ ∎

Consider the problem

$$\frac{4}{x-3} + \frac{x+5}{3-x}$$

How do we add these rational expressions? One method would be to write each fraction with the denominator $(x-3)(3-x)$. However there is an easier way to do this problem. Study the Helpful Hint below.

HELPFUL HINT

> When adding or subtracting fractions whose denominators are opposites (therefore differ only in signs), multiply both numerator *and* denominator of *either one* of the fractions by -1. This will result in both fractions having a common denominator.
>
> $$\frac{x}{a-b} + \frac{y}{b-a} = \frac{x}{a-b} + \frac{-1\,y}{-1\,(b-a)}$$
>
> $$= \frac{x}{a-b} + \frac{-y}{a-b}$$
>
> $$= \frac{x-y}{a-b}$$

EXAMPLE 6   Add $\dfrac{4}{x-3} + \dfrac{x+5}{3-x}$.

Solution: Since the denominators differ only in sign, we will multiply both the numerator and the denominator of the second fraction by -1. This will result in both fractions having a common denominator of $x-3$.

$$\frac{4}{x-3} + \frac{x+5}{3-x} = \frac{4}{x-3} + \frac{-1}{-1} \cdot \frac{(x+5)}{(3-x)}$$

$$= \frac{4}{x-3} + \frac{-x-5}{x-3}$$

$$= \frac{4-x-5}{x-3}$$

$$= \frac{-x-1}{x-3} \quad \text{or} \quad \frac{-(x+1)}{x-3} \quad \blacksquare$$

EXAMPLE 7   Subtract $\dfrac{x+2}{2x-5} - \dfrac{3x+5}{5-2x}$.

Solution: The denominators of the two fractions differ only in sign. If one of the denominators is multiplied by -1, it becomes identical to the other denominator. Since we need identical denominators to add or subtract fractions, we will multiply both numerator and denominator of one of the fractions by -1. In this example we will multiply the numerator and denominator of the second fraction by -1.

$$\frac{x+2}{2x-5} - \frac{3x+5}{5-2x} = \frac{x+2}{2x-5} - \frac{(-1)}{(-1)} \cdot \frac{(3x+5)}{(5-2x)}$$

$$= \frac{x+2}{2x-5} - \frac{-3x-5}{-5+2x}$$

$$= \frac{x+2}{2x-5} - \frac{-3x-5}{2x-5}$$

$$= \frac{x+2-(-3x-5)}{2x-5}$$

$$= \frac{x+2+3x+5}{2x-5}$$

$$= \frac{4x+7}{2x-5} \quad \blacksquare$$

In Example 7 we selected to multiply the numerator and denominator of the second fraction by -1. The same results could be obtained by multiplying the numerator and denominator of the first fraction by -1. Try this now.

EXAMPLE 8   Add $\dfrac{3}{x^2+5x+6} + \dfrac{1}{x^2-x-12}$.

Solution: $\dfrac{3}{(x+2)(x+3)} + \dfrac{1}{(x+3)(x-4)}$

The LCD is $(x+2)(x+3)(x-4)$.

$$\frac{(x-4)}{(x-4)} \cdot \frac{3}{(x+2)(x+3)} + \frac{1}{(x+3)(x-4)} \cdot \frac{(x+2)}{(x+2)}$$

$$= \frac{3x-12}{(x-4)(x+2)(x+3)} + \frac{x+2}{(x-4)(x+2)(x+3)}$$

$$= \frac{3x-12+x+2}{(x-4)(x+2)(x+3)}$$

$$= \frac{4x-10}{(x-4)(x+2)(x+3)} \quad \blacksquare$$

COMMON STUDENT ERROR

A common mistake is to treat an addition or subtraction problem as a multiplication problem. Here is one such example.

Correct	*Wrong*

$$\frac{1}{x} + \frac{x}{1} = \frac{1}{x} + \frac{x^2}{x}$$

$$= \frac{x^2+1}{x}$$

$$\frac{1}{x} + \frac{x}{1} = \frac{1}{x} + \frac{x}{1}$$

$$= 1 + 1 = 2$$

Note that you can only divide out common factors when multiplying expressions.

Exercise Set 6.5 _____

Add or subtract as indicated.

1. $\dfrac{4}{x} + \dfrac{3}{2x}$

2. $\dfrac{1}{3x} + \dfrac{1}{2}$

3. $\dfrac{6}{x^2} + \dfrac{3}{2x}$

4. $3 + \dfrac{5}{x}$

5. $2 - \dfrac{1}{x^2}$

6. $\dfrac{5}{6y} + \dfrac{3}{4y^2}$

7. $\dfrac{1}{x^2} + \dfrac{3}{5x}$

8. $\dfrac{3x}{4y} + \dfrac{5}{6xy}$

9. $\dfrac{3}{4x^2 y} + \dfrac{7}{5xy^2}$

10. $\dfrac{5}{12x^4 y} - \dfrac{1}{5x^2 y^3}$

11. $x + \dfrac{x}{y}$

12. $\dfrac{5}{x} + \dfrac{4}{x^2}$

13. $\dfrac{3x - 1}{x} + \dfrac{2}{3x}$

14. $\dfrac{3}{x} + 4$

15. $\dfrac{5x}{y} + \dfrac{y}{x}$

16. $\dfrac{3}{p} - \dfrac{6}{2p^2}$

17. $\dfrac{4}{5x^2} - \dfrac{6}{y}$

18. $\dfrac{x - 3}{x} - \dfrac{x}{4x^2}$

19. $\dfrac{5}{x} + \dfrac{3}{x - 2}$

20. $6 - \dfrac{3}{x - 3}$

21. $\dfrac{9}{a + 3} + \dfrac{2}{a}$

22. $\dfrac{b}{a - b} + \dfrac{a + b}{b}$

23. $\dfrac{4}{3x} - \dfrac{2x}{3x + 6}$

24. $\dfrac{2}{x - 3} - \dfrac{4}{x - 1}$

25. $\dfrac{3}{x - 2} + \dfrac{1}{2 - x}$

26. $\dfrac{3}{x - 2} - \dfrac{1}{2 - x}$

27. $\dfrac{5}{x + 3} - \dfrac{4}{-x - 3}$

28. $\dfrac{5}{2x - 5} - \dfrac{3}{5 - 2x}$

29. $\dfrac{3}{x + 1} + \dfrac{4}{x - 1}$

30. $\dfrac{x}{2x - 4} + \dfrac{3}{x - 2}$

31. $\dfrac{x + 5}{x - 5} - \dfrac{x - 5}{x + 5}$

32. $\dfrac{x + 7}{x + 3} - \dfrac{x - 3}{x + 7}$

33. $\dfrac{x}{x^2 - 9} + \dfrac{4}{x + 3}$

34. $\dfrac{3}{(x + 5)^2} + \dfrac{2}{x + 5}$

35. $\dfrac{x + 2}{x^2 - 4} - \dfrac{2}{x + 2}$

36. $\dfrac{3}{(x - 2)(x + 3)} + \dfrac{5}{(x + 2)(x + 3)}$

37. $\dfrac{3}{x^2 + 2x - 8} + \dfrac{2}{x^2 - 3x + 2}$

38. $\dfrac{y}{xy - x^2} - \dfrac{x}{y^2 - xy}$

39. $\dfrac{5}{x^2 - 9x + 8} - \dfrac{3}{x^2 - 6x - 16}$

40. $\dfrac{3}{x + 4} + \dfrac{1}{x^2 + 8x + 16}$

41. $\dfrac{1}{x^2 - 4} + \dfrac{3}{x^2 + 5x + 6}$

42. $\dfrac{2x - 5}{6x + 9} - \dfrac{4}{2x^2 + 3x}$

43. $\dfrac{2x + 3}{x^2 - 7x + 12} - \dfrac{2}{x - 3}$

44. $\dfrac{2}{5 - x} + \dfrac{x + 3}{x^2 - 3x - 10}$

45. $\dfrac{2 - x}{x^2 + 2x - 8} - \dfrac{x - 4}{x + 4}$

46. $\dfrac{x - 1}{x^2 - 2x + 1} - \dfrac{x + 1}{x - 1}$

47. $\dfrac{x - 1}{x^2 + 4x + 4} + \dfrac{x - 1}{x + 2}$

JUST FOR FUN

Simplify each expression.

1. $\dfrac{x + 6}{4 - x^2} - \dfrac{x + 3}{x + 2} + \dfrac{x - 3}{2 - x}$

2. $3x + \dfrac{4x}{x - 2} - \dfrac{5}{x + 3} + 1$

3. $\dfrac{3x - 1}{x + 2} + \dfrac{x}{x - 3} - \dfrac{4}{2x + 3}$

6.6
Complex Fractions (Optional)

1️⃣ *Identify complex fractions.*

2️⃣ *Simplify complex fractions using Method 1.*

3️⃣ *Simplify complex fractions using Method 2.*

1️⃣ A complex fraction is one that has a fraction in its numerator or its denominator or both its numerator and denominator. Examples of complex fractions are:

$$\dfrac{\dfrac{2}{3}}{5} \qquad \dfrac{\dfrac{x + 1}{x}}{3x} \qquad \dfrac{\dfrac{x}{y}}{x + 1} \qquad \dfrac{\dfrac{a + b}{a}}{\dfrac{a - b}{b}}$$

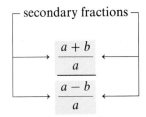

$$\begin{array}{l}\text{numerator of}\\ \text{complex fraction}\end{array}\left\{\dfrac{a + b}{a}\right.$$

← main
fraction line

$$\begin{array}{l}\text{denominator of}\\ \text{complex fraction}\end{array}\left\{\dfrac{a - b}{b}\right.$$

secondary fractions

$$\dfrac{a + b}{a}$$

$$\dfrac{a - b}{a}$$

The expression above the main fraction line is the numerator, and the expression below the main fraction line is the denominator of the complex fraction.

2️⃣ There are two methods to simplify complex fractions. Method 1 reinforces many of the concepts used in this chapter. Many students prefer to use Method 2, because the answer may be obtained more quickly. We will give 3 examples using Method 1 and then work the same 3 examples using Method 2.

Method 1

To Simplify A Complex Fraction

1. Add or subtract each secondary fraction as indicated to obtain a single fraction in both the numerator and denominator.

2. Invert and multiply the denominator of the complex fraction by the numerator of the complex fraction.

3. Simplify when possible.

EXAMPLE 1   Simplify $\dfrac{\dfrac{2}{3}+\dfrac{3}{4}}{\dfrac{3}{4}-\dfrac{2}{3}}.$

Solution: First, simplify the secondary fractions to obtain single fractions in the numerator and the denominator. The LCD of both secondary fractions is 12.

$$\frac{\dfrac{2}{3}+\dfrac{3}{4}}{\dfrac{3}{4}-\dfrac{2}{3}}=\frac{\dfrac{4}{4}\cdot\dfrac{2}{3}+\dfrac{3}{4}\cdot\dfrac{3}{3}}{\dfrac{3}{3}\cdot\dfrac{3}{4}-\dfrac{2}{3}\cdot\dfrac{4}{4}}$$

$$=\frac{\dfrac{8}{12}+\dfrac{9}{12}}{\dfrac{9}{12}-\dfrac{8}{12}}=\frac{\dfrac{17}{12}}{\dfrac{1}{12}}$$

Next, invert and multiply the denominator of the complex fraction.

$$\frac{\dfrac{17}{12}}{\dfrac{1}{12}}=\frac{17}{\cancel{12}}\cdot\frac{\cancel{12}}{1}=\frac{17}{1}=17\qquad\blacksquare$$

EXAMPLE 2   Simplify $\dfrac{a+\dfrac{1}{x}}{x+\dfrac{1}{a}}.$

Solution: Express the secondary fractions as single fractions, then invert and multiply.

$$\frac{\dfrac{x}{x}\cdot a+\dfrac{1}{x}}{\dfrac{a}{a}\cdot x+\dfrac{1}{a}}=\frac{\dfrac{ax}{x}+\dfrac{1}{x}}{\dfrac{ax}{a}+\dfrac{1}{a}}$$

$$=\frac{\dfrac{ax+1}{x}}{\dfrac{ax+1}{a}}$$

$$=\frac{\cancel{(ax+1)}}{x}\cdot\frac{a}{\cancel{(ax+1)}}$$

$$=\frac{a}{x}\qquad\blacksquare$$

EXAMPLE 3   Simplify $\dfrac{a}{\dfrac{1}{a}+\dfrac{1}{b}}$.

Solution: $\dfrac{a}{\dfrac{1}{a}+\dfrac{1}{b}} = \dfrac{a}{\dfrac{b}{b}\cdot\dfrac{1}{a}+\dfrac{1}{b}\cdot\dfrac{a}{a}} = \dfrac{a}{\dfrac{b}{ab}+\dfrac{a}{ab}}$

$$= \dfrac{a}{\dfrac{b+a}{ab}}$$

$$= \dfrac{a}{1}\cdot\dfrac{ab}{b+a}$$

$$= \dfrac{a^2b}{b+a} \quad \text{or} \quad \dfrac{a^2b}{a+b} \quad\blacksquare$$

Here is the second method that may be used to simplify a complex fraction.

Method 2

> **To Simplify A Complex Fraction**
>
> 1. Find the least common denominator of each of the two secondary fractions.
> 2. Next find the LCD of the complex fraction. The LCD of the complex fraction will be the LCD of the two expressions found in Step 1.
> 3. Multiply both secondary fractions by the LCD of the complex fraction found in Step 2.
> 4. Simplify when possible.

We will now rework examples 1, 2, and 3 using Method 2.

EXAMPLE 4   Simplify $\dfrac{\dfrac{2}{3}+\dfrac{3}{4}}{\dfrac{3}{4}-\dfrac{2}{3}}$.

Solution: *Step 1:* The LCD of the numerator of the complex fraction is 12. The LCD of the denominator is 12.

Step 2: The LCD of the complex fraction is the LCD of 12 (from the numerator) and 12 (from the denominator) which is 12.

Step 3: Multiply both secondary fractions by 12.

$$\dfrac{12}{12}\cdot\dfrac{\left(\dfrac{2}{3}+\dfrac{3}{4}\right)}{\left(\dfrac{3}{4}-\dfrac{2}{3}\right)} = \dfrac{12\left(\dfrac{2}{3}\right)+12\left(\dfrac{3}{4}\right)}{12\left(\dfrac{3}{4}\right)-12\left(\dfrac{2}{3}\right)}$$

Step 4: Simplify.

$$= \frac{12\left(\dfrac{2}{3}\right) + 12\left(\dfrac{3}{4}\right)}{12\left(\dfrac{3}{4}\right) - 12\left(\dfrac{2}{3}\right)}$$

$$= \frac{8 + 9}{9 - 8} = \frac{17}{1} = 17 \qquad \blacksquare$$

Notice the answers to Examples 1 and 4 are the same.

EXAMPLE 5   Simplify $\dfrac{a + \dfrac{1}{x}}{x + \dfrac{1}{a}}$

Solution: The LCD of the numerator is x and the LCD of the denominator is a. Therefore the LCD of the complex fraction is ax. Multiply both secondary fractions by ax.

$$\frac{ax \cdot \left(a + \dfrac{1}{x}\right)}{ax \cdot \left(x + \dfrac{1}{a}\right)} = \frac{a^2 x + a}{ax^2 + x}$$

$$= \frac{a(ax + 1)}{x(ax + 1)}$$

$$= \frac{a}{x} \qquad \blacksquare$$

Notice the answers to Examples 2 and 5 are the same.

EXAMPLE 6   Simplify $\dfrac{a}{\dfrac{1}{a} + \dfrac{1}{b}}$

Solution: The LCD of the numerator is 1. The LCD of the denominator is ab. The LCD of the complex fraction is therefore ab. Multiply both secondary fractions by ab.

$$\frac{ab \cdot a}{ab \cdot \left(\dfrac{1}{a} + \dfrac{1}{b}\right)} = \frac{a^2 b}{b + a} \qquad \blacksquare$$

Notice the answers to Examples 3 and 6 are the same. When asked to simplify a complex fraction you may use either method unless you are told by your instructor to use a specific method.

Exercise Set 6.6

Simplify each expression.

1. $\dfrac{1 + \dfrac{3}{5}}{2 + \dfrac{1}{5}}$

2. $\dfrac{1 - \dfrac{9}{16}}{3 + \dfrac{4}{5}}$

3. $\dfrac{2 + \dfrac{3}{8}}{1 + \dfrac{1}{3}}$

4. $\dfrac{\dfrac{3}{5} + \dfrac{2}{7}}{\dfrac{1}{5} + \dfrac{5}{6}}$

5. $\dfrac{\dfrac{4}{9} - \dfrac{3}{8}}{4 - \dfrac{3}{5}}$

6. $\dfrac{1 - \dfrac{x}{y}}{x}$

7. $\dfrac{\dfrac{x^2 y}{4}}{\dfrac{2}{x}}$

8. $\dfrac{\dfrac{15a}{b^2}}{\dfrac{b^3}{5}}$

9. $\dfrac{\dfrac{8x^2 y}{3z^3}}{\dfrac{4xy}{9z^5}}$

10. $\dfrac{\dfrac{36x^4}{5y^4 z^5}}{\dfrac{9xy^2}{15z^5}}$

11. $\dfrac{x + \dfrac{1}{y}}{\dfrac{x}{y}}$

12. $\dfrac{x - \dfrac{x}{y}}{\dfrac{1+x}{y}}$

13. $\dfrac{\dfrac{9}{x} + \dfrac{3}{x^2}}{3 + \dfrac{1}{x}}$

14. $\dfrac{\dfrac{2}{a} + \dfrac{1}{2a}}{a + \dfrac{a}{2}}$

15. $\dfrac{3 - \dfrac{1}{y}}{2 - \dfrac{1}{y}}$

16. $\dfrac{\dfrac{x}{x-y}}{\dfrac{x^2}{y}}$

17. $\dfrac{\dfrac{x}{y} - \dfrac{y}{x}}{\dfrac{x+y}{x}}$

18. $\dfrac{1}{\dfrac{1}{x} + y}$

19. $\dfrac{\dfrac{a^2}{b} - b}{\dfrac{b^2}{a} - a}$

20. $\dfrac{\dfrac{1}{x} + \dfrac{2}{x^2}}{2 + \dfrac{1}{x^2}}$

21. $\dfrac{\dfrac{a}{b} - 2}{\dfrac{-a}{b} + 2}$

22. $\dfrac{\dfrac{x^2 - y^2}{x}}{\dfrac{x+y}{x^3}}$

23. $\dfrac{\dfrac{4x+8}{3x^2}}{\dfrac{4x}{6}}$

24. $\dfrac{\dfrac{1}{a} + \dfrac{1}{b}}{ab}$

25. $\dfrac{\dfrac{1}{a} + \dfrac{1}{b}}{\dfrac{1}{ab}}$

26. $\dfrac{\dfrac{1}{a} + 1}{\dfrac{1}{b} - 1}$

27. $\dfrac{\dfrac{a}{b} + \dfrac{1}{a}}{\dfrac{b}{a} + \dfrac{1}{a}}$

28. $\dfrac{\dfrac{1}{a} + \dfrac{1}{b}}{\dfrac{1}{a}}$

29. $\dfrac{\dfrac{1}{x} - \dfrac{1}{y}}{\dfrac{1}{x} + \dfrac{1}{y}}$

30. $\dfrac{\dfrac{1}{x^2} + \dfrac{1}{x}}{\dfrac{1}{x} + \dfrac{1}{x^2}}$

JUST FOR FUN

1. The efficiency of a jack, E, is given by the formula $E = \dfrac{\frac{1}{2}h}{h + \frac{1}{2}}$

where h is determined by the pitch of the jack's thread. Determine the efficiency of a jack whose value of h is

(a) $\frac{2}{3}$ (b) $\frac{4}{5}$

Pitch

6.7

Solving Equations Containing Rational Expressions

1 Solve equations with rational expressions.

1   In Sections 6.1 through 6.6, we focused on how to add, subtract, multiply, and divide rational expressions. In this section we are ready to solve equations containing fractions.

To Solve Equations Containing Fractions

1. Determine the LCD of all fractions in the equation.
2. Multiply **both** sides of the equation by the LCD. This will result in every term in the equation being multiplied by the LCD.
3. Remove any parentheses and combine like terms on each side of the equation.
4. Solve the equation using the properties discussed in earlier sections.
5. Check your solution in the original equation.

The purpose of multiplying both sides of the equation by the LCD (Step 2) is to eliminate all fractions from the equation. We will omit some of the checks to save space.

EXAMPLE 1   Solve $\dfrac{x}{3} + 2x = 7$.

Solution:

$$3\left(\dfrac{x}{3} + 2x\right) = 7 \cdot 3 \qquad \text{multiply both sides of the equation by the LCD, 3}$$

$$\cancel{3}\left(\dfrac{x}{\cancel{3}}\right) + 3 \cdot 2x = 7 \cdot 3 \qquad \text{distributive property}$$

$$x + 6x = 21$$

$$7x = 21$$

$$x = 3$$

Check:

$$\dfrac{x}{3} + 2x = 7$$

$$\dfrac{3}{3} + 2(3) = 7$$

$$1 + 6 = 7$$

$$7 = 7 \qquad \text{true} \qquad \blacksquare$$

EXAMPLE 2   Solve the equation $\dfrac{3}{4} + \dfrac{5x}{9} = \dfrac{x}{6}$.

Solution: Multiply both sides of the equation by the LCD, 36.

$$36\left(\dfrac{3}{4} + \dfrac{5x}{9}\right) = \dfrac{x}{6} \cdot 36$$

$$\overset{9}{\cancel{36}}\left(\dfrac{3}{\cancel{4}}\right) + \overset{4}{\cancel{36}}\left(\dfrac{5x}{\cancel{9}}\right) = \dfrac{x}{\cancel{6}} \cdot \overset{6}{\cancel{36}}$$

$$27 + 20x = 6x$$

$$27 = -14x$$

$$x = \dfrac{27}{-14} \quad \text{or} \quad \dfrac{-27}{14} \qquad \blacksquare$$

EXAMPLE 3   Solve the equation $\dfrac{x}{4} + 3 = 2(x - 2)$.

Solution: Multiply both sides of the equation by the LCD, 4.

$$\frac{x}{4} + 3 = 2(x - 2)$$

$$4\left(\frac{x}{4} + 3\right) = 4\left[2(x - 2)\right]$$

$$4\left(\frac{x}{4}\right) + 4(3) = 4[2(x - 2)]$$

$$4\left(\frac{x}{4}\right) + 4(3) = 8(x - 2)$$

$$x + 12 = 8(x - 2)$$

$$x + 12 = 8x - 16$$

$$12 = 7x - 16$$

$$28 = 7x$$

$$4 = x \quad \blacksquare$$

EXAMPLE 4   Solve the equation $3 - \dfrac{4}{x} = \dfrac{5}{2}$.

Solution: Multiply both sides of the equation by the LCD, $2x$.

$$2x\left(3 - \frac{4}{x}\right) = \left(\frac{5}{2}\right) \cdot 2x$$

$$2x(3) - 2x\left(\frac{4}{x}\right) = \frac{5}{2} \cdot 2x$$

$$6x - 8 = 5x$$

$$x - 8 = 0$$

$$x = 8$$

Warning: **Whenever a variable appears in any denominator, it is necessary to check your answer in the original equation. If the answer obtained makes any denominator equal to zero, that value is not a solution to the equation.** Such values are called **extraneous roots** or **extraneous solutions.**

Check: $3 - \dfrac{4}{x} = \dfrac{5}{2}$

$$3 - \frac{4}{8} = \frac{5}{2}$$

$$3 - \frac{1}{2} = \frac{5}{2}$$

$$\frac{5}{2} = \frac{5}{2} \qquad \text{true}$$

Since 8 does check, it is the solution to the equation. $\blacksquare$

EXAMPLE 5   Solve the equation $\dfrac{x-7}{x+2}=\dfrac{1}{4}$.

Solution: The LCD is $4(x+2)$. Multiply both sides of the equation by the LCD.

$$4(x+2)\cdot\frac{(x-7)}{(x+2)}=\frac{1}{4}\cdot 4(x+2)$$

$$4(x-7)=1(x+2)$$

$$4x-28=x+2$$

$$3x-28=2$$

$$3x=30$$

$$x=10$$

A check will show that 10 is the solution.   ■

In Section 2.6 we illustrated that proportions of the form

$$\frac{a}{b}=\frac{c}{d}$$

can be cross-multiplied to obtain $a\cdot d=b\cdot c$. Example 5 is a proportion and can also be solved by cross-multiplying.

EXAMPLE 6   Solve the following equation using cross-multiplication $\dfrac{3}{x+4}=\dfrac{4}{x-1}$.

Solution: $3(x-1)=4(x+4)$

$3x-3=4x+16$

$-x-3=16$

$-x=19$

$x=-19$

A check will show that -19 is the solution to the equation.   ■

Now let us examine some examples that involve quadratic equations. Recall from Section 5.6 that quadratic equations are equations of the form $ax^2+bx+c=0$, $a\neq 0$.

EXAMPLE 7   Solve the equation $x+\dfrac{12}{x}=-7$.

Solution: $x\cdot\left(x+\dfrac{12}{x}\right)=-7\cdot x$ multiply both sides of the equation by x

$$x(x)+x\left(\frac{12}{x}\right)=-7x$$

$$x^2+12=-7x$$

$$x^2+7x+12=0$$

$$(x+3)(x+4)=0$$

$$x+3=0 \quad\text{or}\quad x+4=0$$

$$x=-3 \quad\text{or}\quad x=-4$$

Check: $x = -3$ $x = -4$

$$x + \frac{12}{x} = -7 \qquad\qquad x + \frac{12}{x} = -7$$

$$-3 + \frac{12}{-3} = -7 \qquad\qquad -4 + \frac{12}{-4} = -7$$

$$-3 + (-4) = -7 \qquad\qquad -4 + (-3) = -7$$

$$-7 = -7 \quad \text{true} \qquad\qquad -7 = -7 \quad \text{true} \qquad \blacksquare$$

EXAMPLE 8 Solve the equation $\dfrac{x^2}{x-4} = \dfrac{16}{x-4}$.

Solution: $\cancel{(x-4)} \cdot \dfrac{x^2}{\cancel{(x-4)}} = \dfrac{16}{\cancel{(x-4)}} \cdot \cancel{(x-4)}$

$$x^2 = 16$$

$$x^2 - 16 = 0 \qquad \text{this is a difference of two squares}$$

$$(x+4)(x-4) = 0$$

$$x + 4 = 0 \quad \text{or} \quad x - 4 = 0$$

$$x = -4 \qquad\qquad x = 4$$

Check: $x = -4$ $x = 4$

$$\frac{x^2}{x-4} = \frac{16}{x-4} \qquad\qquad \frac{x^2}{x-4} = \frac{16}{x-4}$$

$$\frac{(-4)^2}{-4-4} = \frac{16}{-4-4} \qquad\qquad \frac{(4)^2}{4-4} = \frac{16}{4-4}$$

$$\frac{16}{-8} = \frac{16}{-8} \qquad\qquad \frac{16}{0} = \frac{16}{0}$$

$$-2 = -2 \quad \text{true} \qquad\qquad \text{not a solution}$$

Since 4 results in a denominator of 0, $x = 4$ is *not* a solution to the equation. The 4 is an extraneous root. The only solution to the equation is $x = -4$. $\blacksquare$

Notice that the equation in Example 8 is a proportion. There are many proportions that are difficult to solve by cross-multiplication. What would happen if you attempted to solve the proportion in Example 8 by cross-multiplication?

EXAMPLE 9 Solve the equation $\dfrac{2x}{x^2-4} + \dfrac{1}{x-2} = \dfrac{2}{x+2}$.

Solution: $\dfrac{2x}{(x+2)(x-2)} + \dfrac{1}{x-2} = \dfrac{2}{x+2}$

Multiply both sides of the equation by the LCD, $(x+2)(x-2)$.

$$(x + 2)(x - 2) \cdot \left[\frac{2x}{(x + 2)(x - 2)} + \frac{1}{x - 2} \right] = \frac{2}{x + 2} \cdot (x + 2)(x - 2)$$

$$(x + 2)(x - 2) \cdot \frac{2x}{(x + 2)(x - 2)} + (x + 2)(x - 2) \cdot \frac{1}{(x - 2)} = \frac{2}{(x + 2)} \cdot (x + 2)(x - 2)$$

$$\cancel{(x + 2)}\cancel{(x - 2)} \cdot \frac{2x}{\cancel{(x + 2)}\cancel{(x - 2)}} + (x + 2)\cancel{(x - 2)} \cdot \frac{1}{\cancel{(x - 2)}} = \frac{2}{\cancel{(x + 2)}} \cdot \cancel{(x + 2)}(x - 2)$$

$$2x + (x + 2) = 2(x - 2)$$

$$2x + x + 2 = 2x - 4$$

$$3x + 2 = 2x - 4$$

$$x + 2 = -4$$

$$x = -6$$

A check will show that -6 is the solution to the equation. ∎

HELPFUL HINT

At this point some students confuse adding and subtracting rational expressions with solving rational equations. When adding or subtracting rational equations, we must rewrite each expression with a common denominator. When solving a rational equation, we multiply both sides of the equation by the LCD to eliminate fractions from the equation. Consider the two problems below. Note that the one on the right is an equation because it contains an equal sign. We will work both problems. The LCD for both problems is $x(x + 4)$.

Adding Rational Expressions

$$\frac{x + 2}{x + 4} + \frac{3}{x}$$

$$= \frac{x}{x} \cdot \frac{x + 2}{x + 4} + \frac{3}{x} \cdot \frac{x + 4}{x + 4}$$

$$= \frac{x(x + 2)}{x(x + 4)} + \frac{3(x + 4)}{x(x + 4)}$$

$$= \frac{x^2 + 2x}{x(x + 4)} + \frac{3x + 12}{x(x + 4)}$$

$$= \frac{x^2 + 2x + 3x + 12}{x(x + 4)}$$

$$= \frac{x^2 + 5x + 12}{x(x + 4)}$$

Solving Rational Equations

$$\frac{x + 2}{x + 4} = \frac{3}{x}$$

$$(x)(x + 4) \left(\frac{x + 2}{x + 4} \right) = \frac{3}{x} \, (x)(x + 4)$$

$$x(x + 2) = 3(x + 4)$$

$$x^2 + 2x = 3x + 12$$

$$x^2 - x - 12 = 0$$

$$(x - 4)(x + 3) = 0$$

$$x - 4 = 0 \quad \text{or} \quad x + 3 = 0$$

$$x = 4 \quad \text{or} \quad x = -3$$

The numbers 4 and -3 will both check, and are thus solutions to the equation.

Note that when adding or subtracting rational expressions we end up with an algebraic expression. When solving rational equations the solution will be a numerical value. The equation on the right could also be solved using cross-multiplication.

Exercise Set 6.7

Solve each equation, and check your solution.

1. $\dfrac{2}{5} = \dfrac{x}{10}$

2. $\dfrac{3}{k} = \dfrac{9}{6}$

3. $\dfrac{5}{12} = \dfrac{20}{x}$

4. $\dfrac{x}{8} = \dfrac{-15}{4}$

5. $\dfrac{a}{25} = \dfrac{12}{10}$

6. $\dfrac{9c}{10} = \dfrac{9}{5}$

7. $\dfrac{9}{3b} = \dfrac{-6}{2}$

8. $\dfrac{5}{8} = \dfrac{2b}{80}$

9. $\dfrac{x+4}{9} = \dfrac{5}{9}$

10. $\dfrac{1}{4} = \dfrac{z+1}{8}$

11. $\dfrac{4x+5}{6} = \dfrac{7}{2}$

12. $\dfrac{a}{5} = \dfrac{a-3}{2}$

13. $\dfrac{6x+7}{10} = \dfrac{2x+9}{6}$

14. $\dfrac{n}{10} = 9 - \dfrac{n}{5}$

15. $\dfrac{x}{3} - \dfrac{3x}{4} = \dfrac{1}{12}$

16. $\dfrac{2}{8} + \dfrac{3}{4} = \dfrac{w}{5}$

17. $\dfrac{3}{4} - x = 2x$

18. $\dfrac{2}{y} + \dfrac{1}{2} = \dfrac{5}{2y}$

19. $\dfrac{5}{3x} + \dfrac{3}{x} = 1$

20. $\dfrac{x}{4} - \dfrac{x}{6} = \dfrac{1}{4}$

21. $\dfrac{x-1}{x-5} = \dfrac{4}{x-5}$

22. $\dfrac{2x+3}{x+1} = \dfrac{3}{2}$

23. $\dfrac{5y-3}{7} = \dfrac{15y-2}{28}$

24. $\dfrac{2}{x+1} = \dfrac{1}{x-2}$

25. $\dfrac{5}{-x-6} = \dfrac{2}{x}$

26. $\dfrac{4}{y-3} = \dfrac{6}{y+3}$

27. $\dfrac{x-2}{x+4} = \dfrac{x+1}{x+10}$

28. $\dfrac{x-3}{x+1} = \dfrac{x-6}{x+5}$

29. $\dfrac{2x-1}{3} - \dfrac{3x}{4} = \dfrac{5}{6}$

30. $x + \dfrac{3}{x} = \dfrac{12}{x}$

31. $x + \dfrac{6}{x} = -5$

32. $\dfrac{15}{x} + \dfrac{9x-7}{x+2} = 9$

33. $\dfrac{3y-2}{y+1} = 4 - \dfrac{y+2}{y-1}$

34. $\dfrac{2b}{b+1} = 2 - \dfrac{5}{2b}$

35. $\dfrac{1}{x+3} + \dfrac{1}{x-3} = \dfrac{-5}{x^2-9}$

36. $c - \dfrac{c}{3} + \dfrac{c}{5} = 26$

37. $\dfrac{2}{x-3} - \dfrac{4}{x+3} = \dfrac{8}{x^2-9}$

38. $\dfrac{x+1}{x+3} + \dfrac{x-3}{x-2} = \dfrac{2x^2-15}{x^2+x-6}$

39. $\dfrac{y}{2y+2} + \dfrac{2y-16}{4y+4} = \dfrac{y-3}{y+1}$

40. $\dfrac{3}{x+3} + \dfrac{5}{x+4} = \dfrac{12x+19}{x^2+7x+12}$

JUST FOR FUN

1. A formula frequently used in optics is

$$\dfrac{1}{p} + \dfrac{1}{q} = \dfrac{1}{f}$$

where p represents the distance of the object from a mirror (or lens), q represents the distance of the image from the mirror (or lens), and f represents the focal length of the mirror (or lens). If a mirror has a focal length of 10 centimeters, how far from the mirror will the image appear when the object is 30 centimeters from the mirror?

2. In electronics the total resistance, R_T, of resistors wired in a parallel circuit is determined by the formula

$$\frac{1}{R_T} = \frac{1}{R_1} + \frac{1}{R_2} + \frac{1}{R_3} + \cdots + \frac{1}{R_n}$$

where $R_1, R_2, R_3, \ldots, R_n$ are the resistances of the individual resistors (measured in ohms) in the circuit.

(a) Find the total resistance if two resistors, one of 200 ohms and the other of 300 ohms, are wired in a parallel circuit.

(b) If three identical resistors are to be wired in parallel, what should be the resistance of each resistor if the total resistance of the circuit is to be 300 ohms?

6.8

Applications of Rational Equations

1 *Set up and solve problems containing rational expressions.*

2 *Set up and solve rate problems.*

3 *Set up and solve work problems.*

1 Many applications of algebra involve rational equations. After we represent the application as an equation, we solve the rational equation as we did in Section 6.7.

EXAMPLE 1 The area of a triangle is 27 square feet. Find the base and height if its height is 3 feet less than twice its base.

Solution: Let

$$x = \text{base}$$

$$2x - 3 = \text{height} \quad \text{(See Fig. 6.1.)}$$

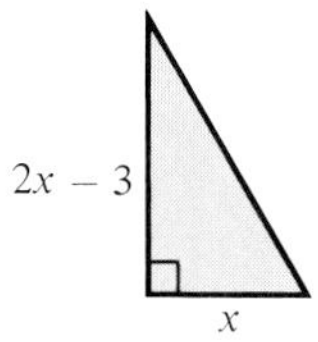

$2x - 3$

x

Figure 6.1

$$\text{area} = \frac{1}{2} \cdot \text{base} \cdot \text{height}$$

$$27 = \frac{1}{2}(x)(2x - 3) \qquad \begin{array}{l}\text{multiply both sides of equation by 2 to} \\ \text{remove fractions}\end{array}$$

$$2(27) = 2\left[\frac{1}{2}(x)(2x - 3)\right]$$

$$54 = x(2x - 3)$$

$$54 = 2x^2 - 3x$$

$$0 = 2x^2 - 3x - 54$$

$$\text{or} \quad 2x^2 - 3x - 54 = 0$$

$$(2x + 9)(x - 6) = 0$$

$$2x + 9 = 0 \qquad \text{or} \quad x - 6 = 0$$

$$2x = -9 \quad \text{or} \qquad x = 6$$

$$x = -\frac{9}{2} \quad \text{or} \qquad x = 6$$

Since the dimensions of a geometric figure cannot be negative, we can eliminate $-\frac{9}{2}$ as an answer to our problem.

$$base = x = 6 \text{ feet}$$
$$height = 2x - 3 = 2(6) - 3 = 9 \text{ feet}$$

Check: $a = \dfrac{1}{2} bh$

$$27 = \frac{1}{2}(6)(9)$$

$$27 = 27 \qquad \text{true} \qquad \blacksquare$$

EXAMPLE 2   One number is 3 times another number. The sum of their reciprocals is 4. Find the numbers.

Solution: Let $x =$ smaller number

then $3x =$ larger number

The sum of their reciprocals is 4, thus

$$\frac{1}{x} + \frac{1}{3x} = 4$$

$$3x\left(\frac{1}{x} + \frac{1}{3x}\right) = 3x(4)$$

$$3x\left(\frac{1}{x}\right) + 3x\left(\frac{1}{3x}\right) = 12x$$

$$3 + 1 = 12x$$

$$4 = 12x$$

$$\frac{4}{12} = x$$

$$\frac{1}{3} = x$$

The smaller number is $\frac{1}{3}$, the larger number is $3x = 3(\frac{1}{3}) = 1$.

Check: $\dfrac{1}{x} + \dfrac{1}{3x} = 4$

$$\frac{1}{\frac{1}{3}} + \frac{1}{3(\frac{1}{3})} = 4$$

$$3 + 1 = 4$$

$$4 = 4 \qquad \text{true} \qquad \blacksquare$$

2   In Chapter 3 we discussed motion problems. Recall that

$$distance = rate \cdot time$$

If we solve this equation for time, we obtain

$$\text{time} = \frac{\text{distance}}{\text{rate}}$$

This equation is useful in solving motion problems when the total time of travel for two objects or the time of travel between two points is known.

EXAMPLE 3   A river has a current of 3 miles per hour. If it takes Jack's motorboat the same time to go 10 miles downstream as 6 miles upstream, find the speed of his boat in still water.

Solution: Let x = speed of boat in still water

then $x + 3$ = speed of boat downstream (with current)

and $x - 3$ = speed of boat upstream (against current)

Boat	d	r	t
Downstream	10	$x + 3$	$\dfrac{10}{x + 3}$
Upstream	6	$x - 3$	$\dfrac{6}{x - 3}$

Since the time it takes to travel 10 miles downstream is the same as the time to travel 6 miles upstream, we set the times equal to each other, then solve the resulting equation.

$$t \text{ downstream} = t \text{ upstream}$$

$$\frac{10}{x + 3} = \frac{6}{x - 3} \qquad \text{now cross-multiply}$$

$$10(x - 3) = 6(x + 3)$$

$$10x - 30 = 6x + 18$$

$$4x - 30 = 18$$

$$4x = 48$$

$$x = 12$$

The speed of the boat in still water is 12 miles per hour.

Check:
$$\frac{10}{x + 3} = \frac{6}{x - 3}$$

$$\frac{10}{12 + 3} = \frac{6}{12 - 3}$$

$$\frac{10}{15} = \frac{6}{9}$$

$$\frac{2}{3} = \frac{2}{3} \qquad \text{true} \qquad ■$$

Work Problems

3   Problems where two or more machines or people work together to complete a certain task are sometimes referred to as **work problems.** Work problems often involve equations containing fractions. Generally, work problems are based on the fact that the fractional part of the work done by person 1 (or machine 1) plus the fractional part of the work done by person 2 (or machine 2) is equal to the total amount of work done by both people (or both machines). We represent the total amount of work done by the number 1, which represents one whole job completed.

$$\boxed{\frac{\text{time together}}{\text{time of } 1^{\text{st}} \text{ person alone}}} + \boxed{\frac{\text{time together}}{\text{time of } 2^{\text{nd}} \text{ person alone}}} = 1$$

fractional part of 1st person + fractional part of 2nd person = 1

EXAMPLE 4   John can mow Mr. Richard's lawn in 3 hours. Pete can mow Mr. Richard's lawn in 4 hours. How long will it take to mow the lawn if both John and Pete work together?

Solution: Let x = time, in hours, for both boys to mow the lawn together.

John can mow the entire lawn in 3 hours. Thus in 1 hour he can mow $\frac{1}{3}$ of the lawn by himself. In two hours he can mow $\frac{2}{3}$ of the lawn and in x hours he can mow $x/3$ of the lawn. Pete can mow the entire lawn in 4 hours. Thus in 1 hour he can mow $\frac{1}{4}$ of the lawn by himself. In 2 hours he can mow $\frac{2}{4}$ of the lawn and in x hours he can mow $x/4$ of the lawn.

To solve this problem we make use of the fact that

$$\binom{\text{part of lawn mowed}}{\text{by John in } x \text{ hours}} + \binom{\text{part of lawn mowed}}{\text{by Pete in } x \text{ hours}} = 1 \text{ (whole lawn mowed)}$$

$$\frac{x}{3} + \frac{x}{4} = 1$$

Multiply both sides of the equation by the LCD, 12, then solve for x.

$$12\left(\frac{x}{3} + \frac{x}{4}\right) = 12 \cdot 1$$

$$12\left(\frac{x}{3}\right) + 12\left(\frac{x}{4}\right) = 12$$

$$4x + 3x = 12$$

$$7x = 12$$

$$x = \frac{12}{7} \quad \text{or} \quad 1\frac{5}{7} \text{ hours}$$

The two boys together can mow the lawn in $1\frac{5}{7}$ hours. Note that the answer is less than the time it takes either boy to mow the lawn by himself (which is what we expect).   ■

EXAMPLE 5   Water pump A can drain an olympic size pool in 20 hours of continuous operation. Water pump B can drain the same pool in 30 hours of continuous operation. How long will it take both water pumps working together to drain the pool?

Solution: Let t = total time for pumps operating together to drain pools.

$$\begin{pmatrix} \text{amount of water} \\ \text{drained by pump A} \\ \text{in } t \text{ hours} \end{pmatrix} + \begin{pmatrix} \text{amount of water} \\ \text{drained by pump B} \\ \text{in } t \text{ hours} \end{pmatrix} = 1 \text{ (whole pool drained)}$$

$$\frac{t}{20} + \frac{t}{30} = 1$$

$$60\left(\frac{t}{20} + \frac{t}{30}\right) = 60 \cdot 1$$

$$60\left(\frac{t}{20}\right) + 60\left(\frac{t}{30}\right) = 60$$

$$3t + 2t = 60$$

$$5t = 60$$

$$t = 12$$

Thus the two pumps working together can drain the pool in 12 hours.   ∎

EXAMPLE 6   A tank can be filled by one pipe in 4 hours, and can be emptied by another pipe in 6 hours. If the valves to both pipes are open, how long will it take to fill the tank?

Solution: As one pipe is filling, the other is emptying the tank. Thus the pipes are working against each other. Let x = amount of time to fill the tank.

$$\begin{pmatrix} \text{amount of water} \\ \text{filled in } x \text{ hours} \end{pmatrix} - \begin{pmatrix} \text{amount of water} \\ \text{emptied in } x \text{ hours} \end{pmatrix} = 1 \text{ (total tank filled)}$$

$$\frac{x}{4} - \frac{x}{6} = 1$$

$$12\left(\frac{x}{4} - \frac{x}{6}\right) = 12 \cdot 1$$

$$12\left(\frac{x}{4}\right) - 12\left(\frac{x}{6}\right) = 12$$

$$3x - 2x = 12$$

$$x = 12$$

The tank will be filled in 12 hours.   ∎

Exercise Set 6.8 _____

Solve each problem.

1. The base of a triangle is 6 centimeters greater than its height. Find the base and height if the area is 80 square centimeters.

2. The height of a triangle is 1 centimeter less than twice its base. Find the base and height if the triangle's area is 33 square centimeters.

3. One number is three times as large as another. The sum of their reciprocals is $\frac{4}{3}$. Find the two numbers.

4. The numerator of the fraction $\frac{3}{4}$ is increased by an amount so that the value of the resulting fraction is $\frac{5}{2}$. Find the amount that the numerator was increased.

5. The reciprocal of 3 plus the reciprocal of 5 is the reciprocal of what number?

6. One number is 4 times as large as another. The sum of their reciprocals is $\frac{5}{8}$. Find the two numbers.

7. Jim can row 4 miles per hour in still water. It takes him as long to row 6 miles upstream as 10 miles downstream. How fast is the current?

8. In the Pixie River a boat can travel 9 miles upstream in the same amount of time it takes to travel 11 miles downstream. If the current of the river is 2 miles per hour, find the speed of the boat in still water.

9. Ms. Duncan took her two sons water skiing in still water. She drove the motor boat one way on the water pulling the younger son at 30 miles per hour. Then she turned around and pulled her older son in the opposite direction the same distance at 30 miles per hour. If the total time spent skiing was $\frac{1}{2}$ hour, how far did each son travel?

10. A business executive traveled 1800 miles by jet and then 300 miles on a private propeller plane. If the rate of the jet is four times the rate of the prop plane and the entire trip took 5 hours, find the speed of each plane.

11. One car travels 30 kilometers per hour faster than another. In the time it takes the slower car to travel 250 kilometers the faster car travels 400 kilometers. Find the speed of both cars.

12. Maria walked at 2 miles per hour and jogged at 4 miles per hour. If she jogged 3 miles further than she walked and the total time for the trip was 3 hours, how far did she walk and how far did she jog?

13. Mario jogs twice as fast as he walks. He jogs for 2 miles and then walks for another 2 miles. If the total time of his outing was 1 hour, find the rates at which he walks and jogs.

14. Marsha can construct a small retaining wall in 3 hours. Her apprentice can complete the same job in 6 hours. How long would it take them to complete the job working together?

15. At the Community Savings Bank it takes one computer 4 hours to complete a certain job and a second computer 5 hours to complete the same job. How long will it take the two computers together to complete the job?

16. Mr. Dell can fertilize the farm in 6 hours. Mrs. Dell can fertilize the farm in 7 hours. How long will it take them to fertilize their farm if they work together?

17. John can paint a house in 8 days. Patti can paint a house in 6 days. How long will it take them to paint the house together?

18. A $\frac{1}{2}$-inch-diameter hose can fill a swimming pool in 8 hours. A $\frac{4}{5}$-inch-diameter hose can fill the same pool in 5 hours. How long will it take to fill the pool when both hoses are used?

19. Jodi can mow a lawn on a rider lawn mower in 4 hours. Kathy can mow the lawn in 6 hours with a push lawn mower. How long will it take them to mow the lawn together?

20. A conveyor belt operating at full speed can fill a tank with topsoil in 3 hours. When a valve at the bottom of the tank is opened, the tank will empty in 4 hours. If the conveyor belt is operating at full speed and the valve at the bottom of the tank is open, how long will it take to fill the tank?

21. When only the cold-water valve is opened, a washtub will fill in 8 minutes. When only the hot-water valve is opened, the washtub will fill in 12 minutes. When the drain of the washtub is open, it will drain completely in 7 minutes. If both the hot- and cold-water valves are open and the drain is open, how long will it take for the washtub to fill?

JUST FOR FUN

1. The reciprocal of 3 less than a certain number is twice the reciprocal of 6 less than twice the number. Find the number(s).

2. If three times a number is added to twice the reciprocal of the number, the answer is 5. Find the number(s).

Summary

Glossary

Algebraic fraction (or rational expression): An algebraic expression of the form $\frac{p}{q}$, where p and q are polynomials and $q \neq 0$.

Complex fraction: A fraction that has a fraction in its numerator or its denominator, or both its numerator and denominator.

Extraneous root or extraneous solution: A number ob-

tained when solving an equation that is not a solution to the original equation.

Reduced to lowest terms: An algebraic fraction is reduced to its lowest terms when the numerator and denominator have no common factors other than 1.

Secondary fractions: The numerator and denominator of a complex fraction are secondary fractions.

Important Facts

To add fractions: $\dfrac{a}{c} + \dfrac{b}{c} = \dfrac{a+b}{c}, c \neq 0$

To subtract fractions: $\dfrac{a}{c} - \dfrac{b}{c} = \dfrac{a-b}{c}, c \neq 0$

To multiply fractions: $\dfrac{a}{b} \cdot \dfrac{c}{d} = \dfrac{ac}{bd}, b \neq 0, d \neq 0$

To divide fractions: $\dfrac{a}{b} \div \dfrac{c}{d} = \dfrac{a}{b} \cdot \dfrac{d}{c} = \dfrac{ad}{bc}, b \neq 0, c \neq 0, d \neq 0$

For any fraction: $-\dfrac{a}{b} = \dfrac{-a}{b} = \dfrac{a}{-b}, b \neq 0$

For any binomial: $a - b = -(b - a)$

$\text{Time} = \dfrac{\text{distance}}{\text{rate}}$

Review Exercises

Reduce each expression to its lowest terms.

1. $\dfrac{x}{x - xy}$

2. $\dfrac{2x}{4x + 8}$

3. $\dfrac{x^3 + 4x^2 + 12x}{x}$

4. $\dfrac{9x^2 + 6xy}{3x}$

5. $\dfrac{x^2 + x - 12}{x - 3}$

6. $\dfrac{x^2 - 4}{x - 2}$

7. $\dfrac{2x^2 - 7x + 3}{3 - x}$

8. $\dfrac{2x^2 - x - 3}{3 - 2x}$

9. $\dfrac{x^2 - 2x - 24}{x^2 + 6x + 8}$

10. $\dfrac{3x^2 - 8x - 16}{x^2 - 8x + 16}$

Multiply as indicated.

11. $\dfrac{4y}{3x} \cdot \dfrac{4x^2 y}{2}$

12. $\dfrac{15x^2 y^3}{3z} \cdot \dfrac{6z^3}{5xy^3}$

13. $\dfrac{40a^3 b^4}{7c^3} \cdot \dfrac{14c^5}{5a^5 b}$

14. $\dfrac{1}{x - 2} \cdot \dfrac{2 - x}{2}$

15. $\dfrac{-x + 2}{3} \cdot \dfrac{6x}{x - 2}$

16. $\dfrac{4x + 4y}{x^2 y} \cdot \dfrac{y^3}{8x}$

17. $\dfrac{a - 2}{a + 3} \cdot \dfrac{a^2 + 4a + 3}{a^2 - a - 2}$

18. $\dfrac{x^2 - y^2}{x - y} \cdot \dfrac{x + y}{xy + x^2}$

Divide as indicated.

19. $\dfrac{6y^3}{x} \div \dfrac{y^3}{6x}$

20. $\dfrac{8xy^2}{z} \div \dfrac{x^4 y^2}{4z^2}$

21. $\dfrac{3x + 3y}{x^2} \div \dfrac{x^2 - y^2}{x^2}$

22. $\dfrac{1}{a^2 + 8a + 15} \div \dfrac{3}{a + 5}$

23. $\dfrac{4x}{a + 2} \div \dfrac{8x^2}{a - 2}$

24. $(x + 3) \div \dfrac{x^2 - 4x - 21}{x - 7}$

25. $\dfrac{x^2 - 3xy - 10y^2}{6x} \div \dfrac{x + 2y}{12x^2}$

26. $\dfrac{4x^2 - 16y^2}{9} \div \dfrac{(x + 2y)^2}{12}$

[6.3] Add or subtract as indicated.

27. $\dfrac{x}{x + 2} + \dfrac{2}{x + 2}$

28. $\dfrac{x}{x + 2} - \dfrac{2}{x + 2}$

29. $\dfrac{4x}{x + 2} + \dfrac{8}{x + 2}$

30. $\dfrac{6x}{3y} - \dfrac{8}{3y}$

31. $\dfrac{9x - 4}{x + 8} + \dfrac{76}{x + 8}$

32. $\dfrac{7x - 3}{x^2 + 7x - 30} - \dfrac{3x + 9}{x^2 + 7x - 30}$

33. $\dfrac{4x^2 - 11x + 4}{x - 3} - \dfrac{x^2 - 4x + 10}{x - 3}$

34. $\dfrac{6x^2 - 4x}{2x - 3} - \dfrac{(-3x + 12)}{2x - 3}$

[6.4] Find the least common denominator.

35. $\dfrac{x}{3} + \dfrac{5x}{8}$

36. $\dfrac{4}{3x} + \dfrac{8}{5x^2}$

37. $\dfrac{6}{x + 1} - \dfrac{3x}{x}$

38. $\dfrac{6x + 3}{x + 2} + \dfrac{4}{x - 3}$

39. $\dfrac{7x - 12}{x^2 + x} - \dfrac{4}{x + 1}$

40. $\dfrac{9x - 3}{x + y} - \dfrac{4x + 7}{x^2 - y^2}$

41. $\dfrac{4x^2}{x - 7} + 8x^2$

42. $\dfrac{19x - 5}{x^2 + 2x - 35} + \dfrac{3x - 2}{x^2 + 9x + 14}$

[6.5] Add or subtract as indicated.

43. $\dfrac{4}{2x} + \dfrac{x}{x^2}$

44. $\dfrac{1}{4x} + \dfrac{6x}{xy}$

45. $\dfrac{5x}{3xy} - \dfrac{4}{x^2}$

46. $6 + \dfrac{x}{x + 2}$

47. $5 - \dfrac{3}{x + 3}$

48. $\dfrac{a + c}{c} - \dfrac{a - c}{a}$

49. $\dfrac{3}{x + 3} + \dfrac{4}{x}$

50. $\dfrac{2}{3x} - \dfrac{3}{3x - 6}$

51. $\dfrac{x + 4}{x + 3} - \dfrac{x - 3}{x + 4}$

52. $\dfrac{4}{x + 5} + \dfrac{6}{(x + 5)^2}$

53. $\dfrac{x + 3}{x^2 - 9} + \dfrac{2}{x + 3}$

54. $\dfrac{4}{(x + 2)(x - 3)} - \dfrac{4}{(x - 2)(x + 2)}$

55. $\dfrac{x + 2}{x^2 - x - 6} + \dfrac{x - 3}{x^2 - 8x + 15}$

56. $\dfrac{x + 5}{x^2 - 15x + 50} - \dfrac{x - 2}{x^2 - 25}$

[6.6] Simplify each complex fraction.

57. $\dfrac{1 + \dfrac{5}{12}}{\dfrac{3}{8}}$

58. $\dfrac{4 - \dfrac{9}{16}}{1 + \dfrac{5}{8}}$

59. $\dfrac{\dfrac{15xy}{6z}}{\dfrac{3x}{z^2}}$

60. $\dfrac{\dfrac{36x^4y^2}{9xy^5}}{4z^2}$

61. $\dfrac{x + \dfrac{1}{y}}{y^2}$

62. $\dfrac{x - \dfrac{x}{y}}{\dfrac{1 + x}{y}}$

63. $\dfrac{\dfrac{4}{x} + \dfrac{2}{x^2}}{6 - \dfrac{1}{x}}$

64. $\dfrac{\dfrac{x}{x + y}}{\dfrac{x^2}{2x + 2y}}$

65. $\dfrac{\dfrac{1}{a}}{\dfrac{1}{a^2}}$

66. $\dfrac{\dfrac{1}{a}+2}{\dfrac{1}{a}+\dfrac{1}{a}}$

67. $\dfrac{\dfrac{1}{x^2}+\dfrac{1}{x}}{\dfrac{1}{x^2}-\dfrac{1}{x}}$

68. $\dfrac{\dfrac{3x}{y}-x}{\dfrac{y}{x}-1}$

[6.7] Solve each equation.

69. $\dfrac{3}{x}=\dfrac{8}{24}$

70. $\dfrac{4}{a}=\dfrac{16}{4}$

71. $\dfrac{x+3}{5}=\dfrac{9}{5}$

72. $\dfrac{x}{6}=\dfrac{x-4}{2}$

73. $\dfrac{3x+4}{5}=\dfrac{2x-8}{3}$

74. $\dfrac{x}{5}+\dfrac{x}{2}=-14$

75. $\dfrac{4}{x}-\dfrac{1}{6}=\dfrac{1}{x}$

76. $\dfrac{1}{x-2}+\dfrac{1}{x+2}=\dfrac{1}{x^2-4}$

77. $\dfrac{x-3}{x-2}+\dfrac{x+1}{x+3}=\dfrac{2x^2+x+1}{x^2+x-6}$

78. $\dfrac{x}{x^2-9}+\dfrac{2}{x+3}=\dfrac{4}{x-3}$

[6.8] Solve each problem.

79. It takes Lee 5 hours to mow Mr. McKane's lawn. It takes Pat 4 hours to mow the same lawn. How long will it take them working together to mow Mr. McKane's lawn?

80. A $\frac{3}{4}$-inch-diameter hose can fill a swimming pool in 7 hours. A $\frac{5}{16}$-inch-diameter hose can fill the pool in 12 hours. How long will it take to fill the pool using both hoses.

81. One number is four times as large as another. The sum of their reciprocals is $\frac{1}{2}$. Find the numbers.

82. A Greyhound bus can travel 400 kilometers in the same time that an Amtrak train can travel 600 kilometers. If the speed of the train is 40 kilometers per hour greater than that of the bus, find the speeds of the bus and train.

Practice Test

Perform the operations indicated.

1. $\dfrac{3x^2y}{4z^2}\cdot\dfrac{8xz^3}{9y^4}$

2. $\dfrac{a^2-9a+14}{a-2}\cdot\dfrac{a^2-4a-21}{(a-7)^2}$

3. $\dfrac{x^2-9y^2}{3x+6y}\div\dfrac{x+3y}{x+2y}$

4. $\dfrac{16}{y^2+2y-15}\div\dfrac{4y}{y-3}$

5. $\dfrac{6x+3}{2y}+\dfrac{x-5}{2y}$

6. $\dfrac{7x^2-4}{x+3}-\dfrac{6x+7}{x+3}$

7. $\dfrac{5}{x}+\dfrac{3}{2x^2}$

8. $5-\dfrac{6x}{x+2}$

9. $\dfrac{x-5}{x^2-16}-\dfrac{x-2}{x^2+2x-8}$

Simplify each expression.

10. $\dfrac{3+\dfrac{5}{8}}{2-\dfrac{3}{4}}$

11. $\dfrac{x+\dfrac{x}{y}}{\dfrac{1}{x}}$

Solve each equation.

12. $\dfrac{x}{3}-\dfrac{x}{4}=5$

13. $\dfrac{x}{x-8}+\dfrac{6}{x-2}=\dfrac{x^2}{x^2-10x+16}$

Solve the problem.

14. Mr. Johnson, on his tractor, can level a 1-acre field in 8 hours. Mr. Hackett, on his tractor, can level a 1-acre field in 5 hours. If they work together, how long will it take them to level a 1-acre field?

7

Graphing Linear Equations

See Section 7.3., Just For Fun 3

7.1

The Cartesian Coordinate System

❶ Use the Cartesian coordinate system.

❷ Plot ordered pairs.

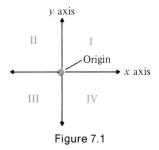

Figure 7.1

❶ In this chapter we discuss procedures that can be used to draw graphs. A **graph** shows the relationship between two variables in an equation. Many algebraic relationships are easier to understand if we can see a picture of them. We draw graphs using the **Cartesian (or rectangular) coordinate system.** The Cartesian coordinate system is named for its developer, the French mathematician and philosopher René Descartes (1596–1650).

Before you learn how to construct a graph, you must understand the Cartesian coordinate system. The Cartesian coordinate system consists of two axes (or number lines) drawn perpendicular to each other. The two intersecting axes form four **quadrants** (see Fig. 7.1).

The horizontal axis is called the *x* **axis.** The vertical axis is called the *y* **axis.** The point of intersection of the two axes is called the **origin.** The origin has an *x* value of 0 and a *y* value of 0. Starting from the origin and moving to the right along the *x* axis, the numbers increase. Starting from the origin and moving to the left, the numbers decrease (see Fig. 7.2). Starting from the origin and moving up the *y* axis, the numbers increase. Starting from the origin and moving down, the numbers decrease.

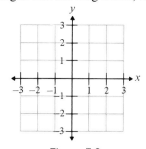

Figure 7.2

❷ To locate a point it is necessary to know both the *x* and *y* values, or **coordinates** of the point. When the *x* and *y* coordinates of a point are placed in parentheses, with the *x* coordinate listed first, we have an **ordered pair.** In the ordered pair (3, 5) the *x* coordinate is 3 and the *y* coordinate is 5. The point representing the ordered pair (3, 5) is plotted in Fig. 7.3.

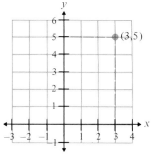

Figure 7.3

244

EXAMPLE 1   Plot each of the following points on the same set of axes.

(a) $A(4, 2)$ (b) $B(2, 4)$ (c) $C(-3, 1)$ (d) $D(4, 0)$ (e) $E(-2, -5)$
(f) $F(0, -3)$ (g) $G(0, 3)$ (h) $H(6, -\frac{7}{2})$ (i) $I(-\frac{3}{2}, -\frac{5}{2})$

Solution: The first number in each ordered pair is the x coordinate and the second number is the y coordinate. The points are plotted in Fig. 7.4.

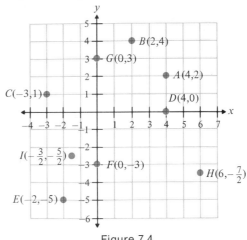

Figure 7.4

Notice that when the x coordinate is 0, as in parts (f) and (g), the point is on the y axis. When the y coordinate is 0, as in part (d), the point is on the x axis.

EXAMPLE 2   List the ordered pairs for each point shown in Fig. 7.5.

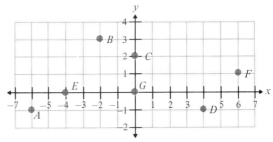

Figure 7.5

Solution: Remember to give the x value first in the ordered pair.

Point	Ordered Pair
A	$(-6, -1)$
B	$(-2, 3)$
C	$(0, 2)$
D	$(4, -1)$
E	$(-4, 0)$
F	$(6, 1)$
G	$(0, 0)$

Exercise Set 7.1

1. List the ordered pairs corresponding to each of the following points.

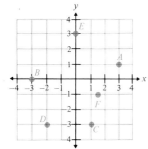

2. List the ordered pairs corresponding to each of the following points.

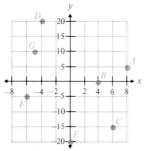

3. Plot each of the following points on the same set of axes.

(a) (4, 2)   (b) (−3, 2)   (c) (0, −3)
(d) (−2, 0)   (e) (−3, −4)   (f) (−4, −2)

4. Plot each of the following points on the same set of axes.

(a) (−3, −1)   (b) (2, 0)   (c) (−3, 2)
(d) ($\frac{1}{2}$, −4)   (e) (−4, 2)   (f) (0, 4)

5. Plot the following points. Then determine if they are all in a straight line.

(a) (1, −1)   (b) (5, 3)   (c) (−3, −5)
(d) (0, −2)   (e) (2, 0)

6. Plot the following points. Then determine if they are all in a straight line.

(a) (1, −2)   (b) (0, −5)   (c) (3, 1)
(d) (−1, −8)   (e) ($\frac{1}{2}$, −$\frac{7}{2}$)

7. In an ordered pair which coordinate is always listed first?

7.2

Graphing Linear Equations

■   *Identify linear equations in two variables.*

▣   *Know that linear equations in two variables have an infinite number of solutions.*

▣   *Graph linear equations by plotting points.*

▣   *Graph linear equations using x and y intercepts.*

▣   *Identify horizontal and vertical lines in equation form.*

■   Most of the equations we have discussed thus far have contained only one variable. Exceptions to this include formulas used in application sections. In this chapter we consider linear equations in two variables.

> A **linear equation in two variables** is an equation that can be put in the form
> $$ax + by = c,$$
> where a, b, and c are real numbers.

Equations of the form $ax + by = c$ will be straight lines when graphed. For this reason such equations are called linear. Linear equations may be written in various forms, as we will show later. A linear equation in the form $ax + by = c$ is said to be in **standard form.**

Examples of linear equations

$$3x - 2y = 4$$
$$y = 5x + 3$$
$$x - 3y + 4 = 0$$

Note that in the examples above only the equation $3x - 2y = 4$ is given in standard form. However the bottom two equations can be written in standard form as illustrated below.

$$y = 5x + 3 \qquad\qquad x - 3y + 4 = 0$$
$$-5x + y = 3 \qquad\qquad x - 3y = -4$$

2 Consider the linear equation in *one* variable, $2x + 3 = 5$. What is its solution?

$$2x + 3 = 5$$
$$2x = 2$$
$$x = 1$$

This equation has only one solution, 1.

Check: $2x + 3 = 5$
$$2(1) + 3 = 5$$
$$5 = 5 \qquad \text{true}$$

Now consider the linear equation in *two* variables, $y = x + 1$. What is the solution? Since the equation contains two variables, its solutions must contain two numbers, one for each variable. One set of numbers that satisfies this equation is $x = 1$ and $y = 2$. To see that this is true, we substitute both values into the equation and see that the equation checks.

$$y = x + 1$$
$$2 = 1 + 1$$
$$2 = 2 \qquad \text{true}$$

We write this answer as an ordered pair by writing the x and y values within parentheses separated by a comma. Remember the x value is always listed first since the form of an ordered pair is (x, y). Therefore, one possible solution to this equation is the ordered pair (1, 2). The equation $y = x + 1$ has other possible solutions, as illustrated below.

Solution	*Solution*	*Solution*
$x = 2, y = 3$	$x = -1, y = 0$	$x = -3, y = -2$
$y = x + 1$	$y = x + 1$	$y = x + 1$
$3 = 2 + 1$	$0 = -1 + 1$	$-2 = -3 + 1$
$3 = 3$ true	$0 = 0$ true	$-2 = -2$ true

Solution written as ordered pair

(2, 3)	(−1, 0)	(−3, −2)

How many possible solutions does the equation $y = x + 1$ have? The equation $y = x + 1$ has an unlimited or *infinite number* of possible solutions. Since it is not possible to list all the specific solutions to the equation, the solutions are illustrated with a graph.

EXAMPLE 1   Determine which of the following ordered pairs satisfy the equation $2x + 3y = 12$.

(a) $(2, 3)$ (b) $(3, 2)$ (c) $(8, -\frac{4}{3})$

Solution: To determine if the ordered pairs are solutions, we substitute them into the equation.

(a) $2x + 3y = 12$ (b) $2x + 3y = 12$ (c) $2x + 3y = 12$

$2(2) + 3(3) = 12$ $2(3) + 3(2) = 12$ $2(8) + 3(-\frac{4}{3}) = 12$

$4 + 9 = 12$ $6 + 6 = 12$ $16 - 4 = 12$

$13 \neq 12$ $12 = 12$ $12 = 12$

$(2, 3)$ is not a solution. $(3, 2)$ is a solution. $(8, -\frac{4}{3})$ is a solution.   ■

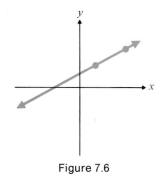

Figure 7.6

3   As mentioned earlier **every linear equation of the form $ax + by = c$ will be a straight line when graphed.** Consider the straight line shown in Fig. 7.6. Note that only two points are needed to draw a straight line.

Since all linear equations are straight lines, only two ordered pairs that satisfy the equation are needed to graph the equation. However, it is always a good idea to use a third ordered pair as a check point. If the three points are not in a straight line, you have made a mistake. A set of points that are in a straight line are said to be **collinear.**

EXAMPLE 2   Determine if the following three points are collinear.

(a) $(2, 7)$, $(0, 3)$, and $(-2, -1)$
(b) $(0, 5)$, $(\frac{5}{2}, 0)$, and $(5, -5)$
(c) $(-2, -5)$, $(0, 1)$, and $(5, 8)$

Solution: We plot the points to determine if they are collinear. The solution is shown in Fig. 7.7.

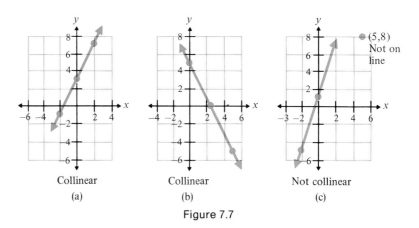

Collinear   Collinear   Not collinear
(a)   (b)   (c)

Figure 7.7   ■

4   One procedure that can be used to graph an equation is illustrated in Example 3.

EXAMPLE 3   Graph the equation $y = 3x + 6$.

Solution: First we determine that this is a linear equation. The graph must therefore be a straight line. Select three values for x, substitute them in the equation, and find the corresponding values for y. We will arbitrarily select the values 0, 2 and -3 for x.

x	$y = 3x + 6$	Ordered pair
0	$y = 3(0) + 6 = 6$	$(0, 6)$
2	$y = 3(2) + 6 = 12$	$(2, 12)$
-3	$y = 3(-3) + 6 = -3$	$(-3, -3)$

It is sometimes convenient to list the x and y values in tabular form.

x	y
0	6
2	12
-3	-3

Now plot the three ordered pairs on the same set of axes (Fig. 7.8).

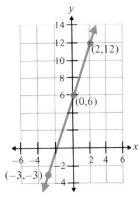

Figure 7.8

Since the three points are collinear, everything appears correct. Connect the three points with a straight line. Place arrows at the ends of the line to show that the line continues infinitely in both directions.   ∎

To graph the equation $y = 3x + 6$, we arbitrarily used the three values $x = 0$, $x = 2$, and $x = -3$. We could have selected three entirely different points and obtained exactly the same graph. When selecting points to substitute for x, use points that make the equation easy to evaluate.

The graph drawn in Example 3 represents the set of *all* ordered pairs that satisfy the equation $y = 3x + 6$. If we select any point on this line, the ordered pair represented by that point will be a solution to the equation $y = 3x + 6$. Similarly, any solution to the equation will be represented by a point on the line. Let us arbitrarily

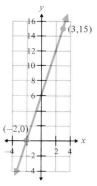

Figure 7.9

select some points on the graph and verify that they are solutions to the equation (see Fig. 7.9).

Points on line selected: (3, 15), (−2, 0)

Check (3, 15): $y = 3x + 6$ *Check* (−2, 0): $y = 3x + 6$

$15 = 3(3) + 6$ $0 = 3(-2) + 6$

$15 = 9 + 6$ $0 = -6 + 6$

$15 = 15$ true $0 = 0$ true

A graph of an equation is an illustration of the set of points whose coordinates satisfy the equation.

EXAMPLE 4   Graph the equation $2y = 4x - 12$.

Solution: By solving the equation for y it will be easier to determine ordered pairs that satisfy the equation. To solve the equation for y divide both sides of the equation by 2.

$$2y = 4x - 12$$

$$y = \frac{4x - 12}{2} = \frac{4x}{2} - \frac{12}{2} = 2x - 6$$

Now select values for x and solve for y in the equation $y = 2x - 6$.

$y = 2x - 6$

		x	y
Let $x = 0$,	$y = 2(0) - 6 = -6$	0	−6
Let $x = 2$,	$y = 2(2) - 6 = -2$	2	−2
Let $x = 3$,	$y = 2(3) - 6 = 0$	3	0

Plot the points and draw the straight line (Fig. 7.10).

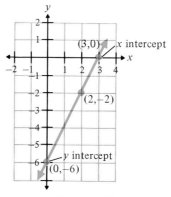

Figure 7.10

4   Let us examine two points on the graph in Fig. 7.10. Note that the graph crosses the x axis at the point (3, 0). Therefore, 3 is called the **x intercept.** The graph crosses the y axis at the point $(0, -6)$. Therefore, -6 is called the **y intercept.** It is often convenient to graph linear equations by finding their x and y intercepts.

---

x and y intercepts

To find the y intercept, set $x = 0$ and solve for y.
To find the x intercept, set $y = 0$ and solve for x.

EXAMPLE 5   Graph the equation $3y = 6x + 12$ by plotting the x and y intercepts.

Solution: To find the y intercept (where the graph crosses the y axis), set $x = 0$ and solve for y.

$$3y = 6x + 12$$
$$3y = 6(0) + 12$$
$$3y = 0 + 12$$
$$3y = 12$$
$$y = \frac{12}{3} = 4$$

The graph crosses the y axis at 4. The ordered pair representing the y intercept is $(0, 4)$.

Check: $3y = 6x + 12$
 $3(4) = 6(0) + 12$
 $12 = 12$ true

To find the x intercept (where the graph crosses the x axis), set $y = 0$ and solve for x.

$$3y = 6x + 12$$
$$3(0) = 6x + 12$$
$$0 = 6x + 12$$
$$-12 = 6x$$
$$\frac{-12}{6} = x$$
$$-2 = x$$

The graph crosses the x axis at -2. The ordered pair representing the x intercept is $(-2, 0)$.

Check: $3y = 6x + 12$
 $3(0) = 6(-2) + 12$
 $0 = 0$ true

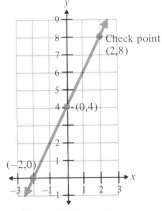

Figure 7.11

Now plot the intercepts (Fig. 7.11).

Before we graph the equation, we will arbitrarily select a value for x, find the corresponding value of y, and make sure that it is collinear with the x and y intercepts. This third point is our **check point.**

$$\text{Let } x = 2$$
$$3y = 6x + 12$$
$$3y = 6(2) + 12$$
$$3y = 12 + 12$$
$$3y = 24$$
$$y = \frac{24}{3} = 8$$

Plot the check point (2, 8). Since the 3 points are collinear draw the straight line through all three points. ■

EXAMPLE 6   Graph the equation $2x + 3y = 9$ by finding the x and y intercepts.

Solution:

Find y intercept	*Find x intercept*	*Check point*
Let $x = 0$	Let $y = 0$	Let $x = 2$
$2x + 3y = 9$	$2x + 3y = 9$	$2x + 3y = 9$
$2(0) + 3y = 9$	$2x + 3(0) = 9$	$2(2) + 3y = 9$
$0 + 3y = 9$	$2x + 0 = 9$	$4 + 3y = 9$
$3y = 9$	$2x = 9$	$3y = 5$
$y = 3$	$x = \dfrac{9}{2}$	$y = \dfrac{5}{3}$

Ordered pairs are (0, 3), $(\frac{9}{2}, 0)$, and $(2, \frac{5}{3})$.
The three points appear to be collinear. Draw the straight line through the three points (Fig. 7.12).

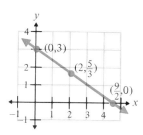

Figure 7.12 ■

EXAMPLE 7   Graph the equation $y = 20x + 60$.

Solution:

Find y intercept	*Find x intercept*	*Check point*
Let $x = 0$	Let $y = 0$	Let $x = 3$
$y = 20x + 60$	$y = 20x + 60$	$y = 20x + 60$
$y = 20(0) + 60$	$0 = 20x + 60$	$y = 20(3) + 60$
$y = 60$	$-60 = 20x$	$y = 60 + 60$
	$-3 = x$	$y = 120$

Since the values of y to be plotted are large, we will let each interval on the y axis be 10 units rather than 1 (Fig. 7.13). In addition, the length of the intervals on the y axis will be made smaller than those on the x axis. Occasionally, you will have to use different scales on the x and y axes, as above, to accommodate the graph.

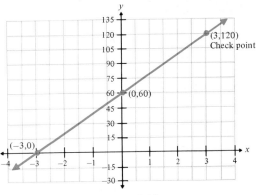

Figure 7.13

When selecting the scales to be used on your axes, you should realize that different scales will result in the same equation having a different appearance. Consider the graphs shown in Fig. 7.14. Both graphs represent the same equation, $y = x$. In Fig. 7.14a both the x and y axes have the same scale. In Fig. 7.14b the x and y axes do not have the same scale. Both graphs are correct in that each represents the graph of $y = x$. The difference in appearance is due solely to the difference in scales on the x axis. When possible, keep the scales on the x and y axis the same, as in Fig. 7.14a.

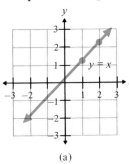   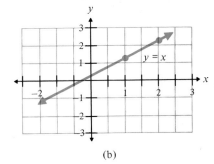

(a)   (b)

Figure 7.14

EXAMPLE 8   Graph the equation $3y + 2x = -4$.

(a) By solving the equation for y, selecting three arbitrary values of x, and finding the corresponding values of y.

(b) By using the x and y intercepts.

Solution: (a) $3y + 2x = -4$

$$3y = -2x - 4$$

$$y = \frac{-2x - 4}{3} = -\frac{2}{3}x - \frac{4}{3}$$

In selecting values for x, we will select those that are multiples of 3 so that the arithmetic will be easier. Let us select 0, 3, and -3.

$$y = -\frac{2}{3}x - \frac{4}{3}$$

Let $x = 0$, $y = -\frac{2}{3}(0) - \frac{4}{3} = 0 - \frac{4}{3} = -\frac{4}{3}.$

Let $x = 3$, $y = -\frac{2}{3}(3) - \frac{4}{3} = -2 - \frac{4}{3} = -\frac{6}{3} - \frac{4}{3} = -\frac{10}{3}.$

Let $x = -3$, $y = -\frac{2}{3}(-3) - \frac{4}{3} = 2 - \frac{4}{3} = \frac{6}{3} - \frac{4}{3} = \frac{2}{3}.$

x	y
0	$-\dfrac{4}{3}$
3	$-\dfrac{10}{3}$
-3	$\dfrac{2}{3}$

The graph is shown in Fig. 7.15. Note that the points do not always come out to be integral values.

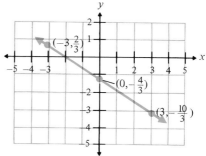

Figure 7.15

(b) $3y + 2x = -4$

Find y intercept	Find x intercept	Check point
Let $x = 0$	Let $y = 0$	Let $x = 3$
$3y + 2x = -4$	$3y + 2x = -4$	$y = -\dfrac{10}{3}$
$3y + 2(0) = -4$	$3(0) + 2x = -4$	from part (a)
$3y = -4$	$2x = -4$	$\left(3, -\dfrac{10}{3}\right)$
$y = -\dfrac{4}{3}$	$x = -2$	

The graph is shown in Fig. 7.16. Note that the graphs in Figs. 7.15 and 7.16 are the same.

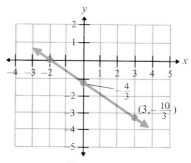

Figure 7.16

EXAMPLE 9   Graph the equation $y = 3$.

Solution: This equation can be written as $y = 3 + 0x$. Thus for any value of x selected, y will be 3. The graph of $y = 3$ is illustrated in Fig. 7.17.

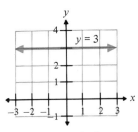

Figure 7.17   ■

The graph of any equation of the form $y = a$ will always be a horizontal line.

EXAMPLE 10   Graph the equation $x = -2$.

Solution: This equation can be written as $x = -2 + 0y$. Thus for any value of y selected, x will have a value of -2. The graph of $x = -2$ is illustrated in Fig. 7.18.

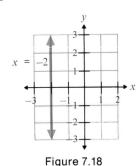

Figure 7.18   ■

The graph of any equation of the form $x = a$ will always be a vertical line.

Exercise Set 7.2

1. Determine which of the following ordered pairs satisfy the equation $y = 2x - 1$.
   (a) $(0, -1)$ **(b)** $(-1, 0)$ **(c)** $(5, 3)$
   (d) $(6, 11)$ **(e)** $(\frac{1}{2}, 0)$

2. Determine which of the following ordered pairs satisfy the equation $y = -3x + 4$.
   (a) $(0, 2)$ **(b)** $(-2, 10)$ **(c)** $(-1, 1)$
   (d) $(5, 11)$ **(e)** $(\frac{1}{2}, \frac{5}{2})$

3. Determine which of the following ordered pairs satisfy the equation $2x + y = -4$.
   (a) $(2, 0)$ **(b)** $(5, -14)$ **(c)** $(-2, 0)$
   (d) $(0, -4)$ **(e)** $(\frac{5}{8}, -\frac{3}{4})$

4. Determine which of the following ordered pairs satisfy the equation $3x - 2y = 6$.
   (a) $(5, 3)$ **(b)** $(7, 4)$ **(c)** $(-2, -3)$
   (d) $(-4, -9)$ **(e)** $(\frac{2}{3}, -2)$

5. Determine which of the following ordered pairs satisfy the equation $5x - 6 = 2y$.
   (a) $(9, 20)$ **(b)** $(-2, -8)$ **(c)** $(0, -3)$
   (d) $(\frac{6}{5}, 0)$ **(e)** $(-\frac{3}{8}, 6)$

6. Determine which of the following ordered pairs satisfy the equation $-3x + 8y = 12$.
   (a) $(4, 3)$ **(b)** $(-3, \frac{21}{8})$ **(c)** $(0, \frac{3}{2})$
   (d) $(\frac{1}{3}, \frac{35}{24})$ **(e)** $(-4, 0)$

Graph each equation by solving the equation for y, selecting three arbitrary values for x, and finding the corresponding values of y.

7. $y = 6$

8. $x = -2$

9. $x = 3$

10. $y = 5$

11. $y = 4x - 2$

12. $y = -x + 3$

13. $y = 6x + 2$

14. $y = x - 4$

15. $y = -\frac{1}{2}x + 3$

16. $2y = 2x + 4$

17. $6x - 2y = 4$

18. $4x - y = 5$

19. $5x - 2y = 8$

20. $-2x + 4y = 8$

21. $6x + 5y = 30$

22. $-2x - 3y = 6$

23. $-4x - y = -2$

24. $8y - 16x = 24$

25. $y = 20x + 40$

26. $2y - 50 = 100x$

27. $y = \frac{2}{3}x$

28. $y = -\frac{3}{5}x$

29. $y = \frac{1}{2}x + 4$

30. $y = -\frac{2}{5}x + 2$

31. $2y = 3x + 6$

32. $4x - 6y = 10$

Graph each equation using x and y intercepts.

33. $y = 2x + 4$

34. $y = -2x + 6$

35. $y = 4x - 3$

36. $y = -3x + 8$

37. $y = -6x + 5$

38. $y = 4x + 16$

39. $2y + 3x = 12$

40. $-2x + 3y = 10$

41. $4x = 3y - 9$

42. $7x + 14y = 21$

43. $\frac{1}{2}x + y = 4$

44. $30x + 25y = 50$

45. $6x - 12y = 24$

46. $25x + 50y = 100$

47. $8y = 6x - 12$

48. $-3y - 2x = -6$

49. $30y + 10x = 45$

50. $120x - 360y = 720$

51. $40x + 6y = 40$

52. $20x - 240 = -60y$

53. $\frac{1}{3}x + \frac{1}{4}y = 12$

54. $\frac{1}{5}x - \frac{2}{3}y = 60$

55. $\frac{1}{2}x = \frac{2}{5}y - 80$

56. $\frac{2}{3}y = \frac{5}{4}x + 120$

57. What does the graph of a linear equation illustrate?

58. Explain how to find the x and y intercepts of a line.

59. How many points are needed to graph a straight line? How many points should be used?

60. What will the graph of $y = a$ look like for any real number a?

61. What will the graph of $x = a$ look like for any real number a?

7.3

Slope-Intercept and Point-Slope Forms of Linear Equations

1   *Learn the meaning of slope.*

2   *Find the slope of a line.*

3   *Write an equation in slope-intercept form.*

4   *Examine the slopes of horizontal and vertical lines.*

5   *Determine if two lines are parallel.*

6   *Write an equation in point-slope form.*

1   The slope of a line is an important concept in many areas of mathematics. A knowledge of slope is helpful in understanding linear equations.

The **slope of a line** is a ratio of the vertical change to the horizontal change between any two selected points on the line. As an example, consider the two points (3, 6) and (1, 2). (See Fig. 7.19a.)

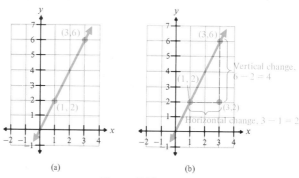

(a)   (b)

Figure 7.19

If we draw a line parallel to the x axis through the point (1, 2) and a line parallel to the y axis through the point (3, 6), the two lines intersect at (3, 2) (see Fig. 7.19b). From Fig. 7.19b we can determine the slope of the line. The vertical change (along the y axis) is $6 - 2$ or 4 units. The horizontal change (along the x axis) is $3 - 1$ or 2 units.

$$\text{slope} = \frac{\text{vertical change}}{\text{horizontal change}} = \frac{4}{2} = 2$$

Thus the slope of the line through these two points is 2. By examining the line connecting these two points, we can see that as the graph moves up two units on the y axis it moves to the right one unit on the x axis (see Fig. 7.20).

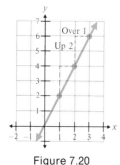

Figure 7.20

2 Let us now determine the procedure to find the slope of a line, symbolized by the letter m, between any two points (x_1, y_1) and (x_2, y_2). Consider Fig. 7.21.

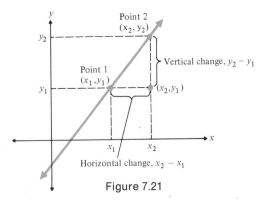

Figure 7.21

The vertical change can be found by subtracting y_1 from y_2. The horizontal change can be found by subtracting x_1 from x_2.

$$\text{slope } (m) = \frac{\text{change in } y \text{ (vertical change)}}{\text{change in } x \text{ (horizontal change)}} = \frac{y_2 - y_1}{x_2 - x_1}$$

It makes no difference which two points are selected when finding the slope of a line. It also makes no difference which point you label (x_1, y_1) or (x_2, y_2). The Greek capital letter delta, Δ, is often used to represent the words "the change in." Thus the slope is sometimes indicated as

$$m = \frac{\Delta y}{\Delta x} = \frac{y_2 - y_1}{x_2 - x_1}$$

In Section 7.2, Example 3, we graphed the equation $y = 3x + 6$. The graph is illustrated in Fig. 7.22.

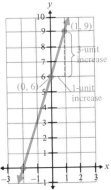

Figure 7.22

Let us find the slope of this line by using the two points (1, 9) and (0, 6).

Let (x_2, y_2) be (1, 9)   Let (x_1, y_1) be (0, 6)

$$m = \frac{y_2 - y_1}{x_2 - x_1} = \frac{9 - 6}{1 - 0} = \frac{3}{1} = 3$$

If we had designated the point $(1, 9)$ to be (x_1, y_1) and $(0, 6)$ to be (x_2, y_2), the slope would not have changed. This is illustrated as follows:

Let (x_2, y_2) be $(0, 6)$ Let (x_1, y_1) be $(1, 9)$

$$m = \frac{y_2 - y_1}{x_2 - x_1} = \frac{6 - 9}{0 - 1} = \frac{-3}{-1} = 3$$

Thus in either case we see that the slope is 3. Note that for each vertical increase of 3 units there is a horizontal increase of 1 unit.

A straight line where the value of y increases as x increases has a **positive slope** (see Fig. 7.23a). A line with a positive slope rises as it moves from left to right. A straight line where the value of y decreases as x increases has a **negative slope** (see Fig. 7.23b). A line with a negative slope falls as it moves from left to right.

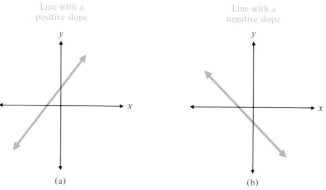

Line with a positive slope   Line with a negative slope

(a)   (b)

Figure 7.23

Consider the graph of the equation $y = -2x + 4$ (Fig. 7.24). Find the slope of the line.

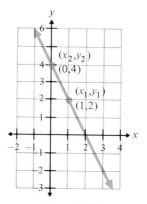

Figure 7.24

Let (x_2, y_2) be $(0, 4)$ Let (x_1, y_1) be $(1, 2)$

$$m = \frac{y_2 - y_1}{x_2 - x_1} = \frac{4 - 2}{0 - 1} = \frac{2}{-1} = -2$$

The slope of the line is -2. Note that the line falls as it goes from left to right.

3   We have shown that the graph of $y = 3x + 6$ (Fig. 7.23) has a slope of **3**, and it has a y intercept of **6**. The graph of $y = -2x + 4$ (Fig. 7.24) has a slope of **-2** and a y intercept of **4**. In general, an equation written in the form $y = mx + b$ will have a **slope of m** and a **y intercept of b.** This form is called the slope-intercept form.

Slope-Intercept Form of a Line
$$y = mx + b$$

where m is the slope, and b is the y intercept of the line.

Examples of equations in slope-intercept form

$$y = 3x - 6, \qquad y = \frac{1}{2}x + \frac{3}{2}, \qquad y = -5x + 3$$

slope $\searrow$ $\curvearrowright$ y intercept
$$y = mx + b$$

Equation	Slope	y intercept
$y = 3x - 6$	3	-6
$y = \frac{1}{2}x + \frac{3}{2}$	$\frac{1}{2}$	$\frac{3}{2}$
$y = -5x + 3$	-5	3
$y = -\frac{2}{3}x - \frac{3}{5}$	$-\frac{2}{3}$	$-\frac{3}{5}$

To write an equation in slope-intercept form, solve the equation for y.

EXAMPLE 1   Write the equation $-3x + 4y = 8$ in slope-intercept form. State the slope and y intercept. Use the slope and the y intercept to draw the graph of the equation.

Solution: To write this equation in slope-intercept form, we solve the equation for y.

$$-3x + 4y = 8$$
$$4y = 3x + 8$$
$$y = \frac{3x + 8}{4}$$
$$y = \frac{3}{4}x + \frac{8}{4}$$
$$y = \frac{3}{4}x + 2$$

The slope is $\frac{3}{4}$, and the y intercept is 2. Starting at the y intercept, 2, we can locate two other points on the graph, as shown in Fig. 7.25.

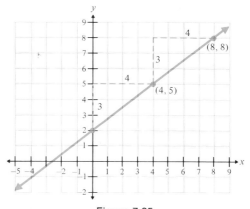

Figure 7.25

The slope is $\frac{3}{4}$. Therefore, the vertical change is 3 and the horizontal change is 4. Since the slope is positive, the line must rise as it moves from left to right!   ▪

In Example 1 if the slope was $-\frac{3}{4}$ we could start at the y intercept and move down 3 units and to the right 4 units to obtain the other points.

EXAMPLE 2   Determine the equation of the graph shown in Fig. 7.26.

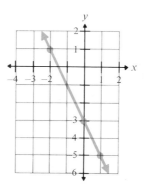

Figure 7.26

Solution: The slope can be determined by selecting any two points on the line. Let us use the point $(-2, 1)$ to represent (x_2, y_2) and the point $(0, -3)$ to represent (x_1, y_1).

$$m = \frac{\Delta y}{\Delta x} = \frac{y_2 - y_1}{x_2 - x_1} = \frac{1 - (-3)}{-2 - 0} = \frac{1 + 3}{-2} = \frac{4}{-2} = -2$$

The slope is -2. The graph shows the y intercept is -3. Recall the slope-intercept form of a line is $y = mx + b$, where m is the slope and b is the y intercept. Thus the equation of this line is $y = -2x - 3$.   ▪

4 Consider the graph of $y = 3$ (see Fig. 7.27). What is its slope?

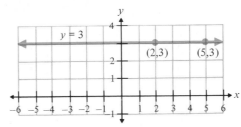

Figure 7.27

The graph is parallel to the x axis and goes through the points (2, 3) and (5, 3). Let the point (5, 3) represent (x_2, y_2) and let (2, 3) represent (x_1, y_1). Then the slope of the line is:

$$m = \frac{y_2 - y_1}{x_2 - x_1} = \frac{3 - 3}{5 - 2} = \frac{0}{3} = 0$$

Since there is no change in y, this graph has a slope of 0. **Any horizontal line has a slope of 0.**

Consider the graph of $x = 3$ (Fig. 7.28). What is its slope?

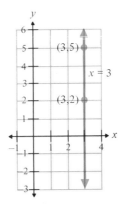

Figure 7.28

The graph is parallel to the y axis and goes through the points (3, 2) and (3, 5). Let the point (3, 5) represent (x_2, y_2) and let (3, 2) represent (x_1, y_1). Then the slope of the line is:

$$m = \frac{y_2 - y_1}{x_2 - x_1} = \frac{5 - 2}{3 - 3} = \frac{3}{0}$$

Since it is meaningless to divide by 0, we say that the slope of this line does not exist. **The slope of any vertical line does not exist.**

EXAMPLE 3 Determine if both of the equations have the same slope.

$$6x + 3y = 8$$
$$-4x - 2y = -3$$

Solution: Solve each of the equations for y to get the equations in slope-intercept form.

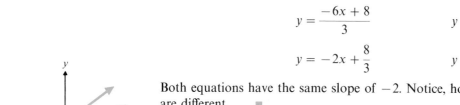

$$6x + 3y = 8 \qquad\qquad -4x - 2y = -3$$
$$3y = -6x + 8 \qquad\qquad -2y = 4x - 3$$
$$y = \frac{-6x + 8}{3} \qquad\qquad y = \frac{4x - 3}{-2}$$
$$y = -2x + \frac{8}{3} \qquad\qquad y = -2x + \frac{3}{2}$$

Both equations have the same slope of -2. Notice, however, that their y intercepts are different.   ■

5   Two lines are **parallel** when they do not intersect no matter how far they are extended. Figure 7.29 illustrates two parallel lines.

Linear equations with the same slope will be parallel (or identical) **lines when graphed.** The graphs of the equations in Example 3 will be parallel lines since they both have a slope of -2. Note that the two equations represent different lines since their y intercepts are different.

Figure 7.29

EXAMPLE 4   Graph the following equations on the same set of axes.

$$y = 2x + 4$$
$$y = 2x - 1$$

Solution: Since both equations have the same slope, 2, they are parallel lines. The equations are graphed in Fig. 7.30.

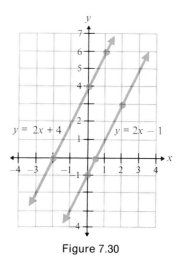

Figure 7.30   ■

6   When the slope of a line and a point on the line are known, we can use the point-slope form to determine the equation of the line. The point-slope form can be developed by beginning with the slope between any two points (x, y) and (x_1, y_1) on a line.

$$m = \frac{y - y_1}{x - x_1} \qquad \text{or} \qquad \frac{m}{1} = \frac{y - y_1}{x - x_1}$$

Now cross-multiply to obtain

$$m(x - x_1) = y - y_1 \qquad \text{or} \qquad y - y_1 = m(x - x_1)$$

Point-Slope Form of a Line

$$y - y_1 = m(x - x_1)$$

where m is the slope of the line and (x_1, y_1) is a point on the line.

EXAMPLE 5 Write an equation of the line that goes through the point (2, 3) and has a slope of 4.

Solution: The slope m is 4. The point on the line is (2, 3); call this point (x_1, y_1). Substitute 4 for m, 2 for x_1, and 3 for y_1, in the point-slope form of a line.

$$y - y_1 = m(x - x_1)$$
$$y - 3 = 4(x - 2)$$
$$y - 3 = 4x - 8$$
$$y = 4x - 5$$

The graph of $y = 4x - 5$ has a slope of 4 and passes through the point (2, 3). ∎

The answer to Example 5 was given in slope-intercept form. The answer could have also been given in standard form. Therefore two other acceptable answers are $-4x + y = -5$ and $4x - y = 5$. Your instructor may specify the form in which the answer is to be given.

EXAMPLE 6 Find an equation of the line through the points $(-1, -3)$ and (4, 2).

Solution: To use the point-slope form, we must first find the slope between the two points. To determine the slope, let us designate $(-1, -3)$ as (x_1, y_1) and (4, 2) as (x_2, y_2).

$$m = \frac{y_2 - y_1}{x_2 - x_1} = \frac{2 - (-3)}{4 - (-1)} = \frac{2 + 3}{4 + 1} = \frac{5}{5} = 1$$

We can use either point (one at a time) in determining the equation of the line. This example will be worked out using both points to show that the solutions obtained are identical.

Use point $(-1, -3)$ as (x_1, y_1):

$$y - y_1 = m(x - x_1)$$
$$y - (-3) = 1[x - (-1)]$$
$$y + 3 = x + 1$$
$$y = x - 2$$

Use point (4, 2) as (x_1, y_1):

$$y - y_1 = m(x - x_1)$$
$$y - 2 = 1(x - 4)$$
$$y - 2 = x - 4$$
$$y = x - 2$$

The solutions are identical. ∎

Exercise Set 7.3 _____

Determine the slope and y intercept of each equation. Graph the equation using the method illustrated in Example 1.

1. $y = 2x - 1$

2. $y = 3x + 2$

3. $y = -x + 5$

4. $y = 2x$

5. $y = -4x$

6. $2x + y = 5$

7. $-2x + y = -3$

8. $3x - y = -2$

9. $3x + 3y = 9$

10. $5x - 2y = 10$

11. $-x + 2y = 8$

12. $5x + 10y = 15$

13. $4x = 6y + 9$

14. $4y = 5x - 12$

15. $-6x = -2y + 8$

16. $6y = 5x - 9$

17. $-3x + 8y = -8$

18. $16y = 8x + 32$

19. $3x = 2y - 4$

20. $15x + 20y = 30$

21. $20x = 80y + 40$

Determine the equation of each line.

22.

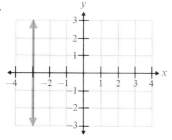

23.

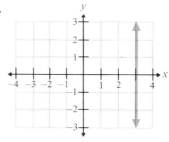

24.

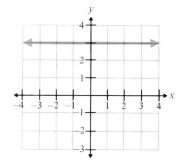

25.

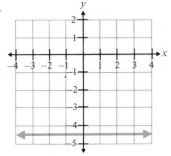

26.

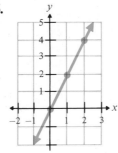

27.

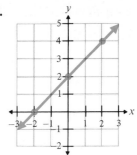

28.

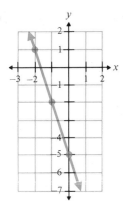

29.

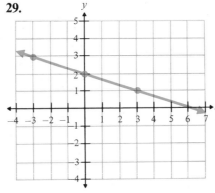

30.

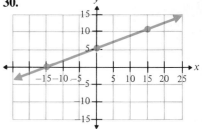

31.

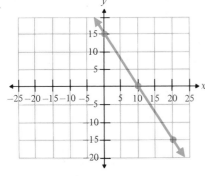

32.

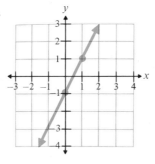

33.

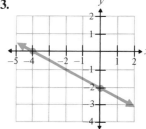

Determine if the two given lines are parallel.

34. $y = 2x - 4$
$y = 2x + 3$

35. $2x + 3y = 8$

$y = \dfrac{-2}{3}x + 5$

36. $4x + 2y = 9$
$8x = 4 - 4y$

37. $3x - 5y = 7$
$5y + 3x = 2$

38. $2x + 5y = 9$
$-x + 3y = 9$

39. $6x + 2y = 8$
$4x - 9 = -y$

40. $y = \dfrac{1}{2}x - 6$

$3y = 6x + 9$

41. $2y - 6 = -5x$

$y = -\dfrac{5}{2}x - 2$

Find an equation of a line with the properties given.

42. Slope $= 4$, through $(2, 3)$
43. Slope $= -2$, through $(-4, 5)$
44. Slope $= -1$, through $(6, 0)$
45. Slope $= \frac{1}{2}$, through $(-1, -5)$
46. Slope $= -\frac{2}{3}$, through $(-1, -2)$
47. Slope $= \frac{3}{5}$, through $(4, -2)$

54. What is the slope of a line?
55. Explain in your own words how to find the slope of a given line.
56. What does it mean when a line has a positive slope?
57. What does it mean when a line has a negative slope?
58. When you are given an equation in a form other than slope-intercept form, how can you change it to slope-intercept form?
59. What are three different forms of a linear equation?

48. Through $(4, 6)$ and $(-1, 1)$
49. Through $(-4, -2)$ and $(-2, 4)$
50. Through $(6, 3)$ and $(5, 2)$
51. Through $(-4, 6)$ and $(4, -6)$
52. Through $(1, 0)$ and $(-2, 4)$
53. Through $(10, 3)$ and $(0, -2)$

Write the equation $4 - 2y = 6x$ in each of the three forms.
60. Write the equation $4y = 6x - 8$ in standard form, slope-intercept form, and point-slope form.
61. Write the equation $5x = 3y + 8$ in standard form, slope-intercept form, and point-slope form.
62. Explain how you can determine if two lines are parallel without actually graphing them.

JUST FOR FUN

1. Write an equation of the line parallel to $3x - 4y = 6$ that passes through the point $(-4, -1)$.
2. Two lines are **perpendicular,** and cross at right angles, when their slopes are negative reciprocals of each other. The negative reciprocal of any number a is $-1/a$. Write an equation of the line perpendicular to $-5x + 2y = -4$ that passes through the point $(2, \frac{1}{2})$.
3. The slope of a hill and the slope of a line both measure steepness. However, there are several important differences.
 (a) Explain how you think the slope of a hill is determined.
 (b) Is the slope of a line, graphed in the Cartesian coordinate system, measured in any specific unit? Is the slope of a hill measured in any specific unit?

7.4

Graphing Linear Inequalities

❶ *Graph linear inequalities in two variables.*

❶   A linear inequality results when the equal sign in a linear equation is replaced with an inequality sign. Examples of linear inequalities in two variables are

$$3x + 2y > 4 \qquad -x + 3y < -2$$
$$-x + 4y \geq 3 \qquad 4x - y \leq 4$$

To Graph A Linear Inequality

1. Replace the inequality symbol with an equal sign.
2. Draw the graph of the equation in Step 1. If the original inequality contained a $\geq$ or $\leq$ symbol, draw the graph using a solid line. If the original inequality contained a $>$ or $<$ symbol, draw the graph using a dashed line.
3. Select any point not on the line and determine if this point is a solution to the original inequality. If the selected point is a solution, shade the region on the side of the line containing this point. If the selected point does not satisfy the inequality, shade the region on the side of the line not containing this point.

EXAMPLE 1   Graph the inequality $y < 2x - 4$.

Solution: Graph the equation $y = 2x - 4$ (Fig. 7.31). Since the original inequality contains a less than sign, $<$, use a dashed line when drawing the graph. The dashed line indicates that the points on this line are not solutions to the inequality $y < 2x - 4$.

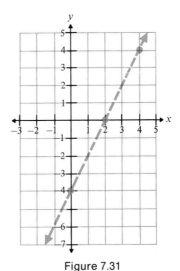

Figure 7.31

Next select a point not on the line and determine if this point satisfies the inequality. Often the easiest point to use is the origin, $(0, 0)$.

$$y < 2x - 4$$
$$0 < 2(0) - 4$$
$$0 < 0 - 4$$
$$0 < -4 \qquad \text{false}$$

Since 0 is not less than -4, the point $(0, 0)$ does not satisfy the inequality. The solution will therefore be all the points on the opposite side of the line from the point $(0, 0)$. Shade in this region (Fig. 7.32).

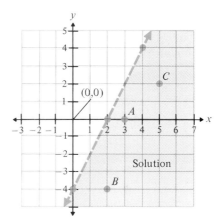

Figure 7.32

Every point in the shaded area satisfies the given inequality. Let us check a few selected points A, B, and C.

Point A	*Point B*	*Point C*
$(3, 0)$	$(2, -4)$	$(5, 2)$
$y < 2x - 4$	$y < 2x - 4$	$y < 2x - 4$
$0 < 2(3) - 4$	$-4 < 2(2) - 4$	$2 < 2(5) - 4$
$0 < 2$ true	$-4 < 0$ true	$2 < 6$ true ■

EXAMPLE 2   Graph the inequality $y \geq -\frac{1}{2}x$.

Solution: Graph the equation $y = -\frac{1}{2}x$. Since the inequality is $\geq$, we use a solid line to indicate that the points on the line are solutions to the inequality (Fig. 7.33).

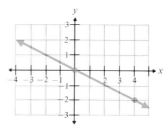

Figure 7.33

Since the point $(0, 0)$ is on the line, we cannot select that point to find the solution. Let us arbitrarily select the point $(3, 1)$.

$$y \geq -\frac{1}{2}x$$

$$1 \geq -\frac{1}{2}(3)$$

$$1 \geq -\frac{3}{2} \quad \text{true}$$

Since the point (3, 1) satisfies the inequality, every point on the same side of the line as (3, 1) will also satisfy the inequality $y \geq -\frac{1}{2}x$. Shade this region as indicated, Fig. 7.34. Every point in the shaded region as well as every point on the line satisfies the inequality.

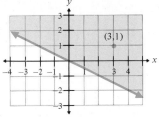

Figure 7.34

Exercise Set 7.4

Graph each inequality.

1. $x > 3$

2. $y < -2$

3. $x \geq \dfrac{5}{2}$

4. $y < x$

5. $y \geq 2x$

6. $y > -2x$

7. $y < 2x + 1$

8. $y \geq 3x - 1$

9. $y < -3x + 4$

10. $y \geq 2x + 4$

11. $y \geq \dfrac{1}{2}x - 4$

12. $y < 3x + 5$

13. $y \leq \dfrac{1}{3}x + 6$

14. $y > 6x + 1$

15. $y \leq -3x + 5$

16. $y \leq \dfrac{2}{3}x + 3$

17. $y > 5x - 9$

18. $y > \dfrac{2}{3}x - 1$

19. $y \leq -x + 4$

20. $y \geq 2x - 3$

21. $y > 3x - 2$

22. $y \leq -2x + 3$

23. $y \geq -4x + 3$

24. $y > -\dfrac{x}{2} + 4$

25. $y < -\dfrac{x}{3} - 2$

26. When graphing inequalities that contain either a $\leq$ or $\geq$, explain why the points on the line will be solutions to the inequality.

27. When graphing inequalities that contain either a $<$ or $>$, explain why the points on the line will not be solutions to the inequality.

Summary

Glossary

Cartesian (or rectangular) coordinate system: Two axes intersecting at right angles that are used when drawing graphs.

Graph: An illustration of the set of points that satisfy an equation or an inequality.

Linear equation in two variables: An equation of the form $ax + by = c$

Negative slope: A line has a negative slope when the values of y decrease as the values of x increase.

Ordered pair: The x and y coordinates of a point listed within parentheses, x coordinate first: (x, y).

Origin: The point of intersection of the x and y axes.

Parallel lines: Lines that never intersect.

Positive slope: A line has a positive slope when the values of y increase as the values of x increase.

Slope of a line: The ratio of the vertical change to the horizontal change between any two selected points on the line.

x axis: The horizontal axis in the Cartesian coordinate system.

x intercept: The value of x at the point where the graph crosses the x axis.

y axis: The vertical axis in the Cartesian coordinate system.

y intercept: The value of y at the point where the graph crosses the y axis.

Important Facts

To find the x intercept: Set $y = 0$ and solve the equation for x.

To find the y intercept: Set $x = 0$ and solve the equation for y.

Slope of line, m, through points (x_1, y_1) and (x_2, y_2)

$$m = \frac{y_2 - y_1}{x_2 - x_1}$$

Standard form of a line: $ax + by = c$.

Slope-intercept form of a line: $y = mx + b$.

Point-slope form of a line: $y - y_1 = m(x - x_1)$.

Review of slope

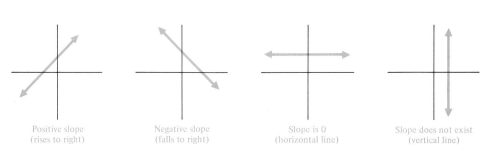

Positive slope
(rises to right)

Negative slope
(falls to right)

Slope is 0
(horizontal line)

Slope does not exist
(vertical line)

Review Exercises _____

[7.1]

1. Determine which of the following ordered pairs satisfy the equation $2x + 3y = 9$.
   (a) $(4, 3)$ **(b)** $(0, 3)$
   (c) $(-1, 4)$ **(d)** $(2, \frac{5}{3})$

2. Determine which of the following ordered pairs satisfy the equation $5x = -3y + 6$.
   (a) $(0, -2)$ **(b)** $(1, \frac{1}{3})$
   (c) $(\frac{6}{5}, 0)$ **(d)** $(15, -23)$

3. Draw the following ordered pairs on the same set of axes.
   (a) $A(5, 3)$ **(b)** $B(0, 6)$ **(c)** $C(5, \frac{1}{2})$
   (d) $D(-4, 3)$ **(e)** $E(-6, -1)$ **(f)** $F(-2, 0)$

4. Determine if the points given below are collinear.
   $$(0, -4), \quad (6, 8), \quad (-2, 0), \quad (4, 4)$$

[7.2] Graph each equation using the method of your choice.

5. $y = 6$

6. $x = -3$

7. $y = 3x$

8. $y = 2x - 1$

9. $y = -3x + 4$

10. $y = -\frac{1}{2}x + 4$

11. $2x + 3y = 6$

12. $3x - 2y = 12$

13. $2y = 3x - 6$

14. $4x - y = 8$

15. $-5x - 2y = 10$

16. $3x = 6y + 9$

17. $25x + 50y = 100$

18. $3x - 2y = 270$

19. $\frac{2}{3}x = \frac{1}{4}y + 20$

[7.3] Determine the slope and y intercept of each equation.

20. $y = -x + 4$

21. $y = 3x + 5$

22. $y = -4x + \frac{1}{2}$

23. $2x + 3y = 8$

24. $3x + 6y = 9$

25. $4y = 6x + 12$

26. $3x + 5y = 12$

27. $9x + 7y = 15$

28. $36x - 72y = 144$

29. $4x - 8 = 0$

30. $3y + 9 = 0$

Write the equation of each line

31.

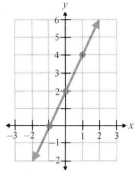

32.

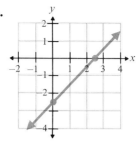

33.

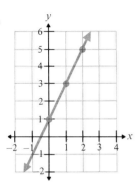

34.

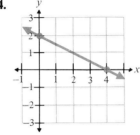

Determine if the two lines are parallel.

35. $y = 3x - 6$
     $6y = 18x + 6$

36. $2x - 3y = 9$
     $3x - 2y = 6$

37. $y = \dfrac{4}{9}x + 5$
     $4x = 9y + 2$

38. $4x = 6y + 3$
     $-2x = -3y + 10$

Find the equation of the line with the properties given.

39. Slope = 2, through (3, 4)
40. Slope = −3, through (−1, 5)
41. Slope = −$\frac{2}{3}$, through (3, 2)
42. Slope = 0, through (4, 2)

43. Slope is undefined, through (3, 5)
44. Through (4, 3) and (2, 1)
45. Through (−2, 3) and (0, −4)
46. Through (−4, −2) and (−4, 3)

[7.5] Graph each inequality.

47. $y \geq -3$

48. $x < 4$

49. $y < 3x$

50. $y > 2x + 1$

51. $y \leq 4x - 3$

52. $y \geq 6x + 5$

53. $y < -x + 4$

54. $y \leq \dfrac{1}{3}x - 2$

Practice Test

1. Determine which of the following ordered pairs satisfy the equation $3y = 5x - 9$.
   (a) $(3, 2)$ (b) $(\frac{9}{5}, 0)$
   (c) $(-2, -6)$ (d) $(0, 3)$
2. Find the slope and y intercept of $4x - 9y = 15$.
3. Write an equation of the graph in the accompanying figure.

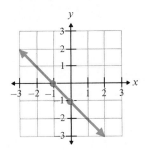

4. Write an equation of the line with a slope of 3 passing through the point $(1, 3)$.
5. Write an equation of the line passing through the points $(3, -1)$ and $(-4, 2)$.
6. Graph $x = -5$.
7. Graph $y = 3x - 2$.
8. Graph $3x + 5y = 15$.
9. Graph $4x = -y + 10$.
10. Graph $3x - 2y = 8$.
11. Graph $y \geq -3x + 5$.
12. Graph $y < 4x - 2$.

8

Systems of Linear Equations

See Section 8.5, Exercise 7

Introduction

1 When we must find the common solution to two or more linear equations, we refer to the equations in this type of problem as **simultaneous linear equations** or as a **system of linear equations.**

$$\left.\begin{array}{ll}(1) & y = x + 5 \\ (2) & y = 2x + 4\end{array}\right\} \text{ system of linear equations}$$

2 The **solution to a system of equations** is the ordered pair or pairs that satisfy all equations. The solution to the system above is (1, 6).

Check: *In equation (1)*   *In equation (2)*

$$\begin{array}{cc} (1, 6) & (1, 6) \\ y = x + 5 & y = 2x + 4 \\ 6 = 1 + 5 & 6 = 2(1) + 4 \\ 6 = 6 \;\; \text{true} & 6 = 6 \;\; \text{true} \end{array}$$

The ordered pair (1, 6) satisfies *both* equations and is a solution to the system of equations. Notice that the ordered pair (2, 7) satisfies the first equation but does not satisfy the second equation.

Check: *In equation (1)*   *In equation (2)*

$$\begin{array}{cc} (2, 7) & (2, 7) \\ y = x + 5 & y = 2x + 4 \\ 7 = 2 + 5 & 7 = 2(2) + 4 \\ 7 = 7 \;\; \text{true} & 7 = 8 \;\; \text{false} \end{array}$$

Since the ordered pair (2, 7) does not satisfy both equations, it is not a solution to the system of equations.

EXAMPLE 1   Determine which of the following ordered pairs satisfy the system of equations.

$$y = 2x - 8$$
$$2x + y = 4$$

(a) (2, −4)   (b) (4, −4)   (c) (3, −2)

Solution: (a) Substitute 2 for x and −4 for y in each of the equations.

$$\begin{array}{cc} y = 2x - 8 & 2x + y = 4 \\ -4 = 2(2) - 8 & 2(2) + (-4) = 4 \\ -4 = 4 - 8 & 4 - 4 = 4 \\ -4 = -4 \;\; \text{true} & 0 = 4 \;\; \text{false} \end{array}$$

Since $(2, -4)$ does not satisfy both equations, it is not a solution to the system of equations.

(b)
$$y = 2x - 8 \qquad\qquad 2x + y = 4$$
$$-4 = 2(4) - 8 \qquad 2(4) + (-4) = 4$$
$$-4 = 8 - 8 \qquad\qquad 8 - 4 = 4$$
$$-4 = 0 \quad \text{false} \qquad\qquad 4 = 4 \quad \text{true}$$

Since $(4, -4)$ does not satisfy both equations, it is not a solution to the system of equations.

(c)
$$y = 2x - 8 \qquad\qquad 2x + y = 4$$
$$-2 = 2(3) - 8 \qquad 2(3) + (-2) = 4$$
$$-2 = 6 - 8 \qquad\qquad 6 - 2 = 4$$
$$-2 = -2 \quad \text{true} \qquad\qquad 4 = 4 \quad \text{true}$$

Since $(3, -2)$ satisfies both equations, it is a solution to the system of linear equations. ■

In this chapter we discuss three methods for finding the solution to a system of equations: the graphic method, the substitution method, and the addition method.

Exercise Set 8.1

Determine which, if any, of the following ordered pairs satisfy each system of linear equations.

1. $y = 3x - 4$
$y = -x + 4$
(a) $(-2, 2)$ **(b)** $(-4, -8)$ **(c)** $(2, 2)$

2. $y = -4x$
$y = -2x + 8$
(a) $(0, 0)$ **(b)** $(-4, 16)$ **(c)** $(2, -8)$

3. $y = 2x - 3$
$y = x + 5$
(a) $(8, 13)$ **(b)** $(4, 5)$ **(c)** $(4, 9)$

4. $x + 2y = 4$
$y = 3x + 3$
(a) $(0, 2)$ **(b)** $(-2, 3)$ **(c)** $(4, 15)$

5. $3x - y = 6$
$2x + y = 9$
(a) $(3, 3)$ **(b)** $(4, -2)$ **(c)** $(-6, 3)$

6. $y = 2x + 4$
$y = 2x - 1$
(a) $(0, 4)$ **(b)** $(3, 10)$ **(c)** $(-2, 0)$

7. $2x - 3y = 6$
$y = \dfrac{2}{3}x - 2$
(a) $(3, 0)$ **(b)** $(3, -2)$ **(c)** $\left(1, -\dfrac{4}{3}\right)$

8. $y = -x + 4$
$2y = -2x + 8$
(a) $(2, 5)$ **(b)** $(0, 4)$ **(c)** $(5, -1)$

9. $3x - 4y = 8$
$2y = \dfrac{3}{2}x - 4$
(a) $(0, -2)$ **(b)** $(1, -6)$ **(c)** $\left(-\dfrac{1}{3}, -\dfrac{9}{4}\right)$

10. $2x + 3y = 6$
$-2x + 5 = y$
(a) $\left(\dfrac{1}{2}, \dfrac{5}{3}\right)$ **(b)** $(2, 1)$ **(c)** $\left(\dfrac{9}{4}, \dfrac{1}{2}\right)$

11. $y = 2x - 3$
$2x - 3y = 4$
(a) $\left(\dfrac{1}{2}, -2\right)$ **(b)** $\left(\dfrac{5}{4}, -\dfrac{1}{2}\right)$ **(c)** $\left(\dfrac{1}{5}, -\dfrac{10}{3}\right)$

8.2

Solving Systems of Equations Graphically

1 *Recognize the three possible types of systems of equations.*

2 *Solve a system of equations graphically.*

1 The **solution to a system of linear equations** is the ordered pair (or pairs) common to all lines when the lines are graphed. When two lines are graphed, three situations are possible, as illustrated in Fig. 8.1.

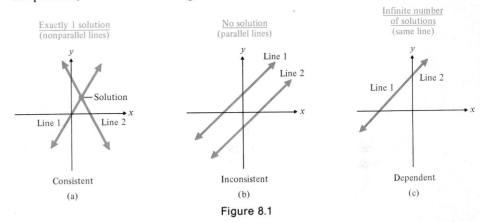

Figure 8.1

In Fig. 8.1a, line 1 and line 2 are nonparallel lines and intersect at exactly one point. This system of equations has *exactly one solution*. This is an example of a **consistent system of equations.** A consistent system of equations is a system of equations that has a solution.

Line 1 and line 2 of Fig. 8.1b are different but parallel lines. The lines do not intersect, and this system of equations has *no solution*. This is an example of an **inconsistent system of equations.** An inconsistent system of equations is a system of equations that has no solution.

In Fig. 8.1c, line 1 and line 2 are actually the same line. In this case every point on the line satisfies both equations and is a solution to the system of equations. This system has *an infinite number of solutions*. This is an example of a **dependent system of equations.** A dependent system of linear equations is a system of equations where both equations represent the same line. *Note that a dependent system is also a consistent system since it has a solution.*

We can determine if a system of linear equations is consistent, inconsistent, or dependent by writing each equation in slope-intercept form and comparing the slopes and *y* intercepts. Note that if the slopes of the lines are different, Fig. 8.1a, the system is consistent. If the slopes are the same but the *y* intercepts are different, Fig. 8.1b, the system is inconsistent, and if both the slopes and the *y* intercepts are the same, Fig. 8.1c, the system is dependent.

EXAMPLE 1   Determine whether the system is consistent, inconsistent, or dependent.

$$3x + 4y = 8$$
$$6x + 8y = 4$$

Solution: Write each equation in slope-intercept form and compare the slopes and y intercepts.

$$3x + 4y = 8 \qquad\qquad 6x + 8y = 4$$
$$4y = -3x + 8 \qquad\qquad 8y = -6x + 4$$
$$y = \frac{-3x + 8}{4} \qquad\qquad y = \frac{-6x + 4}{8}$$
$$y = \frac{-3}{4}x + 2 \qquad\qquad y = \frac{-6}{8}x + \frac{4}{8}$$
$$\qquad\qquad\qquad\qquad y = \frac{-3}{4}x + \frac{1}{2}$$

Since the equations have the same slope, $-\frac{3}{4}$, and different y intercepts the lines are parallel. This system of equations is therefore inconsistent and has no solution. ■

2

> **To obtain the solution to a system of equations graphically,** graph each equation and determine the point or points of intersection.

EXAMPLE 2   Solve the following system of equations graphically.

$$2x + y = 11$$
$$x + 3y = 18$$

Solution: Find the x and y intercepts of each graph, then draw the graphs.

$2x + y = 11$	*Ordered pair*	$x + 3y = 18$	*Ordered pair*
Let $x = 0$, then $y = 11$	$(0, 11)$	Let $x = 0$, then $y = 6$	$(0, 6)$
Let $y = 0$, then $x = \frac{11}{2}$	$\left(\frac{11}{2}, 0\right)$	Let $y = 0$, then $x = 18$	$(18, 0)$

The two graphs (Fig. 8.2) intersect at the point $(3, 5)$. The point $(3, 5)$ is the solution to the system of equations.

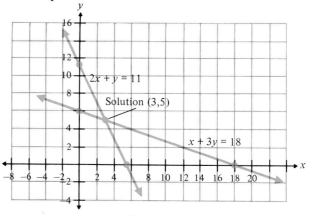

Figure 8.2

Check: $2x + y = 11$ $x + 3y = 18$

   $2(3) + 5 = 11$ $3 + 3(5) = 18$

      $11 = 11$ true $18 = 18$ true   ∎

EXAMPLE 3   Solve the following system of equations graphically.

$$2x + y = 3$$
$$4x + 2y = 12$$

Solution: $2x + y = 3$ *Ordered pair* $4x + 2y = 12$ *Ordered pair*

Let $x = 0$, then $y = 3$ (0, 3)   Let $x = 0$, then $y = 6$ (0, 6)

Let $y = 0$, then $x = \dfrac{3}{2}$ $\left(\dfrac{3}{2}, 0\right)$ Let $y = 0$, then $x = 3$ (3, 0)

The two lines (Fig. 8.3) appear to be parallel.

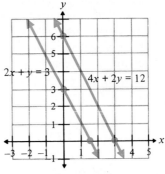

Figure 8.3

To show that the two lines are indeed parallel, write each equation in slope-intercept form.

$$2x + y = 3 \qquad 4x + 2y = 12$$
$$y = -2x + 3 \qquad 2y = -4x + 12$$
$$y = -2x + 6$$

Both equations have the same slope, -2, and different y intercepts; thus the lines must be parallel. Since parallel lines do not intersect, this system of equations has no solution.   ∎

EXAMPLE 4   Solve the following system of equations graphically.

$$x - \frac{1}{2}y = 2$$
$$y = 2x - 4$$

Solution: $x - \dfrac{1}{2}y = 2$ *Ordered pair* $y = 2x - 4$ *Ordered pair*

Let $x = 0$, then $y = -4$ (0, −4)   Let $x = 0$, then $y = -4$ (0, −4)

Let $y = 0$, then $x = 2$ (2, 0)   Let $y = 0$, then $x = 2$ (2, 0)

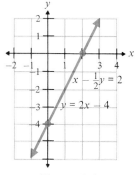

Figure 8.4

Both equations represent the same line (Fig. 8.4). When the equations are changed to slope-intercept form, it becomes clear that the equations are identical, and the system is dependent.

$$x - \frac{1}{2}y = 2 \qquad\qquad y = 2x - 4$$

$$2\left(x - \frac{1}{2}y\right) = 2(2)$$

$$2x - y = 4$$

$$-y = -2x + 4$$

$$y = 2x - 4$$

The solution to this system of equations is all the points on the line.   ■

When graphing a system of equations, the intersection of the lines is not always easy to read on the graph. For example, the true solution to a system of equations may be $(\frac{5}{9}, -\frac{4}{11})$. In cases like this it is not easy to find the exact value of the solution by observation, but you should be able to give an approximate answer. An approximate answer to this system might be $(\frac{1}{2}, -\frac{1}{3})$ or $(0.6, -0.3)$. The accuracy of your answer will depend on how carefully you draw the graphs and on the scale of the graph paper used.

Exercise Set 8.2

Identify each system of linear equations (labeled 1 and 2) as consistent, inconsistent, or dependent. State whether the system has exactly one solution, no solution, or an infinite number of solutions.

1.

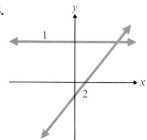

2.

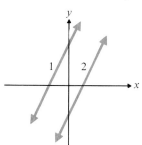

3.

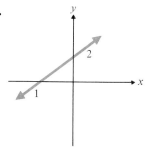

4.

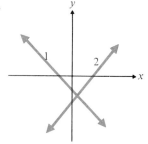

5.

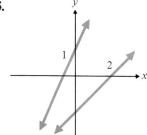

6.

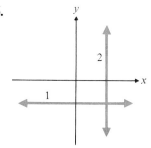

7.

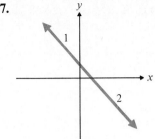

8.

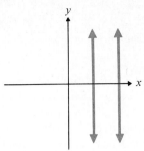

Express each equation in slope-intercept form. Without graphing the equations, state whether the system of equations has exactly one solution, no solution, or an infinite number of solutions.

9. $y = 3x - 2$
$2y = 4x - 6$

10. $x + y = 6$
$x - y = 6$

11. $3y = 2x + 3$
$y = \frac{2}{3}x - 2$

12. $y = \frac{1}{2}x + 4$
$2y = x + 8$

13. $4x = y - 6$
$3x = 4y + 5$

14. $x + 2y = 6$
$2x + y = 4$

15. $2x = 3y + 4$
$6x - 9y = 12$

16. $x - y = 2$
$2x - 2y = -2$

17. $y = \frac{3}{2}x + \frac{1}{2}$
$3x - 2y = -\frac{1}{2}$

18. $x - y = 3$
$\frac{1}{2}x - 2y = -6$

Determine the solution to each system of equations graphically.

19. $y = x + 2$
$y = -x + 2$

20. $y = 2x + 4$
$y = -3x - 6$

21. $y = 3x - 6$
$y = -x + 6$

22. $y = 2x - 1$
$2y = 4x + 6$

23. $2x = 4$
$y = -3$

24. $2x + 3y = 6$
$4x = -6y + 12$

25. $y = x + 2$
$x + y = 4$

26. $2x + y = 6$
$2x - y = -2$

27. $y = -\frac{1}{2}x + 4$
$x + 2y = 6$

28. $x + 2y = -4$
$2x - y = -3$

29. $x + 2y = 8$
$2x - 3y = 2$

30. $4x - y = 5$
$2y = 8x - 10$

31. $x + y = 5$
$2y = x - 2$

32. $2x + 3y = 6$
$2x + y = -2$

33. $y = 3$
$y = 2x - 3$

34. $x = 3$
$y = 2x - 2$

35. $x - 2y = 4$
$2x - 4y = 8$

36. $3x + y = -6$
$2x = 1 + y$

37. $2x + y = -2$
$6x + 3y = 6$

38. $y = 2x - 3$
$y = -x$

39. $4x - 3y = 6$
$2x + 4y = 14$

40. $2x + 6y = 6$
$y = -\frac{1}{3}x + 1$

41. $2x - 3y = 0$
$x + 2y = 0$

42. $2x = 4y - 12$
$-4x + 8y = 8$

43. $6x + 8y = 36$
$-3x - 4y = -9$

44. $x + 4y = -8$
$3x + 2y = 6$

8.3

Solving Systems of Equations by Substitution

❶ *Solve a system of equations by substitution.*

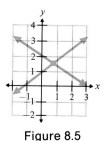

Figure 8.5

Often a graphic solution to a system of equations may be inaccurate since you must estimate the answer by observation. For example, can you determine the answer to the system of equations shown in Fig. 8.5? You may estimate the solution to be $(\frac{7}{10}, \frac{3}{2})$ when it may actually be $(\frac{4}{5}, \frac{8}{5})$. When an exact answer is necessary, the system should be solved algebraically, either by substitution or by addition of equations.

❶ The procedure used to solve a system of equations by substitution is illustrated in Example 1. The procedure used to solve by addition is presented in the next section. Regardless of which of the two algebraic techniques are used to solve a system of equations, our immediate goal remains the same, that is, to obtain one equation containing only one unknown.

EXAMPLE 1 Solve the following system of equations by substitution.

$$2x + y = 11$$
$$x + 3y = 18$$

Solution: Begin by solving for one of the variables in either of the equations. You may solve for any of the variables; however, if you solve for a variable with a numerical coefficient of 1, you may avoid working with fractions. In this system the y term in $2x + y = 11$ and the x term in $x + 3y = 18$ both have a numerical coefficient of 1.

Let us elect to solve for y in $2x + y = 11$

$$2x + y = 11$$
$$y = -2x + 11$$

Next, substitute $-2x + 11$ for y in the *other equation*, $x + 3y = 18$, and solve for the remaining variable, x.

$$x + 3y = 18$$
$$x + 3(\overbrace{-2x + 11}) = 18$$
$$x - 6x + 33 = 18$$
$$-5x + 33 = 18$$
$$-5x = -15$$
$$x = 3$$

Finally, substitute $x = 3$ in the equation solved for y and find the value of y.

$$y = -2x + 11$$
$$y = -2(3) + 11$$
$$y = -6 + 11$$
$$y = 5$$

The solution is the ordered pair (3, 5). ■

Note that this solution is identical to the graphic solution obtained in Example 2 of Section 8.2.

> **To Solve A System of Equations by Substitution**
>
> **1.** Solve for a variable in either equation. (If possible solve for a variable with a numerical coefficient of 1 to avoid working with fractions.)
> **2.** Substitute the expression found for the variable in Step 1 into the other equation.
> **3.** Solve the equation determined in Step 2 to find the value of one variable.
> **4.** Substitute the value found in Step 3 into the equation obtained in Step 1 to find the remaining variable.

EXAMPLE 2   Solve the following system of equations by substitution.

$$2x + y = 3$$
$$4x + 2y = 12$$

Solution: Solve for y in $2x + y = 3$.

$$2x + y = 3$$
$$y = -2x + 3$$

Now substitute the expression $-2x + 3$ for y in the *other equation*, $4x + 2y = 12$, and solve for x.

$$4x + 2y = 12$$
$$4x + 2(-2x + 3) = 12$$
$$4x - 4x + 6 = 12$$
$$6 = 12 \quad \text{false}$$

Since the statement $6 = 12$ is false, the system has no solution. (Therefore, the lines will be parallel and the system is inconsistent.) ∎

Note that the solution in Example 2 is identical to the graphic solution obtained in Example 3 of Section 8.2.

EXAMPLE 3   Solve the following system of equations by substitution.

$$x - \frac{1}{2}y = 2$$
$$y = 2x - 4$$

Solution: The equation $y = 2x - 4$ is already solved for y. Substitute $2x - 4$ for y in the other equation, $x - \frac{1}{2}y = 2$, and solve for the remaining variable, x.

$$x - \frac{1}{2}y = 2$$
$$x - \frac{1}{2}(2x - 4) = 2$$
$$x - x + 2 = 2$$
$$2 = 2 \quad \text{true}$$

Since the statement $2 = 2$ is true, this system has an infinite number of solutions. (Therefore, the lines will be the same when graphed and the system is dependent.) ∎

Note that the solution in Example 3 is identical to the solution obtained graphically in Example 4 of Section 8.2.

EXAMPLE 4   Solve the following system of equations by substitution.

$$2x + 4y = 6$$
$$4x - 2y = -8$$

Solution: None of the variables in either equation has a numerical coefficient of 1. However, since the numbers 4 and 6 are both divisible by 2, if we solve the first equation for x, we will avoid having to work with fractions.

$$2x + 4y = 6$$
$$2x = -4y + 6$$
$$\frac{2x}{2} = \frac{-4y + 6}{2}$$
$$x = -2y + 3$$

Now substitute $-2y + 3$ for x in the other equation, $4x - 2y = -8$, and solve for the remaining variable, y.

$$4x - 2y = -8$$
$$4(-2y + 3) - 2y = -8$$
$$-8y + 12 - 2y = -8$$
$$-10y + 12 = -8$$
$$-10y = -20$$
$$y = 2$$

Finally, solve for x by substituting $y = 2$ in the equation previously solved for x, $x = -2y + 3$.

$$x = -2y + 3$$
$$x = -2(2) + 3$$
$$x = -4 + 3$$
$$x = -1$$

The solution is $(-1, 2)$.   ∎

COMMON STUDENT ERROR

A student will often successfully solve for one of the variables and forget to solve for the other. Remember that a solution must contain *both* an x and a y value.

EXAMPLE 5   Solve the following system of equation by substitution.

$$4x + 4y = 3$$
$$2x = 2y + 5$$

Solution: We will elect to solve for x in the second equation.

$$2x = 2y + 5$$

$$x = \frac{2y + 5}{2}$$

$$x = y + \frac{5}{2}$$

Now substitute $y + \frac{5}{2}$ for x in the other equation.

$$4x + 4y = 3$$

$$4\left(y + \frac{5}{2}\right) + 4y = 3$$

$$4y + 10 + 4y = 3$$

$$8y + 10 = 3$$

$$8y = -7$$

$$y = -\frac{7}{8}$$

Finally, find x.

$$x = y + \frac{5}{2}$$

$$x = -\frac{7}{8} + \frac{5}{2}$$

$$x = -\frac{7}{8} + \frac{20}{8}$$

$$x = \frac{13}{8}$$

The solution is $\left(\frac{13}{8}, -\frac{7}{8}\right)$.   ■

Exercise Set 8.3

Find the solution to each system of equations using substitution.

1. $x + 2y = 4$
   $2x - 3y = 1$
4. $2x + y = 3$
   $2y = 6 - 4x$

7. $x = 4$
   $x + y + 5 = 0$

10. $2x + 3y = 7$
   $6x - y = 1$

2. $y = x + 3$
   $y = -x - 5$
5. $2x + y = 3$
   $2x + y + 5 = 0$

8. $y = \frac{1}{3}x - 2$
   $x - 3y = 6$
11. $3x + y = -1$
   $y = 3x + 5$

3. $x + y = -2$
   $x - y = 0$
6. $y = 2x + 4$
   $y = -2$

9. $x - \frac{1}{2}y = 2$
   $y = 2x - 4$
12. $y = -2x + 5$
   $x + 3y = 0$

13. $y = 2x - 13$
$-4x - 7 = 9y$

14. $x = y + 4$
$3x + 7y = -18$

15. $2x + 3y = 7$
$6x - 2y = 10$

16. $4x - 3y = 6$
$2x + 4y = 5$

17. $3x - y = 14$
$6x - 2y = 10$

18. $5x - 2y = -7$
$5 = y - 3x$

19. $2x - 7y = 6$
$5x - 8y = -4$

20. $4x - 5y = -4$
$3x = 2y - 3$

21. $3x + 4y = 10$
$4x + 5y = 14$

22. $5x + 4y = -7$

$x - \dfrac{5}{3}y = -2$

23. When solving the system of equations

$$3x + 6y = 9$$
$$4x + 3y = 5$$

by substitution, which variable, in which equation, would you choose to solve for to make the solution easier? Explain your answer.

24. When solving a system of linear equations by substitution, how will you know if the system is inconsistent?

25. When solving a system of linear equations by substitution, how will you know if the system is dependent?

8.4

Solving Systems of Equations by the Addition Method

☐ Solve a system of equations by addition.

☐ A third, and often the easiest, method of solving a system of equations is by the addition (or elimination) method. The object of this process is to obtain two equations whose sum will be an equation containing only one variable. Always keep in mind that our immediate goal is to obtain one equation containing only one unknown.

EXAMPLE 1   Solve the following system of equations using the addition method.

$$x + y = 6$$
$$2x - y = 3$$

Solution: Note that one equation contains $+y$ and the other contains $-y$. By adding the equations, we can eliminate the variable y and obtain one equation containing only one unknown.

$$\begin{array}{r} x + y = 6 \\ 2x - y = 3 \\ \hline 3x = 9 \end{array}$$

Now solve for the remaining variable, x.

$$\frac{3x}{3} = \frac{9}{3}$$

$$x = 3$$

Finally, solve for y by inserting $x = 3$ in either of the original equations.

$$x + y = 6$$
$$3 + y = 6$$
$$y = 3$$

The solution is (3, 3).

Check answer in *both* equations.

$$x + y = 6 \qquad\qquad 2x - y = 3$$
$$3 + 3 = 6 \qquad\qquad 2(3) - 3 = 3$$
$$6 = 6 \quad \text{true} \qquad\quad 6 - 3 = 3$$
$$3 = 3 \quad \text{true} \quad \blacksquare$$

EXAMPLE 2   Solve the following system of equations using the addition method.

$$-x + 3y = 8$$
$$x + 2y = -13$$

Solution: By adding the equations we can eliminate the variable x.

$$-x + 3y = 8$$
$$\underline{x + 2y = -13}$$
$$5y = -5$$
$$\frac{5y}{5} = \frac{-5}{5}$$
$$y = -1$$

Now solve for x by substituting $y = -1$ in either of the original equations.

$$x + 2y = -13$$
$$x + 2(-1) = -13$$
$$x - 2 = -13$$
$$x = -11$$

The solution is $(-11, -1)$. $\blacksquare$

To Solve A System of Equations by the Addition (or Elimination) Method

1. If necessary, rewrite each equation so that the terms containing variables appear on the left side of the equal sign and any constants appear on the right side of the equal sign.
2. If necessary, multiply one or both equations by a constant(s) so that when the equations are added the resulting sum will contain only one variable.
3. Add the equations. This will result in a single equation containing only one variable.
4. Solve for the variable in the equation in Step 3.
5. Substitute the value found in Step 4 into either of the original equations. Solve that equation to find the value of the remaining variable.

EXAMPLE 3   Solve the following system of equations using the addition method.

$$2x + y = 6$$
$$3x + y = 5$$

Solution: The object of the addition process is to obtain two equations whose sum will be an equation containing only one variable. If we add these two equations, none of the variables will be eliminated. However, if we multiply either equation by -1, and then add, we will accomplish our goal. In this text we will use brackets, [ ], to indicate multiplication of *an entire equation* by a real number. Thus 2[ ] means multiply the entire equation within the brackets by 2, and -3[ ] means multiply the entire equation within the brackets by -3.

$$-1[2x + y = 6] \qquad \text{gives} \qquad -2x - y = -6$$
$$3x + y = 5 \qquad\qquad\qquad 3x + y = 5$$

Remember that both sides of the equation must be multiplied by the -1. This process has the effect of changing the sign of each term in the equation being multiplied without changing the solution to the system of equations. Now add the last two equations.

$$
\begin{array}{r}
-2x - y = -6 \\
3x + y = 5 \\
\hline
x = -1
\end{array}
$$

Solve for y in either of the original equations.

$$2x + y = 6$$
$$2(-1) + y = 6$$
$$-2 + y = 6$$
$$y = 8$$

The solution is $(-1, 8)$.   ■

EXAMPLE 4   Solve the following system of equations using the addition method.

$$2x + y = 11$$
$$x + 3y = 18$$

Solution: To eliminate the variable x, we multiply the second equation by -2 and add the two equations.

$$2x + y = 11 \qquad \text{gives} \qquad 2x + y = 11$$
$$-2[x + 3y = 18] \qquad\qquad -2x - 6y = -36$$

Now add:

$$
\begin{array}{r}
2x + y = 11 \\
-2x - 6y = -36 \\
\hline
-5y = -25 \\
y = 5
\end{array}
$$

Solve for x.

$$2x + y = 11$$
$$2x + 5 = 11$$
$$2x = 6$$
$$x = 3$$

The solution $(3, 5)$ is identical to the one obtained graphically in Example 2 of Section 8.2 and by substitution in Example 1 of Section 8.3.   ■

In Example 4, we could have multiplied the first equation by -3 to eliminate the variable y.

$$-3[2x + y = 11] \quad \text{gives} \quad -6x - 3y = -33$$
$$x + 3y = 18 \quad\quad\quad\quad\quad x + 3y = 18$$

Now add:

$$
\begin{array}{r}
-6x - 3y = -33 \\
x + 3y = 18 \\
\hline
-5x = -15 \\
x = 3
\end{array}
$$

Solve for y.

$$2x + y = 11$$
$$2(3) + y = 11$$
$$6 + y = 11$$
$$y = 5$$

The solution remains the same, $(3, 5)$.

EXAMPLE 5   Solve the following system of equations using the addition method.

$$4x + 2y = -18$$
$$-2x - 5y = 10$$

Solution: To eliminate the variable x, we can multiply the second equation by 2, and then add.

$$4x + 2y = -18 \quad \text{gives} \quad 4x + 2y = -18$$
$$2[-2x - 5y = 10] \quad\quad\quad\quad -4x - 10y = 20$$

$$
\begin{array}{r}
4x + 2y = -18 \\
-4x - 10y = 20 \\
\hline
-8y = 2 \\
y = -\dfrac{1}{4}
\end{array}
$$

Solve for x:

$$4x + 2y = -18$$
$$4x + 2\left(-\frac{1}{4}\right) = -18$$
$$4x - \frac{1}{2} = -18$$
$$2\left(4x - \frac{1}{2}\right) = 2(-18) \quad\quad \text{multiply both sides of equation}$$
$$\text{by 2 to remove fractions}$$
$$8x - 1 = -36$$
$$8x = -35$$
$$x = -\frac{35}{8}$$

The solution is $(-\frac{35}{8}, -\frac{1}{4})$.

Check the solution $(-\frac{35}{8}, -\frac{1}{4})$ in both equations.

$$4x + 2y = -18 \qquad\qquad -2x - 5y = 10$$

$$4\left(-\frac{35}{8}\right) + 2\left(-\frac{1}{4}\right) = -18 \qquad -2\left(-\frac{35}{8}\right) - 5\left(-\frac{1}{4}\right) = 10$$

$$-\frac{35}{2} - \frac{1}{2} = -18 \qquad\qquad \frac{35}{4} + \frac{5}{4} = 10$$

$$-\frac{36}{2} = -18 \qquad\qquad \frac{40}{4} = 10$$

$$-18 = -18 \quad \text{true} \qquad\qquad 10 = 10 \quad \text{true} \qquad \blacksquare$$

Note that the solution to Example 5 contains fractions. You should not always expect to get integers as answers.

EXAMPLE 6   Solve the following system of equations using the addition method.

$$2x + 3y = 6$$
$$5x - 4y = -8$$

Solution: The variable x can be eliminated by multiplying the first equation by -5 and the second by 2 and then adding the equations.

$$-5[2x + 3y = 6] \qquad \text{gives} \qquad -10x - 15y = -30$$
$$2[5x - 4y = -8] \qquad\qquad\qquad 10x - 8y = -16$$

$$\begin{array}{r} -10x - 15y = -30 \\ 10x - 8y = -16 \\ \hline -23y = -46 \\ y = 2 \end{array}$$

The same value could be obtained for y by multiplying the first equation by 5 and the second by -2 and then adding. Try it now and see.

Solve for x.

$$2x + 3y = 6$$
$$2x + 3(2) = 6$$
$$2x + 6 = 6$$
$$2x = 0$$
$$x = 0$$

The solution to this system is $(0, 2)$. $\blacksquare$

EXAMPLE 7   Solve the following system of equations using the addition method.

$$2x + y = 3$$
$$4x + 2y = 12$$

Solution: The variable y can be eliminated by multiplying the first equation by -2 and then adding the two equations.

$$-2[2x + y = 3] \quad \text{gives} \quad -4x - 2y = -6$$
$$4x + 2y = 12 \qquad\qquad 4x + 2y = 12$$

$$\begin{array}{r} -4x - 2y = -6 \\ 4x + 2y = 12 \\ \hline 0 = 6 \quad \text{false} \end{array}$$

Since $0 = 6$ is a false statement, this system has no solution. The system is inconsistent and the lines will be parallel when graphed.

This solution is identical to the solutions obtained by graphing in Example 3 of Section 8.2 and by substitution in Example 2 of Section 8.3. ∎

EXAMPLE 8   Solve the following system of equations using the addition method.

$$x - \frac{1}{2}y = 2$$
$$y = 2x - 4$$

Solution: First align the x and y terms on the left side of the equation.

$$x - \frac{1}{2}y = 2$$
$$-2x + y = -4$$

Now proceed as in the previous examples.

$$2\left[x - \frac{1}{2}y = 2\right] \quad \text{gives} \quad 2x - y = 4$$
$$-2x + y = -4 \qquad\qquad -2x + y = -4$$

$$\begin{array}{r} 2x - y = 4 \\ -2x + y = -4 \\ \hline 0 = 0 \quad \text{true} \end{array}$$

Since $0 = 0$ is a true statement, the system is dependent and has an infinite number of solutions. When graphed, both equations will be the same line. This solution is identical to the solutions obtained by graphing in Example 4 of Section 8.2 and by substitution in Example 3 of Section 8.3. ∎

EXAMPLE 9   Solve the following system of linear equations using the addition method.

$$2x + 3y = 7$$
$$5x - 7y = -3$$

Solution: We can eliminate the variable x by multiplying the first equation by -5 and the second by 2.

$$-5[2x + 3y = 7] \qquad \text{gives} \qquad -10x - 15y = -35$$
$$2[5x - 7y = -3] \qquad\qquad\qquad 10x - 14y = -6$$

$$
\begin{array}{r}
-10x - 15y = -35 \\
10x - 14y = -6 \\
\hline
-29y = -41
\end{array}
$$

$$y = \frac{41}{29}$$

We can now find x by substituting $y = \frac{41}{29}$ into one of the original equations and solving for x. If you try this, you will see that although it can be done, it gets pretty messy. An easier method that can be used to solve for x is to go back to the original equations and eliminate the variable y.

$$7[2x + 3y = 7] \qquad \text{gives} \qquad 14x + 21y = 49$$
$$3[5x - 7y = -3] \qquad\qquad\qquad 15x - 21y = -9$$

$$
\begin{array}{r}
14x + 21y = 49 \\
15x - 21y = -9 \\
\hline
29x = 40
\end{array}
$$

$$x = \frac{40}{29}$$

Thus the solution is $(\frac{40}{29}, \frac{41}{29})$.

Check by substituting $(\frac{40}{29}, \frac{41}{29})$ in both equations.

$$2x + 3y = 7 \qquad\qquad\qquad 5x - 7y = -3$$

$$2\left(\frac{40}{29}\right) + 3\left(\frac{41}{29}\right) = 7 \qquad\qquad 5\left(\frac{40}{29}\right) - 7\left(\frac{41}{29}\right) = -3$$

$$\frac{80}{29} + \frac{123}{29} = 7 \qquad\qquad\qquad \frac{200}{29} - \frac{287}{29} = -3$$

$$\frac{203}{29} = 7 \qquad\qquad\qquad\qquad \frac{-87}{29} = -3$$

$$7 = 7 \quad \text{true} \qquad\qquad\qquad -3 = -3 \quad \text{true} \qquad \blacksquare$$

We have illustrated three methods that can be used to solve a system of linear equations: graphing, substitution, and addition. When you are given a system of equations, which method should you use to solve the system? When you need an exact solution, graphing should not be used. Of the two algebraic methods, the addition method may be easier to use if there are no numerical coefficients of 1 in the system. If one or more of the equations has a coefficient of 1, you may wish to use either method.

method may be easier to use if there are no numerical coefficients of 1 in the system. If one or more of the equations has a coefficient of 1, you may use either method.

Exercise Set 8.4

Solve each system of equations using the addition method.

1. $x + y = 8$
$x - y = 4$

2. $x - y = 6$
$x + y = 4$

3. $-x + y = 5$
$x + y = 1$

4. $x + y = 10$
$-x + y = -2$

5. $x + 2y = 15$
$x - 2y = -7$

6. $3x + y = 10$
$4x - y = 4$

7. $4x + y = 6$
$-8x - 2y = 20$

8. $5x + 3y = 30$
$3x + 3y = 18$

9. $-5x + y = 14$
$-3x + y = -2$

10. $2x - y = 7$
$3x + 2y = 0$

11. $3x + y = 10$
$3x - 2y = 16$

12. $-4x + 3y = 0$
$5x - 6y = 9$

13. $4x - 3y = 8$
$2x + y = 14$

14. $2x - 3y = 4$
$2x + y = -4$

15. $5x + 3y = 6$
$2x - 4y = 5$

16. $6x - 4y = 9$
$2x - 8y = 3$

17. $4x - 2y = 6$
$y = 2x - 3$

18. $5x - 2y = -4$
$-3x - 4y = -34$

19. $3x - 2y = -2$
$3y = 2x + 4$

20. $5x + 4y = 10$
$-3x - 5y = 7$

21. $5x - 4y = 20$
$-3x + 2y = -15$

22. $5x = 2y - 4$
$3x - 5y = 6$

23. $6x + 2y = 5$
$3y = 5x - 8$

24. $4x - 3y = -4$
$3x - 5y = 10$

25. $4x + 5y = 0$
$3x = 6y + 4$

26. $4x - 3y = 8$
$-3x + 4y = 9$

27. $x - \frac{1}{2}y = 4$
$3x + y = 6$

28. $2x - \frac{1}{3}y = 6$
$5x - y = 4$

29. When solving a system of linear equations by the addition method, how will you know if the system is inconsistent?

30. When solving a system of linear equations by the addition method, how will you know if the system is dependent?

JUST FOR FUN

Solve each system of equations using the addition method. (*Hint:* First remove all fractions.)

1. $\dfrac{x + 2}{2} - \dfrac{y + 4}{3} = 4$

$\dfrac{x + y}{2} = \dfrac{1}{2} + \dfrac{x - y}{3}$

2. $\dfrac{5x}{2} + 3y = \dfrac{9}{2} + y$

$\dfrac{1}{4}x - \dfrac{1}{2}y = 6x + 12$

8.5

Applications of Systems of Equations

1 *Use systems equations to solve practical problems.*

1 The method you use to solve a system of equations may depend on such things as whether you wish to see "the entire picture" or are interested in finding the exact answer. If you are interested in the trend as the variable changes, you might decide to graph the equations. If you just want the answer, that is the ordered pair common to both equations, you might use one of the two algebraic methods to find the common solution.

Many of the applications solved in earlier chapters using only one variable can now be solved using two variables. The following example illustrates how Example 3 of Section 3.3 can be solved using two variables.

EXAMPLE 1   The sum of two numbers is 17. Find the two numbers if one number is 5 more than twice the other number.

Solution: Let x = one number

y = second number

Statement	Equation
The sum of two numbers is 17	$x + y = 17$
One number is 5 more than twice the other number	$y = 2x + 5$

$$\text{system of equations} \begin{cases} x + y = 17 \\ y = 2x + 5 \end{cases}$$

Substitute $2x + 5$ for y in first equation.

$$x + y = 17$$
$$x + (2x + 5) = 17$$
$$3x + 5 = 17$$
$$3x = 12$$
$$x = 4$$

Second number:

$$y = 2x + 5$$
$$y = 2(4) + 5 = 13$$

Thus the two numbers are 4 and 13.   ■

EXAMPLE 2   The Delicious Juice Company sells cans of apple juice for 40 cents and cans of apple drink for 16 cents each. The company wishes to market and sell, for 25 cents each, cans of juice-drink that are part juice and part drink. How many ounces of each should be used if the juice drink is to be sold in 8-ounce cans?

Solution: Let x = number of ounces of juice

y = number of ounces of drink

The value of the juice (or drink or juice drink) is found by multiplying the number of ounces by the cost per ounce.

Statement (or reason) for equation	Equation
The juice drink is to be sold in 8-ounce cans	$x + y = 8$
Value of juice + value of drink = value of juice drink	$0.40x + 0.16y = 0.25(x + y)$

$$\text{system of equations} \begin{cases} x + y = 8 \\ 0.40x + 0.16y = 0.25(x + y) \end{cases}$$

Solve the first equation for y.

$$x + y = 8$$
$$y = -x + 8$$

Substitute $-x + 8$ for *each* y in the second equation.

$$0.40x + 0.16y = 0.25(x + y)$$
$$0.40x + 0.16(-x + 8) = 0.25[x + (-x + 8)]$$
$$0.40x - 0.16x + 1.28 = 0.25(8)$$
$$0.24x + 1.28 = 2$$
$$0.24x = 0.72$$
$$x = \frac{0.72}{0.24} = 3 \text{ ounces of juice}$$

Find the number of ounces of drink.

$$y = -x + 8$$
$$y = -3 + 8$$
$$y = 5 \text{ ounces of drink}$$

Thus the mixture should contain 3 ounces of juice and 5 ounces of drink. ∎

EXAMPLE 3   A plane can travel 600 miles per hour with the wind and 450 miles per hour against the wind. Find the speed of the wind and the speed of the plane in still air.

Solution: Let $x =$ speed of plane in still air

$y =$ speed of wind

Speed of plane going with wind: $\left. \begin{array}{l} x + y = 600 \\ x - y = 450 \end{array} \right\}$ *system of equations*
Speed of plane going against wind:

$$
\begin{array}{rl}
x + y = & 600 \\
x - y = & 450 \\
\hline
2x \quad\;\; = & 1050 \\
x = & 525
\end{array}
$$

The plane's speed is 525 miles per hour in still air.

$$x + y = 600$$
$$525 + y = 600$$
$$y = 75$$

The wind's speed is 75 miles per hour.   ∎

EXAMPLE 4   It requires 3 hours for a motorboat to make a trip downstream when the water current is 2 miles per hour. The motorboat requires 4 hours to make the return trip upstream against the current. Find

(a) The speed of the motorboat in still water.
(b) The one-way distance.

Solution: (a) To solve this problem we use the fact that the distance traveled downstream is equal to the distance traveled upstream. Figure 8.6 may be helpful. Recall from earlier sections that distance = rate · time.

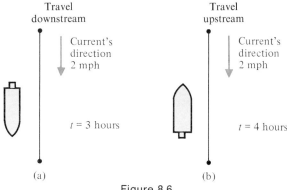

Figure 8.6

Let x = speed (or rate) of the boat in still water

then $x + 2$ = speed of the boat with the current (downstream)

and $x - 2$ = speed of the boat against the current (upstream)

Downstream	*Upstream*
distance = rate · time	distance = rate · time
$= (x + 2)3$	$= (x - 2)4$
$= 3x + 6$	$= 4x - 8$

If we represent the one-way distance by the letter d, the system of equations is,

$$d = 3x + 6$$
$$d = 4x - 8$$

Since the distance traveled downstream equals the distance traveled upstream, we set the two distances equal to each other (or we substitute $3x + 6$ for d in the second equation).

$$3x + 6 = 4x - 8$$
$$-x + 6 = -8$$
$$-x = -14$$
$$x = 14$$

The speed of the motorboat in still water is 14 miles per hour.

(b) The distance can now be found by substituting $x = 14$ in either of the distance equations.

$d = 3x + 6$	or	$d = 4x - 8$
$= 3(14) + 6$		$= 4(14) - 8$
$= 42 + 6$		$= 56 - 8$
$= 48$		$= 48$

Thus the one-way distance is 48 miles.   ■

EXAMPLE 5   A 50% saltwater solution is to be mixed with a 75% saltwater solution to get 60 liters of a 60% saltwater solution. How many liters of the 50% solution and the 75% solution should be mixed?

Solution: Let x = number of liters of 50% solution

y = number of liters of 75% solution

Statement	*Equation*
Total volume of combination is 60 liters	$x + y = 60$
Salt content of 50% solution + salt content of 75% solution = salt content of mixture	$0.5x + 0.75y = 0.6(60)$

$$\text{System of equations} \quad \begin{cases} x + y = 60 \\ 0.5x + 0.75y = 0.6(60) \end{cases}$$

Solve for y in first equation.

$$x + y = 60$$
$$y = 60 - x$$

Substitute $60 - x$ for y in the second equation.

$$0.5x + 0.75y = 0.6(60)$$
$$0.5x + 0.75(60 - x) = 36$$
$$0.5x + 45 - 0.75x = 36$$
$$-0.25x + 45 = 36$$
$$-0.25x = -9$$
$$x = \frac{-9}{-0.25} = 36$$

Now solve for y.

$$y = 60 - x$$
$$y = 60 - 36$$
$$y = 24$$

Thus 36 liters of 50% solution should be mixed with 24 liters of 75% solution to obtain 60 liters of 60% solution.   ■

Sometimes you may be interested in more than just the answer to a system of equations. For example, you may wish to make a comparison of the two equations under different types of situations. To get this comparison, you may wish to graph the system of equations. We will solve the system in Example 6 graphically.

EXAMPLE 6   A major university is doing research to determine the most cost-efficient method to set up a number of computer terminals throughout the university. The university is considering two options. Option 1 is an $80,000 minicomputer whose terminals cost $1000 each. Option 2 is a $20,000 network system whose terminals cost $2500 each. How many terminals would the university have to install to make the total cost of the network system equal to the total cost of the minicomputer?

Solution: The network system has a much smaller initial cost ($20,000 versus $80,000); however, their terminals cost more ($2500 versus $1000).

Let n = number of terminals.

Option 1: total cost = cost of minicomputer + cost of n terminals

$$c = 80{,}000 + 1000n$$

Option 2: total cost = cost of network + cost of n terminals

$$c = 20{,}000 + 2500n$$

$$\text{system of equations} \begin{cases} c = 80{,}000 + 1000n \\ c = 20{,}000 + 2500n \end{cases}$$

Now graph each equation.

$c = 80{,}000 + 1000n$

		x	y
Let $n = 0$,	$c = 80{,}000 + 1000(0) = 80{,}000$	0	80,000
Let $n = 30$,	$c = 80{,}000 + 1000(30) = 110{,}000$	30	110,000
Let $n = 50$,	$c = 80{,}000 + 1000(50) = 130{,}000$	50	130,000

$c = 20{,}000 + 2500n$

		x	y
Let $n = 0$,	$c = 20{,}000 + 2500(0) = 20{,}000$	0	20,000
Let $n = 30$,	$c = 20{,}000 + 2500(30) = 95{,}000$	30	95,000
Let $n = 50$,	$c = 20{,}000 + 2500(50) = 145{,}000$	50	145,000

The graph (Fig. 8.7) shows that the total cost of the network system equals the cost of the minicomputer when 40 terminals are used. If the university plans to have fewer than 40 terminals, the network system would be less expensive.   ■

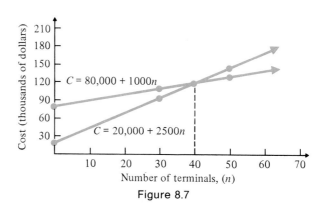

Figure 8.7

Exercise Set 8.5

Express each problem as a system of linear equations, and then find the solution.

1. The sum of two integers is 37. Find the numbers if one number is 1 greater than twice the other.
2. The sum of two consecutive even integers is 54. Find the two numbers.
3. The sum of two consecutive odd integers is 76. Find the two numbers.
4. The difference of two integers is 25. Find the two numbers if the larger is 1 less than three times the smaller.
5. The difference of two integers is 28. Find the two numbers if the larger is three times the smaller.
6. A plane can travel 540 miles per hour with the wind and 490 against the wind. Find the speed of the plane in still air and the speed of the wind.
7. Carlos can paddle a kayak 4.5 miles per hour with the current and 3.2 miles per hour against the current. Find the speed of the kayak in still water and the speed of the current.
8. Maria can make a weekly salary of $200 plus 5% commission on sales, or a weekly salary consisting of a straight 15% commission of sales. Determine the amount of sales necessary for the 15% straight commission salary to equal the $200 plus 5% commission salary.
9. A Hertz Automobile Rental Agency charges a daily fee of $36 plus 12 cents a mile. A National Automobile Rental Agency charges a daily fee of $30 plus 15 cents a mile for the same car. What distance would you have to drive in 1 day to make the cost of renting from Hertz equal to the cost of renting from National.
10. A druggist needs 1000 milliliters of a 10% phenobarbital solution. She has only 5% and 25% phenobarbital solutions available. How many milliliters of each solution should she mix to obtain the desired solution?
11. The ABC Printing Company charges a set up fee of $1600, plus $6 per book it prints. The XYZ Printing Company has a set up fee of $1200, plus an $8 fee per book it prints. How many books would have to be ordered to make the ABC Company's total charges equal the XYZ Company's total charges?

See Exercise 14.

12. The total cost of printing a certain booklet consists of a fixed charge for typesetting and an additional charge for each booklet. If the total cost for 1000 booklets is $550 and the total cost of 2000 booklets is $800, find the fixed charge and the charge for each booklet.
13. Janet wishes to mix 30 pounds of coffee to sell for a total cost of $100. To obtain the mixture, she will mix coffee that sells for $3 per pound with coffee that sells for $5 per pound. How many pounds of each type coffee should she use?
14. In chemistry class, Steve has an 80% acid solution and a 50% acid solution. How much of each solution must he mix to get 100 liters of a 75% acid solution?
15. Jason has milk that is 5% butterfat and skim milk without butterfat. How much 5% milk and how much skim milk should he mix to make 100 gallons of milk that is 3.5% butterfat?
16. Pierre's recipe for Quiche Lorraine calls for 2 cups (16 ounces) of light cream which is 20% milk fat. It is often difficult to find light cream with 20% milk fat at the supermarket. What is commonly found is heavy cream which is 36% milk fat and half and half which is 10.5% milk fat. How many ounces of the heavy cream and how much of the half and half should be mixed to obtain the 2 cups of light cream which is 20% milkfat?

17. Mr. and Mrs. McAdams invest a total of $8000 in two savings accounts. One account gives 10% interest and the other 8%. Find the amount placed in each account if they receive a total of $750 in interest after 1 year.

18. The Webers wish to invest a total $12,500 in two savings accounts. One account is giving 10% interest and the other $5\frac{1}{4}$%. The Webers wish for their interest from the two accounts to be at least $1200 at the end of the year. Find the minimum amount that can be placed in the account giving 10% interest.

19. If the minicomputer system in Example 6 costs $60,000 plus $1500 per terminal, and the network system cost $20,000 plus $3000 per terminal, how many terminals would have to be ordered for the cost of the minicomputer system to equal the cost of the network system?

20. Two cars start at the same point and travel in opposite directions. One car travels at 80 kilometers per hour and the other at 65 kilometers per hour. In how many hours will they be 435 kilometers apart?

21. Two trains start at the same point going in the same direction on parallel tracks. One train travels at 70 miles per hour and the other at 42 miles per hour. In how many hours will they be 154 miles apart?

22. An airplane takes 4 hours to fly a certain distance with a 20 mile per hour headwind. The return trip against the headwind takes $4\frac{3}{4}$ hours.
(a) Find the speed of the plane in still air.
(b) Find the one-way distance.

JUST FOR FUN

1. Two brothers jog to school daily. The older jogs at 9 miles per hour, the younger at 5 miles per hour. When the older brother reaches the school, the younger brother is $\frac{1}{2}$ mile from the school. How far is the school from the boys' house?

2. By weight, an alloy of brass is 70% copper and 30% zinc. Another alloy of brass is 40% copper and 60% zinc. How many grams of each of these alloys need to be melted and combined to obtain 300 grams of a brass alloy that is 60% copper and 40% zinc?

8.6 _____

Systems of Linear Inequalities (Optional)

1 *Solve systems of linear inequalities graphically.*

---

1 In Section 7.4 we showed how to graph linear inequalities in two variables. In Section 8.2 we learned how to solve systems of equations graphically. In this section we discuss how to solve systems of linear inequalities graphically. The solution to a system of linear inequalities is the set of points that satisfy all inequalities in the system. Although a system of linear inequalities may contain more than two inequalities, in this book we will consider only systems with two inequalities.

> **To Solve A System of Linear Inequalities**
>
> Graph each inequality on the same set of axes. The solution is the set of points that satisfy all the inequalities in the system.

EXAMPLE 1   Determine the solution to the system of inequalities.

$$x + 2y \le 6$$
$$y > 2x - 4$$

Solution: First graph the inequality $x + 2y \le 6$, see Fig. 8.8.

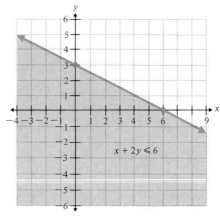

Figure 8.8

Now on the same set of axes graph the inequality $y > 2x - 4$, see Fig. 8.9.

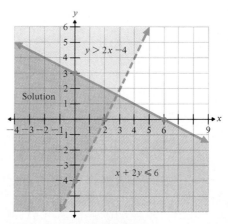

Figure 8.9

The solution is the set of points common to both inequalities. It is the part of the graph that contains both shadings. The dashed line is not part of the solution. However the part of the solid line that satisfies both inequalities is part of the solution.   ■

EXAMPLE 2   Determine the solution to the system of inequalities.

$$2x + 3y \geq 4$$
$$2x - y > -6$$

Solution: Graph $2x + 3y \geq 4$, see Fig. 8.10. Graph $2x - y > -6$ on the same set of axes, Fig. 8.11. The solution is the part of the graph with both shadings and the part of the solid line which satisfies both inequalities.

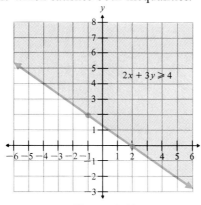

Figure 8.10

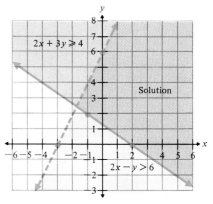

Figure 8.11

EXAMPLE 3   Determine the solution to the system of inequalities.

$$y < 2$$
$$x > -3$$

Solution: Graph both inequalities on the same set of axes, Fig. 8.12.

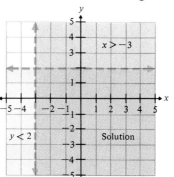

Figure 8.12

Exercise Set 8.6

Determine the solution to each system of inequalities.

1. $x + y > 2$
$\quad x - y < 2$

2. $y \leq 3x - 2$
$\quad y > -4x$

3. $y \leq x$
$\quad y < -2x + 4$

4. $2x + 3y < 6$
   $4x - 2y \geq 8$

5. $y > x + 1$
   $y \geq 3x + 2$

6. $x + 3y \geq 6$
   $2x - y > 4$

7. $x - 2y < 6$
   $y \leq -x + 4$

8. $y \leq 3x + 4$
   $y > 2$

9. $4x + 5y < 20$
   $x \geq -3$

10. $3x - 4y \leq 12$
   $y > -x + 4$

11. $x \leq 4$
   $y \geq -2$

12. $x \geq 0$
   $y \leq 0$

13. $x > -3$
   $y > 1$

14. $4x + 2y > 8$
   $y \leq 2$

15. $-2x + 3y \geq 6$
   $x + 4y \geq 4$

Summary

Glossary

Consistent system of linear equations: A system of linear equations that has a solution.
Dependent system of linear equations: A system of linear equations where both equations represent the same line.
Inconsistent system of linear equations: A system of linear equations that has no solution.

Simultaneous linear equations or a system of linear equations: Two or more linear equations considered together.
Solution to a system of equations: The ordered pair or pairs that satisfy all equations in the system.

Important Facts

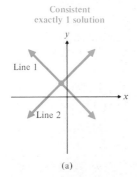

Consistent
exactly 1 solution

(a)

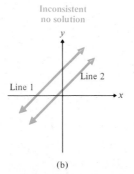

Inconsistent
no solution

(b)

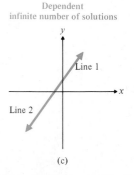

Dependent
infinite number of solutions

(c)

Three methods that can be used to solve a system of linear equations are (a) graphical solution, (b) substitution method, (c) addition (or elimination) method.

Review Exercises

[8.1] Determine which, if any, of the ordered pairs satisfy each system of equations.

1. $y = 3x - 2$
   $2x + 3y = 5$
   (a) $(0, -2)$ **(b)** $(2, 4)$ **(c)** $(1, 1)$

2. $y = -x + 4$
   $3x + 5y = 15$
   (a) $\left(\dfrac{5}{2}, \dfrac{3}{2}\right)$ **(b)** $(0, 4)$ **(c)** $\left(\dfrac{1}{2}, \dfrac{3}{5}\right)$

[8.2] Identify each system of linear equations as consistent, inconsistent, or dependent. State whether the system has exactly one solution, no solution, or an infinite number of solutions.

3. 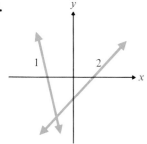   **4.** 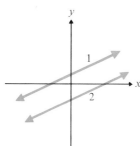   **5.** 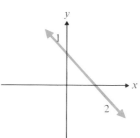   **6.**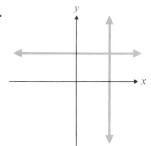

Express each equation in slope-intercept form. Without graphing or solving the system of equations, state whether the system of linear equations has exactly one solution, no solution, or an infinite number of solutions.

7. $x + 2y = 8$
   $3x + 6y = 12$

8. $y = -3x - 6$
   $2x + 3y = 8$

9. $y = \dfrac{1}{2}x + 4$
   $x + 2y = 8$

10. $6x = 4y - 8$
   $4x = 6y + 8$

Determine the solution to each system of equations graphically.

11. $y = x + 3$
    $y = 2x + 5$

12. $x = -2$
    $y = 3$

13. $y = 3$
    $y = -2x + 5$

14. $x + 3y = 6$
    $y = 2$

15. $x + 2y = 8$
    $2x - y = -4$

16. $y = x - 3$
    $2x - 2y = 6$

17. $2x + y = 0$
    $4x - 3y = 10$

18. $x + 2y = 4$
    $\dfrac{1}{2}x + y = -2$

[8.3] Find the solution to each system of equations using substitution.

19. $y = 2x - 8$
    $2x - 5y = 0$

20. $x = 3y - 9$
    $x + 2y = 1$

21. $2x + y = 5$
    $3x + 2y = 8$

22. $2x - y = 6$
    $x + 2y = 13$

23. $3x + y = 17$
    $2x - 3y = 4$

24. $x = -3y$
    $x + 4y = 6$

25. $2x + 3y = 10$
    $2x + 4y = 12$

26. $2x + 4y = 8$
    $4x + 8y = 16$

27. $2x - 3y = 8$
    $6x + 5y = 10$

28. $4x - y = 6$
    $x + 2y = 8$

[8.4] Find the solution to each system of equations using the addition method.

29. $x + y = 6$
    $x - y = 10$

30. $x + 2y = -3$
    $2x - 2y = 6$

31. $2x + 3y = 4$
    $x + 2y = -6$

32. $x + y = 12$
    $2x + y = 5$

33. $4x - 3y = 8$
    $2x + 5y = 8$

34. $-2x + 3y = 15$
    $3x + 3y = 10$

35. $5x + 2y = 9$
    $2x - 3y = -4$

36. $2x + 2y = 8$
    $y = 4x - 3$

37. $3x + 4y = 10$
    $4x + 5y = 14$

38. $2x - 5y = 12$
    $3x - 4y = -6$

[8.5] Express the problem as a system of linear equations, and then find the solution.

39. The sum of two integers is 48. Find the two numbers if the larger is 3 less than twice the smaller.

40. The difference of two integers is 18. Find the two numbers if the larger is four times the smaller.

41. Each year an employee has the option of selecting one of two retirement plans. Option 1 is a yearly rate of 8% of the employee's salary. Option 2 is $500 plus 3% of the salary. At what salary is it beneficial for the employee to select option 1?

42. A chemist has a 30% acid solution and a 50% acid solution. How much of each must be mixed to get 6 liters of a 40% acid solution?

43. A plane can travel 600 miles per hour with the wind and 530 miles per hour against the wind. Find the speed of the wind and the speed of the plane in still air.

44. It takes Joan 2 hours in her motorboat to travel downstream to a store when the current is 2 miles per hour. The return trip against the current takes her $2\frac{1}{2}$ hours. Find
   (a) The speed of the boat in still water.
   (b) The one-way distance.

Determine the solution to each system of inequalities.

45. $x + y > 2$
    $2x - y \le 4$

46. $2x - 3y \le 6$
    $x + 4y > 4$

47. $2x - 6y > 6$
    $x > -2$

48. $x < 2$
    $y \ge -1$

Practice Test

1. Determine which, if any, of the ordered pairs satisfy the system of equations.

$$x + 2y = -6$$
$$3x + 2y = -12$$

(a) $(0, -6)$ **(b)** $\left(-3, -\dfrac{3}{2}\right)$ **(c)** $(2, -4)$

Identify each system as consistent, inconsistent, or dependent. State whether the system has exactly one solution, no solution, or an infinite number of solutions.

2.

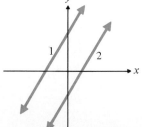

3.

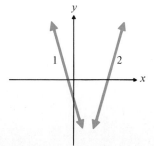

4.

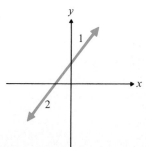

Express each equation in slope-intercept form. Then determine, without solving the system, whether the system of equations has exactly one solution, no solution, or an infinite number of solutions.

5. $3y = 6x - 9$
   $2x - y = 6$

6. $3x + 2y = 10$
   $3x - 2y = 10$

Solve each system of equations graphically.

7. $y = 3x - 2$
   $y = -2x + 8$

8. $3x - 2y = -3$
   $3x + y = 6$

Solve each system of equations using substitution.

9. $3x + y = 8$
   $x - y = 6$

10. $4x - 3y = 9$
   $2x + 4y = 10$

Solve each system of equations using the addition method.

11. $2x + y = 5$
   $x + 3y = -10$

12. $3x + 2y = 12$
   $-2x + 5y = 8$

Express the problem as a system of linear equations, and then find the solution.

13. Budget Rent a Car Agency charges $40 per day plus 3 cents per mile to rent a certain car. Hertz charges $45 per day plus 8 cents per mile to rent the same car. How many miles will have to be driven for the cost of Budget's car to equal the cost of Hertz's car?

14. The Modern Grocery has cashews that sell for $4.00 a pound and peanuts that sell for $2.50 a pound.

How much of each must Albert, the grocer, mix to get 20 pounds of a mixture that he can sell for $3.00 per pound?

*15. Determine the solution to the system of inequalities.

$$2x + 4y < 8$$
$$x - 3y \geq 6$$

9 Roots and Radicals

See Section 9.6, Exercise 16.

▣ *Evaluate square roots.*

▣ *Know square roots of negative numbers are not real numbers.*

▣ *Recognize perfect square numbers, rational and irrational numbers.*

▣ You will use the square root in many areas of mathematics and science. We will begin with what a square root looks like:

$$\sqrt{x} \text{ is read "the square root of } x\text{."}$$

The $\sqrt{}$ is called the **radical sign.** The number or expression inside the radical sign is called the **radicand.**

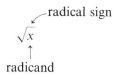

The entire expression, including the radical sign and radicand, is called the **radical expression.**

Another part of a radical expression is its **index.** The index tells the "root" of the expression. Square roots have an index of 2. The index of square roots are generally not written.

$$\underset{\text{index}}{\downarrow}$$

$$\sqrt{x} \qquad \text{means} \qquad \sqrt[2]{x}$$

Other types of radical expressions have different indexes. For example, $\sqrt[3]{x}$ is the third or cube root of x. The index of cube roots is 3. Cube roots are discussed in optional Section 9.7.

Examples of square roots	*How read*	*Radicand*
$\sqrt{8}$	the square root of 8	8
$\sqrt{5x}$	the square root of 5x	5x
$\sqrt{\dfrac{x}{2y}}$	the square root of x over 2y	$\dfrac{x}{2y}$

Every positive number has two square roots, a principal or positive square root and a negative square root.

> The **principal or positive square root** of a positive real number x, written $\sqrt{x}$, is that *positive* number whose square equals x.

Examples

$$\sqrt{25} = 5 \qquad \text{since } 5^2 = 5 \cdot 5 = 25$$

$$\sqrt{36} = 6 \qquad \text{since } 6^2 = 6 \cdot 6 = 36$$

$$\sqrt{\frac{1}{4}} = \frac{1}{2} \qquad \text{since } \left(\frac{1}{2}\right)^2 = \left(\frac{1}{2}\right)\left(\frac{1}{2}\right) = \frac{1}{4}$$

$$\sqrt{\frac{4}{9}} = \frac{2}{3} \qquad \text{since } \left(\frac{2}{3}\right)^2 = \left(\frac{2}{3}\right)\left(\frac{2}{3}\right) = \frac{4}{9}$$

The **negative square root** of a positive real number x, written $-\sqrt{x}$, is the additive inverse or opposite of the principal square root. For example, $-\sqrt{25} = -5$ and $-\sqrt{36} = -6$. **Whenever we use the term "square root" in this book we will be referring to the principal or positive square root.**

EXAMPLE 1   Evaluate: (a) $\sqrt{64}$ (b) $\sqrt{100}$

Solution: (a) $\sqrt{64} = 8$ since $8^2 = (8)(8) = 64$

(b) $\sqrt{100} = 10$ since $(10)^2 = 100$ ∎

EXAMPLE 2   Evaluate: (a) $-\sqrt{64}$ (b) $-\sqrt{100}$

Solution: (a) $-\sqrt{64} = -8$

(b) $-\sqrt{100} = -10$ ∎

2   An important point you must understand is that **square roots of negative numbers are not real numbers.** Consider $\sqrt{-4}$; to what is $\sqrt{-4}$ equal? To evaluate $\sqrt{-4}$ we must find some number whose square equals -4. But we know that the square of any nonzero number must be a positive number. Therefore no number squared gives -4, and $\sqrt{-4}$ has no real value. Numbers like $\sqrt{-4}$, or the square root of any negative number, are called **imaginary numbers.** The study of imaginary numbers is beyond the scope of this book.

Now suppose we have an expression like $\sqrt{x}$ where x represents some number. To make sure that the radical $\sqrt{x}$ is a real number, and not imaginary, we will always assume that x is a nonnegative number.

In this chapter, unless stated otherwise, we will assume that all variables that appear in a radicand represent nonnegative numbers.

We will end this section with a discussion of rational and irrational numbers. To help in our discussion, we will define perfect square numbers.

3   The numbers 1, 4, 9, 16, 25, 36, 49, ... are called **perfect square numbers (or perfect square factors)** because each number is *the square of a natural number.*

1, 2, 3, 4, 5, 6, 7, ...	natural numbers
$1^2, 2^2, 3^2, 4^2, 5^2, 6^2, 7^2, ...$	the squares of the natural number
1, 4, 9, 16, 25, 36, 49, ...	perfect square number

What are the next two perfect square numbers? Note that the square root of a perfect square number is an integer. That is $\sqrt{1} = 1$, $\sqrt{4} = 2$, $\sqrt{9} = 3$, $\sqrt{16} = 4$, and so on.

A **rational number** is one that can be written in the form $\frac{a}{b}$, where a and b are integers, $b \neq 0$. Examples of rational numbers are $\frac{1}{2}$, $\frac{3}{5}$, $-\frac{9}{2}$, 4, and 0. All integers are rational numbers since they can be expressed with a denominator of 1. For example, $4 = \frac{4}{1}$, and $0 = \frac{0}{1}$. The square roots of perfect square numbers are also rational numbers since each is an integer. When a rational number is written as a decimal, it will be either a terminating or repeating decimal.

Terminating decimals	Repeating decimals
$\frac{1}{2} = 0.5$	$\frac{1}{3} = 0.333\ldots$
$\frac{1}{4} = 0.25$	$\frac{2}{3} = 0.666\ldots$
$\sqrt{4} = 2.0$	$\frac{1}{6} = .1666\ldots$

Real numbers that are not rational numbers are called **irrational numbers.** Irrational numbers when written as decimals are nonterminating, nonrepeating decimals. The square root of every nonperfect square number is an irrational number. For example, $\sqrt{2}$ and $\sqrt{3}$ are irrational numbers.

Appendix B gives the square roots of numbers from 1 to 100 rounded to the nearest thousandth. The symbol $\approx$ means "is approximately equal to." From Appendix B we can determine that $\sqrt{2} \approx 1.414$. The actual value of the square root of 2, and every other irrational number, when written as a decimal, will continue indefinitely in a nonrepeating pattern. Thus whenever we give a value of a square root for an irrational number it is only an approximation.

Calculator Corner

The square root key, $\boxed{\sqrt{x}}$, on calculators can be used to find square roots of nonnegative numbers. For example, to find the square root of 4, we press

$$\boxed{c}\ \boxed{4}\ \boxed{\sqrt{x}}\ \boxed{=}\ \boxed{2}$$

The calculator would display the answer 2. What would the calculator display if we evaluate $\sqrt{7}$?

$$\boxed{c}\ \boxed{7}\ \boxed{\sqrt{x}}\ \boxed{=}\ 2.6457513$$

Note that $\sqrt{7}$ is an irrational number since 7 is not a perfect square. $\sqrt{7}$ is therefore a nonrepeating, nonterminating decimal. The exact decimal value of $\sqrt{7}$, or any other irrational number, can never be given exactly. The answers given on a calculator for irrational numbers are only close approximations of their value.

Suppose that we tried to evaluate $\sqrt{-4}$ on a calculator. What would the calculator give as an answer?

$$\boxed{c}\ \boxed{4}\ \boxed{\pm}\ \boxed{\sqrt{x}}\ \boxed{=}\ \text{Error}$$

The calculator would give some type of error message since the square root of -4, or the square root of any other negative number, is not a real number.

Exercise Set 9.1 _____

Evaluate each expression.

1. $\sqrt{1}$
2. $\sqrt{4}$
3. $\sqrt{0}$
4. $\sqrt{64}$

5. $-\sqrt{81}$
6. $\sqrt{9}$
7. $\sqrt{121}$
8. $\sqrt{100}$

9. $-\sqrt{16}$
10. $-\sqrt{36}$
11. $\sqrt{144}$
12. $\sqrt{49}$

13. $\sqrt{169}$
14. $\sqrt{225}$
15. $-\sqrt{1}$
16. $-\sqrt{100}$

17. $\sqrt{81}$
18. $-\sqrt{25}$
19. $-\sqrt{121}$
20. $-\sqrt{169}$

21. $\sqrt{\dfrac{1}{4}}$
22. $\sqrt{\dfrac{4}{9}}$
23. $\sqrt{\dfrac{9}{16}}$
24. $\sqrt{\dfrac{25}{64}}$

25. $-\sqrt{\dfrac{4}{25}}$
26. $-\sqrt{\dfrac{100}{144}}$

Use Appendix B or your calculator to evaluate each expression to the nearest thousandth.

27. $\sqrt{8}$
28. $\sqrt{2}$
29. $\sqrt{15}$
30. $\sqrt{30}$

31. $\sqrt{80}$
32. $\sqrt{79}$
33. $\sqrt{81}$
34. $\sqrt{52}$

35. $\sqrt{97}$
36. $\sqrt{5}$
37. $\sqrt{3}$
38. $\sqrt{40}$

39. Whenever we see a variable in a radical what assumption do we make about the variable? Why do we make this assumption?

Answer True or False

40. $\sqrt{6}$ is a rational number
41. $\sqrt{25}$ is a rational number
42. $\sqrt{-4}$ is not a real number
43. $\sqrt{-5}$ is not a real number

44. $\sqrt{5}$ is an irrational number
45. $\sqrt{9}$ is an irrational number
46. $\sqrt{\frac{1}{4}}$ is a rational number
47. $\sqrt{\frac{4}{9}}$ is a rational number

9.2 _____

Multiplying and Simplifying Square Roots

1. Use the product rule to simplify radicals containing numbers.
2. Use the product rule to simplify radicals containing variables.

To simplify square roots in this section we will make use of the Product Rule for Radicals.

Product Rule for Radicals

$$\sqrt{a} \cdot \sqrt{b} = \sqrt{a \cdot b} \qquad a \geq 0, b \geq 0 \qquad \text{Rule 1}$$

The product rule says that the product of two square roots is equal to the square root of the product. The product rule applies only when both a and b are nonnegative, for the square roots of negative numbers are not real numbers.

Example of the Product Rule

$$\left.\begin{array}{c}\sqrt{1}\cdot\sqrt{60}\\\sqrt{2}\cdot\sqrt{30}\\\sqrt{3}\cdot\sqrt{20}\\\sqrt{4}\cdot\sqrt{15}\\\sqrt{6}\cdot\sqrt{10}\end{array}\right\}=\sqrt{60}$$

Note $\sqrt{60}$ can be factored into any of these forms.

To Simplify A Square Root Containing Only A Numerical Value

1. Write the numerical value as a product of the largest perfect square factor and another factor.
2. Use the product rule to write the expression as a product of square roots, with each square root containing one of the factors.
3. Find the square root of the perfect square factor.

EXAMPLE 1   Simplify $\sqrt{60}$.

Solution: The only perfect square factor of 60 is 4.

$$\sqrt{60} = \sqrt{4 \cdot 15}$$
$$= \sqrt{4} \cdot \sqrt{15}$$
$$= 2\sqrt{15}$$

$2\sqrt{15}$ is read "two times the square root of fifteen" or "two radical fifteen."   ∎

EXAMPLE 2   Simplify $\sqrt{75}$.

Solution: $\sqrt{75} = \sqrt{25 \cdot 3}$
$$= \sqrt{25} \cdot \sqrt{3}$$
$$= 5\sqrt{3}$$ ∎

EXAMPLE 3   Simplify $\sqrt{80}$.

Solution: $\sqrt{80} = \sqrt{16 \cdot 5}$
$$= \sqrt{16} \cdot \sqrt{5}$$
$$= 4\sqrt{5}$$ ∎

HELPFUL HINT

It is not uncommon for students to use a perfect square factor that is not the largest perfect square factor of the number. Let us consider Example 3 again. Four is also a perfect square factor of 80.

$$80 = \sqrt{4 \cdot 20} = \sqrt{4} \cdot \sqrt{20} = 2\sqrt{20}$$

Since 20 itself contains a perfect square factor of 4, the problem is not complete. Rather than starting the entire problem again, you can continue the simplification process as follows:

$$\sqrt{80} = 2\sqrt{20} = 2\sqrt{4 \cdot 5} = 2\sqrt{4} \cdot \sqrt{5} = 2 \cdot 2\sqrt{5} = 4\sqrt{5}$$

Now the answer checks with the answer in Example 3.

EXAMPLE 4   Simplify $\sqrt{180}$.

Solution: $\sqrt{180} = \sqrt{36 \cdot 5}$
$= \sqrt{36} \cdot \sqrt{5}$
$= 6\sqrt{5}$ ∎

EXAMPLE 5   Simplify $\sqrt{156}$.

Solution: $\sqrt{156} = \sqrt{4 \cdot 39}$
$= \sqrt{4} \cdot \sqrt{39}$
$= 2\sqrt{39}$ ∎

Although 39 can be factored into $3 \cdot 13$, neither of these factors is a perfect square. Thus the answer can be simplified no further.   ∎

2   Now we will simplify radicals that contain variables in the radicand.

In Section 9.1 we noted that certain numbers were **perfect square factors.** When a radical contains a variable (or number) raised to an **even** exponent, that variable (or number) and exponent together also form a perfect square factor. For example, in the expression $\sqrt{x^4}$, the x^4 is a perfect square factor since the exponent, 4, is even. In the expression $\sqrt{x^5}$, the x^5 is not a perfect square factor since the exponent is odd.

To evaluate square roots when the radicand is a perfect square factor we use the following rule.

$$\sqrt{a^{2 \cdot n}} = a^n \qquad a \geq 0 \qquad \text{Rule 2}$$

This rule states that **the square root of a variable raised to an even power equals the variable raised to one-half that power.**

Examples

$$\sqrt{x^2} = x$$
$$\sqrt{y^4} = y^2$$
$$\sqrt{x^{12}} = x^6$$
$$\sqrt{x^{20}} = x^{10}$$

A special case of the above rule is

$$\sqrt{a^2} = a, \qquad a \geq 0$$

EXAMPLE 6   Simplify: (a) $\sqrt{x^{54}}$ (b) $\sqrt{x^4 y^6}$ (c) $\sqrt{x^8 y^2}$ (d) $\sqrt{y^8 z^{12}}$

Solution: (a) $\sqrt{x^{54}} = x^{27}$
(b) $\sqrt{x^4 y^6} = \sqrt{x^4}\sqrt{y^6} = x^2 y^3$
(c) $\sqrt{x^8 y^2} = x^4 y$
(d) $\sqrt{y^8 z^{12}} = y^4 z^6$ ∎

> **To Evaluate the Square Root of A Radicand Containing A Variable Raised To An Odd Power**
>
> **1.** Express the variable as the product of two factors, one of which has an exponent of 1 (the other will therefore be a perfect square factor).
> **2.** Use the product rule to simplify.

Examples 7 and 8 illustrate this procedure.

EXAMPLE 7 Simplify each expression.

(a) $\sqrt{x^3}$ (b) $\sqrt{y^{11}}$ (c) $\sqrt{x^{99}}$

Solution: (a) $\sqrt{x^3} = \sqrt{x^2 \cdot x} = \sqrt{x^2} \cdot \sqrt{x}$
$$= x \cdot \sqrt{x} \quad \text{or} \quad x\sqrt{x}$$

(Remember that x is the same as x^1.)

(b) $\sqrt{y^{11}} = \sqrt{y^{10} \cdot y} = \sqrt{y^{10}} \cdot \sqrt{y}$
$$= y^5\sqrt{y}$$

(c) $\sqrt{x^{99}} = \sqrt{x^{98} \cdot x} = \sqrt{x^{98}} \cdot \sqrt{x}$
$$= x^{49}\sqrt{x} \quad \blacksquare$$

More complex radicals can be simplified using the product rule for radicals and the principles discussed in this section.

EXAMPLE 8 Simplify each expression.

(a) $\sqrt{16x^3}$ (b) $\sqrt{32x^2}$ (c) $\sqrt{32x^3}$

Solution: Write each expression as the product of square roots, one of which has a perfect square radicand.

(a) $\sqrt{16x^3} = \sqrt{16x^2} \cdot \sqrt{x}$
$$= 4x\sqrt{x}$$

(b) $\sqrt{32x^2} = \sqrt{16x^2} \cdot \sqrt{2}$
$$= 4x\sqrt{2}$$

(c) $\sqrt{32x^3} = \sqrt{16x^2} \cdot \sqrt{2x}$
$$= 4x\sqrt{2x} \quad \blacksquare$$

EXAMPLE 9 Simplify each expression.

(a) $\sqrt{50x^2y}$ (b) $\sqrt{48x^3y^2}$ (c) $\sqrt{98x^9y^7}$

Solution: (a) $\sqrt{50x^2y} = \sqrt{25x^2} \cdot \sqrt{2y}$
$$= 5x\sqrt{2y}$$

(b) $\sqrt{48x^3y^2} = \sqrt{16x^2y^2} \cdot \sqrt{3x}$
$$= 4xy\sqrt{3x}$$

(c) $\sqrt{98x^9y^7} = \sqrt{49x^8y^6} \cdot \sqrt{2xy}$
$= 7x^4y^3\sqrt{2xy}$ ∎

The radicand of your simplified answer should not contain any perfect square factors or any variables with an exponent greater than 1.

Now let us look at an example where we use the product rule to multiply two radicals together before simplifying.

EXAMPLE 10   Multiply and simplify where possible.

(a) $\sqrt{2} \cdot \sqrt{8}$ (b) $\sqrt{2x} \cdot \sqrt{8}$ (c) $\sqrt{2x} \cdot \sqrt{8x}$

Solution: (a) $\sqrt{2} \cdot \sqrt{8} = \sqrt{2 \cdot 8} = \sqrt{16} = 4$

(b) $\sqrt{2x} \cdot \sqrt{8} = \sqrt{16x} = \sqrt{16} \cdot \sqrt{x} = 4\sqrt{x}$

(c) $\sqrt{2x} \cdot \sqrt{8x} = \sqrt{16x^2} = 4x$ ∎

EXAMPLE 11   Multiply and simplify where possible

(a) $\sqrt{8x^3y} \cdot \sqrt{4xy^5}$ (b) $\sqrt{5xy^6}\sqrt{6x^3y}$

Solution: (a) $\sqrt{8x^3y} \cdot \sqrt{4xy^5} = \sqrt{32x^4y^6} = \sqrt{16x^4y^6} \cdot \sqrt{2}$
$= 4x^2y^3\sqrt{2}$

(b) $\sqrt{5xy^6} \cdot \sqrt{6x^3y} = \sqrt{30x^4y^7} = \sqrt{x^4y^6} \cdot \sqrt{30y}$
$= x^2y^3\sqrt{30y}$

Note: In part (b), 30 can be factored in many ways. However, none of the factors are perfect squares, so we leave the answer as given. ∎

Exercise Set 9.2

Simplify each expression.

1. $\sqrt{16}$	2. $\sqrt{64}$	3. $\sqrt{8}$	4. $\sqrt{75}$
5. $\sqrt{96}$	6. $\sqrt{125}$	7. $\sqrt{32}$	8. $\sqrt{52}$
9. $\sqrt{160}$	10. $\sqrt{28}$	11. $\sqrt{48}$	12. $\sqrt{27}$
13. $\sqrt{108}$	14. $\sqrt{128}$	15. $\sqrt{156}$	16. $\sqrt{180}$
17. $\sqrt{256}$	18. $\sqrt{212}$	19. $\sqrt{900}$	20. $\sqrt{x^4}$
21. $\sqrt{y^6}$	22. $\sqrt{x^9}$	23. $\sqrt{x^2y^4}$	24. $\sqrt{x^2y}$
25. $\sqrt{x^9y^{12}}$	26. $\sqrt{x^4y^5z^6}$	27. $\sqrt{a^2b^4c}$	28. $\sqrt{a^3b^9c^{11}}$
29. $\sqrt{3x^3}$	30. $\sqrt{12x^4y^2}$	31. $\sqrt{50x^2y^3}$	
32. $\sqrt{125x^3y^5}$	33. $\sqrt{200y^5z^{12}}$	34. $\sqrt{64xyz^5}$	
35. $\sqrt{243q^2b^3c}$	36. $\sqrt{500ab^4c^3}$	37. $\sqrt{128x^3yz^5}$	
38. $\sqrt{112x^6y^8}$	39. $\sqrt{250x^4yz}$	40. $\sqrt{98x^4y^4z}$	

Simplify each expression.

41. $\sqrt{8} \cdot \sqrt{3}$	42. $\sqrt{5} \cdot \sqrt{5}$	43. $\sqrt{18} \cdot \sqrt{3}$
44. $\sqrt{60} \cdot \sqrt{5}$	45. $\sqrt{75} \cdot \sqrt{6}$	46. $\sqrt{30} \cdot \sqrt{5}$

47. $\sqrt{3x}\sqrt{5x}$

48. $\sqrt{4x^3}\sqrt{4x}$

49. $\sqrt{5x^2}\sqrt{8x^3}$

50. $\sqrt{15x^2}\sqrt{6x^5}$

51. $\sqrt{12x^2y}\sqrt{6xy^3}$

52. $\sqrt{20xy^4}\sqrt{6x^5}$

53. $\sqrt{18xy^4}\sqrt{3x^2y}$

54. $\sqrt{40x^2y^4}\sqrt{6x^3y^5}$

55. $\sqrt{15xy^6}\sqrt{6xyz}$

56. $\sqrt{14xyz^5}\sqrt{3xy^2z^6}$

57. $\sqrt{9x^4y^6}\sqrt{4x^2y^4}$

58. $\sqrt{3x^3yz^6}\sqrt{6x^4y^5z^6}$

59. $(\sqrt{4x})^2$

60. $(\sqrt{6x^2})^2$

61. $(\sqrt{13x^4y^6})^2$

9.3
Dividing and Simplifying Square Roots

1 Understand what it means for a radical to be simplified.

2 Use the quotient rule to simplify radicals.

3 Rationalize denominators.

1 In this section we will use a new rule, the Quotient Rule, to simplify radicals containing fractions. However, before we do that, we need to discuss what it means for a square root to be simplified.

We Will Consider A Square Root Simplified When

1. There are no perfect square factors in any radicand.

2. No radicand contains a fraction.

3. There are no square roots in any denominator.

All three criteria must be met for an expression to be simplified. Let us look at some radical expressions that *are not simplified*.

Radical	Reason not simplified
$\sqrt{8}$	(1) contains perfect square factor, 4. (note $\sqrt{8} = \sqrt{4} \cdot \sqrt{2} = 2\sqrt{2}$)
$\sqrt{x^3}$	(1) contains perfect square factor, x^2. (note $\sqrt{x^3} = \sqrt{x^2} \cdot \sqrt{x} = x\sqrt{x}$)
$\sqrt{\dfrac{1}{2}}$	(2) radicand contains a fraction.
$\dfrac{1}{\sqrt{2}}$	(3) square root in the denominator.

2 The quotient rule for radicals states that the quotient of the square root of a divided by the square root of b is equal to the square root of the quotient of a divided by b.

Quotient Rule for Radicals

$$\frac{\sqrt{a}}{\sqrt{b}} = \sqrt{\frac{a}{b}} \qquad a \geq 0, b > 0 \qquad \text{Rule 3}$$

Examples 1 through 4 illustrate how the quotient rule is used to simplify square roots.

EXAMPLE 1   Simplify each expression.

(a) $\sqrt{\dfrac{8}{2}}$ (b) $\sqrt{\dfrac{25}{5}}$ (c) $\sqrt{\dfrac{9}{4}}$

Solution: When the square root contains a fraction, divide out any factor common to both the numerator and denominator. If the square root still contains a fraction, use the quotient rule for radicals to simplify.

(a) $\sqrt{\dfrac{8}{2}} = \sqrt{4} = 2$

(b) $\sqrt{\dfrac{25}{5}} = \sqrt{5}$

(c) $\sqrt{\dfrac{9}{4}} = \dfrac{\sqrt{9}}{\sqrt{4}} = \dfrac{3}{2}$ ■

EXAMPLE 2   Simplify each expression.

(a) $\sqrt{\dfrac{16x^2}{8}}$ (b) $\sqrt{\dfrac{64x^4y}{2x^2y}}$ (c) $\sqrt{\dfrac{3x^2y^4}{27x^4}}$ (d) $\sqrt{\dfrac{15xy^5z^2}{3x^5yz}}$

Solution: First divide out any common factors to both the numerator and denominator, then use the quotient rule for radicals to simplify.

(a) $\sqrt{\dfrac{16x^2}{8}} = \sqrt{2x^2} = \sqrt{2}\sqrt{x^2} = \sqrt{2}x$ or $x\sqrt{2}$.

(b) $\sqrt{\dfrac{64x^4y}{2x^2y}} = \sqrt{32x^2} = \sqrt{16x^2}\sqrt{2} = 4x\sqrt{2}$

(c) $\sqrt{\dfrac{3x^2y^4}{27x^4}} = \sqrt{\dfrac{y^4}{9x^2}} = \dfrac{\sqrt{y^4}}{\sqrt{9x^2}} = \dfrac{y^2}{3x}$

(d) $\sqrt{\dfrac{15xy^5z^2}{3x^5yz}} = \sqrt{\dfrac{5y^4z}{x^4}} = \dfrac{\sqrt{5y^4z}}{\sqrt{x^4}} = \dfrac{\sqrt{y^4}\sqrt{5z}}{\sqrt{x^4}} = \dfrac{y^2\sqrt{5z}}{x^2}$ ■

When given a fraction containing radical expressions in both the numerator and denominator we use the quotient rule to simplify as illustrated in Examples 3 and 4.

EXAMPLE 3   Simplify each expression.

(a) $\dfrac{\sqrt{2}}{\sqrt{8}}$ (b) $\dfrac{\sqrt{75}}{\sqrt{3}}$

Solution: (a) $\dfrac{\sqrt{2}}{\sqrt{8}} = \sqrt{\dfrac{2}{8}} = \sqrt{\dfrac{1}{4}} = \dfrac{\sqrt{1}}{\sqrt{4}} = \dfrac{1}{2}$

(b) $\dfrac{\sqrt{75}}{\sqrt{3}} = \sqrt{\dfrac{75}{3}} = \sqrt{25} = 5$ ■

EXAMPLE 4   Simplify each expression.

(a) $\dfrac{\sqrt{32x^4y^3}}{\sqrt{8xy}}$ (b) $\dfrac{\sqrt{75x^8y^4}}{\sqrt{3x^5y^8}}$

Solution: (a) $\dfrac{\sqrt{32x^4y^3}}{\sqrt{8xy}} = \sqrt{\dfrac{32x^4y^3}{8xy}} = \sqrt{4x^3y^2} = \sqrt{4x^2y^2} \cdot \sqrt{x} = 2xy\sqrt{x}$

(b) $\dfrac{\sqrt{75x^8y^4}}{\sqrt{3x^5y^8}} = \sqrt{\dfrac{75x^8y^4}{3x^5y^8}} = \sqrt{\dfrac{25x^3}{y^4}} = \dfrac{\sqrt{25x^3}}{\sqrt{y^4}} = \dfrac{\sqrt{25x^2}\sqrt{x}}{\sqrt{y^4}} = \dfrac{5x\sqrt{x}}{y^2}$ ∎

3   When the denominator of a fraction contains the square root of a nonperfect square number, we generally simplify the expression by **rationalizing the denominator. To rationalize a denominator means to remove all radicals from the denominator.** We rationalize the denominator because it is easier (without a calculator) to obtain the approximate value of a number like $\sqrt{2}/2$ than a number like $1/\sqrt{2}$.

> **To rationalize a denominator** multiply *both* the numerator and the denominator of the fraction by the square root that appears in the denominator, or by the square root of a number that makes the denominator a perfect square.

EXAMPLE 5   Simplify $\dfrac{1}{\sqrt{2}}$.

Solution: Since $\sqrt{2} \cdot \sqrt{2} = \sqrt{4} = 2$, we multiply both numerator and denominator by $\sqrt{2}$.

$$\frac{1}{\sqrt{2}} = \frac{1}{\sqrt{2}} \cdot \frac{\sqrt{2}}{\sqrt{2}} = \frac{\sqrt{2}}{\sqrt{4}} = \frac{\sqrt{2}}{2}$$

The answer $\dfrac{\sqrt{2}}{2}$ is simplified since it satisfies the three requirements stated earlier.   ∎

In Example 5, multiplying both the numerator and denominator by $\sqrt{2}$ is equivalent to multiplying the fraction by 1, which does not change the value of the original fraction.

EXAMPLE 6   Simplify each expression.

(a) $\sqrt{\dfrac{2}{3}}$ (b) $\sqrt{\dfrac{x^2}{18}}$

Solution: (a) $\sqrt{\dfrac{2}{3}} = \dfrac{\sqrt{2}}{\sqrt{3}} = \dfrac{\sqrt{2}}{\sqrt{3}} \cdot \dfrac{\sqrt{3}}{\sqrt{3}} = \dfrac{\sqrt{6}}{3}$

(b) $\sqrt{\dfrac{x^2}{18}} = \dfrac{\sqrt{x^2}}{\sqrt{18}} = \dfrac{x}{\sqrt{9} \cdot \sqrt{2}} = \dfrac{x}{3\sqrt{2}}$

Now rationalize the denominator.

$$\frac{x}{3\sqrt{2}} \cdot \frac{\sqrt{2}}{\sqrt{2}} = \frac{x\sqrt{2}}{3\sqrt{4}} = \frac{x\sqrt{2}}{3\cdot 2} = \frac{x\sqrt{2}}{6}$$

Part (b) can also be rationalized as illustrated below.

$$\sqrt{\frac{x^2}{18}} = \frac{\sqrt{x^2}}{\sqrt{18}} = \frac{x}{\sqrt{18}} \cdot \frac{\sqrt{2}}{\sqrt{2}} = \frac{x\sqrt{2}}{\sqrt{36}} = \frac{x\sqrt{2}}{6}$$

Note that $\dfrac{x}{\sqrt{18}} \cdot \dfrac{\sqrt{18}}{\sqrt{18}}$ will also result in the correct answer when simplified.   ■

COMMON STUDENT ERROR

A number within a square root *cannot* be divided by a number not within the square root.

Correct	*Wrong*

$\dfrac{\sqrt{2}}{2}$ cannot be simplified any further

$\dfrac{\sqrt{2}^{\,1}}{\underset{1}{\cancel{2}}} = \sqrt{1} = 1$

$\dfrac{\sqrt{6}}{3}$ cannot be simplified any further

$\dfrac{\sqrt{6}^{\,2}}{\underset{1}{\cancel{3}}} = \sqrt{2}$

$\dfrac{\sqrt{x^3}}{x} = \dfrac{\sqrt{x^2}\sqrt{x}}{x} = \dfrac{\cancel{x}\sqrt{x}}{\cancel{x}} = \sqrt{x}$

$\dfrac{\sqrt{x^3}^{\,2}}{\underset{1}{\cancel{x}}} = \sqrt{x^2} = x$

Each of the following simplifications is correct because the numbers divided out are not within square roots.

Correct	*Correct*

$\dfrac{\overset{2}{\cancel{6}}\sqrt{2}}{\underset{1}{\cancel{3}}} = 2\sqrt{2}$ $\dfrac{\cancel{x}\sqrt{2}}{\cancel{x}} = \sqrt{2}$

$\dfrac{\overset{1}{\cancel{4}}\sqrt{3}}{\underset{2}{\cancel{8}}} = \dfrac{\sqrt{3}}{2}$ $\dfrac{3x^2\sqrt{5}}{\cancel{x}} = 3x\sqrt{5}$

Exercise Set 9.3

Simplify each expression.

1. $\sqrt{\dfrac{12}{3}}$ 2. $\sqrt{\dfrac{8}{2}}$ 3. $\sqrt{\dfrac{27}{3}}$ 4. $\sqrt{\dfrac{16}{4}}$

5. $\dfrac{\sqrt{18}}{\sqrt{2}}$ 6. $\dfrac{\sqrt{3}}{\sqrt{27}}$ 7. $\sqrt{\dfrac{1}{25}}$ 8. $\sqrt{\dfrac{16}{25}}$

9. $\sqrt{\dfrac{9}{49}}$ **10.** $\sqrt{\dfrac{4}{81}}$ **11.** $\dfrac{\sqrt{10}}{\sqrt{490}}$ **12.** $\sqrt{\dfrac{16x^3}{4x}}$

13. $\sqrt{\dfrac{40x^3}{2x}}$ **14.** $\sqrt{\dfrac{45x^2}{16x^2y^4}}$ **15.** $\sqrt{\dfrac{9xy^4}{3y^3}}$ **16.** $\sqrt{\dfrac{50x^3y^6}{10x^3y^8}}$

17. $\sqrt{\dfrac{25x^6y}{45x^6y^3}}$ **18.** $\sqrt{\dfrac{14xyz^5}{56x^3y^3z^4}}$ **19.** $\sqrt{\dfrac{72xy}{72x^3y^5}}$ **20.** $\dfrac{\sqrt{16x^4}}{\sqrt{8x}}$

21. $\dfrac{\sqrt{32x^5}}{\sqrt{8x}}$ **22.** $\dfrac{\sqrt{60x^2y^2}}{\sqrt{6x^2y^4}}$ **23.** $\dfrac{\sqrt{16x^4y}}{\sqrt{25x^6y^3}}$ **24.** $\dfrac{\sqrt{72}}{\sqrt{36x^2y^6}}$

25. $\dfrac{\sqrt{45xy^6}}{\sqrt{9xy^4z^2}}$ **26.** $\dfrac{\sqrt{24x^2y^6}}{\sqrt{8x^4z^4}}$

Simplify each expression.

27. $\dfrac{3}{\sqrt{2}}$ **28.** $\dfrac{2}{\sqrt{3}}$ **29.** $\dfrac{4}{\sqrt{8}}$ **30.** $\dfrac{6}{\sqrt{6}}$

31. $\dfrac{5}{\sqrt{10}}$ **32.** $\dfrac{9}{\sqrt{50}}$ **33.** $\sqrt{\dfrac{2}{5}}$ **34.** $\sqrt{\dfrac{7}{12}}$

35. $\sqrt{\dfrac{3}{15}}$ **36.** $\sqrt{\dfrac{3}{10}}$ **37.** $\sqrt{\dfrac{x^2}{2}}$ **38.** $\sqrt{\dfrac{x^2}{7}}$

39. $\sqrt{\dfrac{x^2}{8}}$ **40.** $\sqrt{\dfrac{x^3}{18}}$ **41.** $\sqrt{\dfrac{x^4}{5}}$ **42.** $\sqrt{\dfrac{x^3}{11}}$

43. $\sqrt{\dfrac{x^6}{15y}}$ **44.** $\sqrt{\dfrac{x^5y}{12y^2}}$ **45.** $\sqrt{\dfrac{8x^4y^2}{32x^2y^3}}$ **46.** $\sqrt{\dfrac{27xz^4}{6y^4}}$

47. $\sqrt{\dfrac{18yz}{75x^4y^5z^3}}$ **48.** $\dfrac{\sqrt{25x^5}}{\sqrt{100xy^5}}$ **49.** $\dfrac{\sqrt{90x^4y}}{\sqrt{2x^5y^5}}$ **50.** $\dfrac{\sqrt{120xyz^2}}{\sqrt{9xy^2}}$

9.4

Addition and Subtraction of Square Roots

▮ *Add and subtract like and unlike square roots.*

▮ *Rationalize a binomial denominator.*

▮ **Like square roots** are square roots having the same radicands. Like square roots are added in much the same manner that like terms are added. **To add like square roots** add their coefficients and then multiply that sum by the like square root.

Examples of adding like terms

$$2x + 3x = (2 + 3)x = 5x$$

$$4x + x = 4x + 1x = (4 + 1)x = 5x$$

Examples of adding like square roots

$$2\sqrt{7} + 3\sqrt{7} = (2 + 3)\sqrt{7} = 5\sqrt{7}$$

$$4\sqrt{x} + \sqrt{x} = 4\sqrt{x} + 1\sqrt{x} = (4 + 1)\sqrt{x} = 5\sqrt{x}$$

Note that adding like square roots is an application of the distributive property.

$$2\sqrt{7} + 3\sqrt{7} = (2 + 3)\sqrt{7}$$
$$= 5\sqrt{7}$$

Other examples of adding like square roots

$$2\sqrt{5} - 3\sqrt{5} = (2 - 3)\sqrt{5} = -1\sqrt{5} = -\sqrt{5}$$

$$\sqrt{x} + \sqrt{x} = 1\sqrt{x} + 1\sqrt{x} = (1 + 1)\sqrt{x} = 2\sqrt{x}$$

$$6\sqrt{2} + 3\sqrt{2} - \sqrt{2} = (6 + 3 - 1)\sqrt{2} = 8\sqrt{2}$$

$$\frac{2\sqrt{3}}{5} + \frac{1\sqrt{3}}{5} = \left(\frac{2}{5} + \frac{1}{5}\right)\sqrt{3} = \frac{3}{5}\sqrt{3} \quad \text{or} \quad \frac{3\sqrt{3}}{5}$$

EXAMPLE 1   Simplify each expression, if possible.

(a) $4\sqrt{3} + 2\sqrt{3} - 2$ (b) $\sqrt{5} - 4\sqrt{5} + 5$

(c) $5 + 3\sqrt{2} - \sqrt{2} + 3$ (d) $2\sqrt{3} + 5\sqrt{2}$

Solution: (a) $4\sqrt{3} + 2\sqrt{3} - 2 = 6\sqrt{3} - 2$

(b) $\sqrt{5} - 4\sqrt{5} + 5 = -3\sqrt{5} + 5$

(c) $5 + 3\sqrt{2} - \sqrt{2} + 3 = 8 + 2\sqrt{2}$

(d) Cannot be simplified since the radicands are different.   ∎

EXAMPLE 2   Simplify each expression.

(a) $2\sqrt{x} - 3\sqrt{x} + 4\sqrt{x}$ (b) $3\sqrt{x} + x + 4\sqrt{x}$

(c) $x + \sqrt{x} + 2\sqrt{x} + 3$ (d) $x\sqrt{x} + 3\sqrt{x} + x$

(e) $\sqrt{xy} + 2\sqrt{xy} - \sqrt{x}$

Solution: (a) $2\sqrt{x} - 3\sqrt{x} + 4\sqrt{x} = 3\sqrt{x}$

(b) $3\sqrt{x} + x + 4\sqrt{x} = x + 7\sqrt{x}$ (only $3\sqrt{x}$ and $4\sqrt{x}$ can be combined)

(c) $x + \sqrt{x} + 2\sqrt{x} + 3 = x + 3\sqrt{x} + 3$ (only the $\sqrt{x}$ and $2\sqrt{x}$ can be combined)

(d) $x\sqrt{x} + 3\sqrt{x} + x = (x + 3)\sqrt{x} + x$ (only the $x\sqrt{x}$ and $3\sqrt{x}$ can be combined)

(e) $\sqrt{xy} + 2\sqrt{xy} - \sqrt{x} = 3\sqrt{xy} - \sqrt{x}$ (only the $\sqrt{xy}$ and $2\sqrt{xy}$ can be combined)   ∎

Unlike square roots are square roots having different radicands. It is sometimes possible to change unlike square roots into like square roots, as illustrated in Examples 3, 4, and 5.

EXAMPLE 3   Simplify $\sqrt{2} + \sqrt{18}$.

Solution: Since 18 has a perfect square factor, 9, we write it as a product of a perfect square factor and another factor.

$$\sqrt{2} + \sqrt{18} = \sqrt{2} + \sqrt{9 \cdot 2}$$
$$= \sqrt{2} + \sqrt{9}\sqrt{2}$$
$$= \sqrt{2} + 3\sqrt{2}$$
$$= 4\sqrt{2}$$ ∎

EXAMPLE 4   Simplify $\sqrt{24} - \sqrt{54}$.

Solution: Write each radicand as a product of a perfect square factor and another factor.

$$\sqrt{24} - \sqrt{54} = \sqrt{4 \cdot 6} - \sqrt{9 \cdot 6}$$
$$= \sqrt{4}\sqrt{6} - \sqrt{9}\sqrt{6}$$
$$= 2\sqrt{6} - 3\sqrt{6} = -\sqrt{6} \quad \blacksquare$$

EXAMPLE 5   Simplify each expressions.

    (a) $2\sqrt{8} - \sqrt{32}$ (b) $3\sqrt{12} + 5\sqrt{27} + 2$ (c) $\sqrt{120} - \sqrt{75}$

Solution: (a) $2\sqrt{8} - \sqrt{32} = 2\sqrt{4 \cdot 2} - \sqrt{16 \cdot 2}$

$$= 2\sqrt{4}\sqrt{2} - \sqrt{16}\sqrt{2}$$
$$= 2 \cdot 2\sqrt{2} - 4\sqrt{2}$$
$$= 4\sqrt{2} - 4\sqrt{2}$$
$$= 0$$

(b) $3\sqrt{12} + 5\sqrt{27} + 2 = 3\sqrt{4 \cdot 3} + 5\sqrt{9 \cdot 3} + 2$

$$= 3\sqrt{4}\sqrt{3} + 5\sqrt{9}\sqrt{3} + 2$$
$$= 3 \cdot 2\sqrt{3} + 5 \cdot 3\sqrt{3} + 2$$
$$= 6\sqrt{3} + 15\sqrt{3} + 2$$
$$= 21\sqrt{3} + 2$$

(c) $\sqrt{120} - \sqrt{75} = \sqrt{4 \cdot 30} - \sqrt{25 \cdot 3}$

$$= \sqrt{4}\sqrt{30} - \sqrt{25}\sqrt{3}$$
$$= 2\sqrt{30} - 5\sqrt{3}$$

Since 30 has no perfect square factors and since the radicands are different, the expression $2\sqrt{30} - 5\sqrt{3}$ cannot be simplified any further. $\blacksquare$

COMMON STUDENT ERROR

The product rule presented in Section 9.2 was $\sqrt{a} \cdot \sqrt{b} = \sqrt{a \cdot b}$. The same principle **does not apply to** addition.

Wrong

$$\cancel{\sqrt{a} + \sqrt{b} = \sqrt{a + b}}$$

For example, to evaluate $\sqrt{9} + \sqrt{16}$,

           Correct   *Wrong*

$$\sqrt{9} + \sqrt{16} = 3 + 4 \qquad \sqrt{9} + \sqrt{16} = \sqrt{9 + 16}$$
$$= 7 \qquad\qquad\qquad\qquad = \sqrt{25}$$
$$= 5$$

2   When the denominator of a rational expression is a binomial with a square root term, we again **rationalize the denominator.** We do this by multiplying both the numerator and the denominator of the fraction by the **conjugate** of the denominator. The conjugate of a binomial is a binomial having the same two terms with the sign of the second term changed.

Binomial	*Its conjugate*
$3 + \sqrt{2}$	$3 - \sqrt{2}$
$\sqrt{5} - 3$	$\sqrt{5} + 3$
$2\sqrt{3} - \sqrt{5}$	$2\sqrt{3} + \sqrt{5}$
$x + \sqrt{3}$	$x - \sqrt{3}$

When a binomial is multiplied by its conjugate using the FOIL method, the outer and inner terms will add to zero.

EXAMPLE 6   Multiply $(2 + \sqrt{3})(2 - \sqrt{3})$ using the FOIL method.

Solution:

$$(2 + \sqrt{3})(2 - \sqrt{3})$$

$$\begin{array}{cccc} \text{F} & \text{O} & \text{I} & \text{L} \\ 2(2) + 2(-\sqrt{3}) + 2(\sqrt{3}) + \sqrt{3}(-\sqrt{3}) \end{array}$$
$$= \quad 4 \ - \ 2\sqrt{3} \ + 2\sqrt{3} \ - \quad \sqrt{9}$$
$$= 4 - \sqrt{9}$$
$$= 4 - 3 = 1 \quad \blacksquare$$

EXAMPLE 7   Multiply $(\sqrt{3} - \sqrt{5})(\sqrt{3} + \sqrt{5})$ using the FOIL method.

Solution:

$$(\sqrt{3} - \sqrt{5})(\sqrt{3} + \sqrt{5})$$

$$\begin{array}{cccc} \text{F} & \text{O} & \text{I} & \text{L} \\ \sqrt{3} \cdot \sqrt{3} + \sqrt{3} \cdot \sqrt{5} + (-\sqrt{5})(\sqrt{3}) + (-\sqrt{5})(\sqrt{5}) \end{array}$$
$$= \quad \sqrt{9} \ + \ \sqrt{15} \ - \quad \sqrt{15} \ - \quad \sqrt{25}$$
$$= \sqrt{9} - \sqrt{25}$$
$$= 3 - 5 = -2 \quad \blacksquare$$

Now let us do some problems where we rationalize the denominator where the denominator is a binomial with one or more terms a radical.

EXAMPLE 8   Simplify $\dfrac{5}{2 + \sqrt{3}}$.

Solution: To rationalize the denominator multiply both the numerator and the denominator by $2 - \sqrt{3}$, which is the conjugate of $2 + \sqrt{3}$.

$$\frac{5}{2 + \sqrt{3}} \cdot \frac{2 - \sqrt{3}}{2 - \sqrt{3}} = \frac{5(2 - \sqrt{3})}{(2 + \sqrt{3})(2 - \sqrt{3})}$$

$$= \frac{5(2 - \sqrt{3})}{4 - 3}$$

$$= \frac{5(2 - \sqrt{3})}{1}$$

$$= 5(2 - \sqrt{3}) \quad \text{or} \quad 10 - 5\sqrt{3}$$

Note that $-5\sqrt{3} + 10$ is also an acceptable answer.   ■

EXAMPLE 9   Simplify $\dfrac{6}{\sqrt{5} - \sqrt{2}}$.

Solution: Multiply both the numerator and the denominator of the fraction by $\sqrt{5} + \sqrt{2}$, the conjugate of $\sqrt{5} - \sqrt{2}$.

$$\frac{6}{\sqrt{5} - \sqrt{2}} \cdot \frac{\sqrt{5} + \sqrt{2}}{\sqrt{5} + \sqrt{2}} = \frac{6(\sqrt{5} + \sqrt{2})}{5 - 2}$$

$$= \frac{\overset{2}{\cancel{6}}(\sqrt{5} + \sqrt{2})}{\underset{1}{\cancel{3}}}$$

$$= 2(\sqrt{5} + \sqrt{2}) \quad \text{or} \quad 2\sqrt{5} + 2\sqrt{2} \qquad ■$$

EXAMPLE 10   Simplify $\dfrac{\sqrt{3}}{2 - \sqrt{6}}$.

Solution: Multiply both the numerator and the denominator of the fraction by $2 + \sqrt{6}$, the conjugate of $2 - \sqrt{6}$.

$$\frac{\sqrt{3}}{2 - \sqrt{6}} \cdot \frac{2 + \sqrt{6}}{2 + \sqrt{6}} = \frac{\sqrt{3}(2 + \sqrt{6})}{4 - 6}$$

$$= \frac{2\sqrt{3} + \sqrt{3} \cdot \sqrt{6}}{-2}$$

$$= \frac{2\sqrt{3} + \sqrt{18}}{-2}$$

$$= \frac{2\sqrt{3} + \sqrt{9} \cdot \sqrt{2}}{-2}$$

$$= \frac{2\sqrt{3} + 3\sqrt{2}}{-2}$$

$$= \frac{-2\sqrt{3} - 3\sqrt{2}}{2} \qquad ■$$

EXAMPLE 11   Simplify $\dfrac{x}{x + \sqrt{y}}$.

Solution: Multiply both the numerator and the denominator of the fraction by the conjugate of the denominator, $x - \sqrt{y}$.

$$\frac{x}{x + \sqrt{y}} \cdot \frac{x - \sqrt{y}}{x - \sqrt{y}} = \frac{x(x - \sqrt{y})}{x^2 - y} = \frac{x^2 - x\sqrt{y}}{x^2 - y}$$

Remember that you cannot divide out the x^2 terms because they are not factors. ■

Exercise Set 9.4

Simplify each expression.

1. $4\sqrt{3} - 2\sqrt{3}$

2. $\sqrt{5} + 2\sqrt{5}$

3. $6\sqrt{7} - 8\sqrt{7}$

4. $4\sqrt{10} + 6\sqrt{10} - \sqrt{10} + 2$

5. $2\sqrt{3} - 2\sqrt{3} - 4\sqrt{3} + 5$

6. $12\sqrt{15} + 5\sqrt{15} - 8\sqrt{15}$

7. $4\sqrt{x} + \sqrt{x}$

8. $-2\sqrt{x} - 3\sqrt{x}$

9. $-\sqrt{x} + 6\sqrt{x} - 2\sqrt{x}$

10. $3\sqrt{y} - 6\sqrt{y}$

11. $3\sqrt{y} - \sqrt{y} + 3$

12. $3\sqrt{5} - \sqrt{x} + 4\sqrt{5} + 3\sqrt{x}$

13. $\sqrt{x} + \sqrt{y} + x + 3\sqrt{y}$

14. $2 + 3\sqrt{y} - 6\sqrt{y} + 5$

15. $3 + 4\sqrt{x} - 6\sqrt{x}$

16. $4\sqrt{x} + 6\sqrt{x} - 3\sqrt{x} + 2x$

Simplify each expression.

17. $\sqrt{8} - \sqrt{12}$

18. $\sqrt{27} + \sqrt{45}$

19. $\sqrt{200} - \sqrt{72}$

20. $\sqrt{75} + \sqrt{108}$

21. $\sqrt{125} + \sqrt{20}$

22. $\sqrt{60} - \sqrt{135}$

23. $4\sqrt{50} - \sqrt{72} + \sqrt{8}$

24. $-4\sqrt{90} + 3\sqrt{40} + 2\sqrt{10}$

25. $-6\sqrt{75} + 4\sqrt{125}$

26. $4\sqrt{80} - \sqrt{75}$

27. $5\sqrt{250} - 9\sqrt{80}$

28. $7\sqrt{108} - 6\sqrt{180}$

29. $8\sqrt{64} - \sqrt{96}$

30. $3\sqrt{250} + 5\sqrt{160}$

Multiply as indicated.

31. $(3 + \sqrt{2})(3 - \sqrt{2})$

32. $(\sqrt{6} + 3)(\sqrt{6} - 3)$

33. $(6 - \sqrt{5})(6 + \sqrt{5})$

34. $(\sqrt{8} - 3)(\sqrt{8} + 3)$

35. $(\sqrt{x} + 3)(\sqrt{x} - 3)$

36. $(\sqrt{x} + 5)(\sqrt{x} - 5)$

37. $(\sqrt{6} + x)(\sqrt{6} - x)$

38. $(\sqrt{y} - 3)(\sqrt{y} + 3)$

39. $(\sqrt{x} + y)(\sqrt{x} - y)$

40. $(\sqrt{x} + \sqrt{y})(\sqrt{x} - \sqrt{y})$

41. $(\sqrt{y} - x)(\sqrt{y} + x)$

42. $(x + \sqrt{y})(x - \sqrt{y})$

Simplify each expression.

43. $\dfrac{4}{2 + \sqrt{3}}$

44. $\dfrac{3}{\sqrt{6} - 5}$

45. $\dfrac{3}{\sqrt{2} + 5}$

46. $\dfrac{4}{\sqrt{2} - 7}$

47. $\dfrac{2}{\sqrt{2} + \sqrt{3}}$

48. $\dfrac{5}{\sqrt{5} - \sqrt{6}}$

49. $\dfrac{8}{\sqrt{5} - \sqrt{8}}$

50. $\dfrac{1}{\sqrt{17} - \sqrt{8}}$

51. $\dfrac{2}{6 + \sqrt{x}}$

52. $\dfrac{5}{\sqrt{x} - 3}$

53. $\dfrac{6}{4 - \sqrt{y}}$

54. $\dfrac{5}{3 + \sqrt{x}}$

55. $\dfrac{4}{\sqrt{x} - y}$

56. $\dfrac{9}{x + \sqrt{y}}$

57. $\dfrac{x}{\sqrt{x} + \sqrt{y}}$

58. $\dfrac{\sqrt{3}}{\sqrt{x} - \sqrt{3}}$

59. $\dfrac{\sqrt{x}}{\sqrt{5} + \sqrt{x}}$

60. $\dfrac{x}{\sqrt{x} - y}$

9.5

Solving Radical Equations

> **1** Solve radical equations with only one square root term.
>
> **2** Solve radical equations with two square root terms.

1 A **radical equation** is an equation that contains a variable in a radicand. Some examples of radical equations are

$$\sqrt{x} = 3, \qquad \sqrt{x + 4} = 6, \qquad \sqrt{x - 2} = x - 6$$

To Solve A Radical Equation Containing Only One Square Root Term

1. Use the appropriate properties to rewrite the equation with the square root term by itself on one side of the equation.
2. Combine like terms.
3. Square both sides of the equation to remove the square root.
4. Solve the equation for the variable.
5. Check the solution in the original equation for extraneous roots.

The following examples illustrate this procedure.

EXAMPLE 1 Solve the equation $\sqrt{x} = 6$.

Solution: The square root containing the variable is already by itself on one side of the equation. Square both sides of the equation.

$$\sqrt{x} = 6$$
$$(\sqrt{x})^2 = (6)^2$$
$$x = 36$$

Check: $\sqrt{x} = 6$
$$\sqrt{36} = 6$$
$$6 = 6 \qquad \text{true} \qquad \blacksquare$$

EXAMPLE 2 Solve the equation $\sqrt{x + 4} = 6$.

Solution: The square root containing the variable is already by itself on one side of the equation. Square both sides of the equation.

$$\sqrt{x + 4} = 6$$
$$(\sqrt{x + 4})^2 = 6^2$$
$$x + 4 = 36$$
$$x + 4 - 4 = 36 - 4$$
$$x = 32$$

Check:
$$\sqrt{x + 4} = 6$$
$$\sqrt{32 + 4} = 6$$
$$\sqrt{36} = 6$$
$$6 = 6 \qquad \text{true} \qquad ■$$

EXAMPLE 3   Solve the equation $\sqrt{x} + 4 = 6$.

Solution: Since the 4 is outside the square root sign, we first subtract 4 from both sides of the equation to isolate the square root term.

$$\sqrt{x} + 4 = 6$$
$$\sqrt{x} + 4 - 4 = 6 - 4$$
$$\sqrt{x} = 2$$

Now square both sides of the equation.

$$(\sqrt{x})^2 = 2^2$$
$$x = 4$$

Check:
$$\sqrt{x} + 4 = 6$$
$$\sqrt{4} + 4 = 6$$
$$2 + 4 = 6$$
$$6 = 6 \qquad \text{true} \qquad ■$$

HELPFUL HINT

When you square both sides of an equation you may introduce extraneous roots. An **extraneous root** is a number obtained when solving an equation that is not a solution to the original equation. Therefore equations that are squared in the process of finding their solutions should always be checked for extraneous roots by substituting the numbers found back in the **original** equation.

Consider the equation

$$x = 5$$

Now square both sides

$$x^2 = 25$$

Note that the equation $x = 5$ is only true when x is 5. However the equation $x^2 = 25$ is true for both 5 and -5. When we squared $x = 5$ we introduced the extraneous root -5.

EXAMPLE 4   Solve the equation $\sqrt{x} = -5$.

Solution:
$$\sqrt{x} = -5$$
$$(\sqrt{x})^2 = (-5)^2$$
$$x = 25$$

Check:
$$\sqrt{x} = -5$$
$$\sqrt{25} = -5$$
$$5 = -5 \quad \text{false}$$

Since the statement $5 = -5$ is false, the number 25 is an extraneous root and is not a solution to the given equation. Thus the equation $\sqrt{x} = -5$ has no real solutions.   ■

EXAMPLE 5   Solve the equation $\sqrt{2x - 3} = x - 3$.

Solution: $\sqrt{2x - 3} = x - 3$

Square both sides of the equation.

$$(\sqrt{2x - 3})^2 = (x - 3)^2$$
$$2x - 3 = (x - 3)(x - 3)$$
$$2x - 3 = x^2 - 6x + 9$$

Now solve the quadratic equation as explained earlier. Move the $2x$ and -3 to the right side of the equation to obtain

$$0 = x^2 - 8x + 12 \quad \text{or} \quad x^2 - 8x + 12 = 0$$

Now factor.

$$x^2 - 8x + 12 = 0$$
$$(x - 6)(x - 2) = 0$$

$$x - 6 = 0 \quad \text{or} \quad x - 2 = 0$$
$$x = 6 \qquad\qquad\qquad x = 2$$

Check:

$x = 6$	$x = 2$
$\sqrt{2x - 3} = x - 3$	$\sqrt{2x - 3} = x - 3$
$\sqrt{2(6) - 3} = 6 - 3$	$\sqrt{2(2) - 3} = 2 - 3$
$\sqrt{9} = 3?$	$\sqrt{1} = -1?$
$3 = 3 \quad \text{true}$	$1 = -1 \quad \text{false}$

The 6 is a solution, but 2 is not a solution to the equation.   ■

EXAMPLE 6   Solve the equation $2x - 5\sqrt{x} - 3 = 0$.

Solution: First write the equation with the square root containing the variable by itself on one side of the equation.

$$2x - 5\sqrt{x} - 3 = 0$$
$$-5\sqrt{x} = -2x + 3$$
$$\text{or} \quad 5\sqrt{x} = 2x - 3$$

Now square both sides of the equation.

$$(5\sqrt{x})^2 = (2x - 3)^2$$
$$25x = (2x - 3)(2x - 3)$$
$$25x = 4x^2 - 12x + 9$$
$$0 = 4x^2 - 37x + 9$$
$$0 = (4x - 1)(x - 9)$$

$$4x - 1 = 0 \quad \text{or} \quad x - 9 = 0$$
$$4x = 1 \qquad\qquad x = 9$$
$$x = \frac{1}{4}$$

Check: $x = \dfrac{1}{4}$ $\qquad\qquad\qquad\qquad x = 9$

$$2x - 5\sqrt{x} - 3 = 0 \qquad\qquad 2x - 5\sqrt{x} - 3 = 0$$
$$2\left(\frac{1}{4}\right) - 5\sqrt{\frac{1}{4}} - 3 = 0 \qquad 2(9) - 5\sqrt{9} - 3 = 0$$
$$\frac{1}{2} - 5\left(\frac{1}{2}\right) - 3 = 0 \qquad\qquad 18 - 5(3) - 3 = 0$$
$$\frac{1}{2} - \frac{5}{2} - 3 = 0 \qquad\qquad\qquad 18 - 15 - 3 = 0$$
$$\qquad\qquad\qquad\qquad\qquad\qquad\qquad 0 = 0 \quad \text{true}$$
$$\frac{1}{2} - \frac{5}{2} - \frac{6}{2} = 0$$
$$-\frac{10}{2} = 0$$
$$-5 = 0 \quad \text{false}$$

The solution is 9; $\frac{1}{4}$ is not a solution.   ∎

2 Consider the radical equations

$$\sqrt{x + 1} = \sqrt{x - 3}, \qquad \sqrt{x + 5} - \sqrt{2x + 4} = 0$$

These equations are different from those previously discussed because they have two square root terms containing the variable x. To solve equations of this type, rewrite the equation, when necessary, so that there is only one square root term on each side of the equation. Then square both sides of the equation. Examples 7 and 8 illustrate this procedure.

EXAMPLE 7   Solve the equation $\sqrt{2x + 2} = \sqrt{3x - 5}$.

Solution: Since each side of the equation already contains one square root term, it is not necessary to rewrite the equation. Square both sides of the equation.

$$(\sqrt{2x + 2})^2 = (\sqrt{3x - 5})^2$$
$$2x + 2 = 3x - 5$$

Now solve for x.

$$2x + 2 = 3x - 5$$
$$2 = x - 5$$
$$7 = x$$

Check:
$$\sqrt{2x + 2} = \sqrt{3x - 5}$$
$$\sqrt{2(7) + 2} = \sqrt{3(7) - 5}$$
$$\sqrt{16} = \sqrt{16}$$
$$4 = 4 \quad \text{true}$$

The solution is 7. ■

EXAMPLE 8   Solve the equation $3\sqrt{x - 2} - \sqrt{7x + 4} = 0$.

Solution: Add $\sqrt{7x + 4}$ to both sides of the equation to get one square root term on each side of the equation.

$$3\sqrt{x - 2} - \sqrt{7x + 4} + \sqrt{7x + 4} = 0 + \sqrt{7x + 4}$$
$$3\sqrt{x - 2} = \sqrt{7x + 4}$$
$$(3\sqrt{x - 2})^2 = (\sqrt{7x + 4})^2$$
$$9(x - 2) = 7x + 4$$
$$9x - 18 = 7x + 4$$
$$2x - 18 = 4$$
$$2x = 22$$
$$x = 11$$

Check:
$$3\sqrt{x - 2} - \sqrt{7x + 4} = 0$$
$$3\sqrt{11 - 2} - \sqrt{7(11) + 4} = 0$$
$$3\sqrt{9} - \sqrt{77 + 4} = 0$$
$$3(3) - \sqrt{81} = 0$$
$$9 - 9 = 0 \quad \text{true} \quad ■$$

Exercise Set 9.5

Solve each equation.

1. $\sqrt{x} = 8$

2. $\sqrt{x} = 5$

3. $\sqrt{x} = -3$

4. $\sqrt{x - 3} = 6$

5. $\sqrt{x + 5} = 3$

6. $\sqrt{2x - 4} = 2$

7. $\sqrt{2x + 4} = -6$

8. $\sqrt{x - 4} = 8$

9. $\sqrt{x + 3} = 5$

10. $4 + \sqrt{x} = 9$

11. $6 = 4 + \sqrt{x}$

12. $2 = 8 - \sqrt{x}$

13. $4 + \sqrt{x} = 2$

14. $\sqrt{3x + 4} = x - 2$

15. $\sqrt{2x - 5} = x - 4$

16. $\sqrt{x^2 + 8} = x + 2$

17. $\sqrt{2x - 6} = \sqrt{5x - 27}$

18. $2\sqrt{x + 3} = 10$

19. $\sqrt{3x + 3} = \sqrt{5x - 1}$

20. $\sqrt{2x - 5} = \sqrt{x + 2}$

21. $\sqrt{3x + 9} = 2\sqrt{x}$

22. $x - 6 = \sqrt{3x}$

23. $\sqrt{4x - 5} = \sqrt{x + 9}$

24. $\sqrt{x^2 + 3} = x + 1$

25. $3\sqrt{x} = \sqrt{x + 8}$

26. $x - 5 = \sqrt{x^2 - 35}$

27. $4\sqrt{x} = x + 3$

28. $2x - 1 = -\sqrt{x}$

29. $\sqrt{2x - 3} = 2\sqrt{3x - 2}$

30. $6 - 2\sqrt{3x} = 0$

31. $\sqrt{x^2 + 5} = x + 5$ **32.** $2\sqrt{4x - 3} = 10$ **33.** $x - 4\sqrt{x} + 3 = 0$

34. $\sqrt{4x + 5} + 5 = 2x$

35. Why is it necessary to always check solutions to radical equations?

JUST FOR FUN

Solve each equation. (*Hint:* It will be necessary to square both sides of the equation twice—good luck.)

1. $\sqrt{x} + 2 = \sqrt{x + 16}$

2. $\sqrt{x + 1} = 2 - \sqrt{x}$

3. $\sqrt{x + 7} = 5 - \sqrt{x - 8}$

9.6

Applications of Radicals

1. Use the Pythagorean Theorem.
2. Use the distance formula.
3. Learn some scientific applications of radicals.

In this section we will focus on some of the many important applications of radicals. We will discuss the Pythagorean Theorem, the distance formula, and then give a few additional applications of radicals.

Pythagorean Theorem

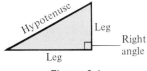

Figure 9.1

1. A **right triangle** is a triangle that contains a right, or 90°, angle. A right triangle is illustrated in Fig. 9.1.

The two smaller sides of a right triangle are called the **legs** and the side opposite the right angle is called the **hypotenuse.** The Pythagorean theorem expresses the relationship between the legs of a right triangle and its hypotenuse.

Pythagorean Theorem

The square of the hypotenuse of a right triangle is equal to the sum of the squares of the two legs.

If a and b represent the legs, and c represents the hypotenuse, then

$$a^2 + b^2 = c^2$$

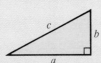

EXAMPLE 1   Find the hypotenuse of the right triangle whose legs are 3 feet and 4 feet.

Solution: Draw a picture of the problem (Fig. 9.2). When drawing the picture it makes no difference which leg is called a and which leg is called b.

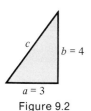

Figure 9.2

$$a^2 + b^2 = c^2$$
$$3^2 + 4^2 = c^2$$
$$9 + 16 = c^2$$
$$25 = c^2$$
$$\sqrt{25} = \sqrt{c^2} \qquad \text{take square root of both sides of equation}$$
$$5 = c$$

The hypotenuse is 5 feet.

Check: $a^2 + b^2 = c^2$
$$3^2 + 4^2 = 5^2$$
$$9 + 16 = 25$$
$$25 = 25 \qquad \text{true} \qquad \blacksquare$$

Note that the length of a side of any geometric figure must always be a positive number, since it makes no sense to talk about negative lengths. Since the hypotenuse, c, is positive we can write $\sqrt{c^2} = c$ by Rule 2 in Section 9.2.

EXAMPLE 2   The hypotenuse of a right triangle is 12 inches. Find the second leg if one leg is 8 inches.

Solution: Draw the triangle (Fig. 9.3).

Figure 9.3

$$a^2 + b^2 = c^2$$
$$8^2 + b^2 = (12)^2$$
$$64 + b^2 = 144$$
$$b^2 = 80$$
$$\sqrt{b^2} = \sqrt{80}$$
$$b = \sqrt{80} \quad \text{or} \quad \text{approximately 8.94 inches} \qquad \blacksquare$$

EXAMPLE 3   A regulation baseball diamond is a square with 90 feet between bases. How far is second base from home plate?

Solution: Draw the baseball diamond (Fig. 9.4). We are asked to find the distance from second base to home plate. This distance is the hypotenuse of the triangle, as illustrated in Fig. 9.5.

Figure 9.4

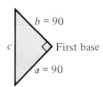

Figure 9.5

$$a^2 + b^2 = c^2$$
$$(90)^2 + (90)^2 = c^2$$
$$8100 + 8100 = c^2$$
$$16{,}200 = c^2$$
$$c = \sqrt{16{,}200} \quad \text{or} \quad \text{approximately 127.28 feet}$$

The $\sqrt{16{,}200}$ was evaluated on a calculator using the $\boxed{\sqrt{}}$ key. ■

Distance Formula

2 The distance formula can be used to find the distance between two points, (x_1, y_1) and (x_2, y_2), in the Cartesian coordinate system.

Distance Formula

$$d = \sqrt{(x_2 - x_1)^2 + (y_2 - y_1)^2}$$

EXAMPLE 4   Find the length of the straight line between the points $(-1, -4)$ and $(5, -2)$.

Solution: The two points are illustrated in Fig. 9.6. It makes no difference which points are labeled (x_1, y_1), and (x_2, y_2). Let $(5, -2)$ be (x_2, y_2) and $(-1, -4)$ be (x_1, y_1). Thus $x_2 = 5$, $y_2 = -2$, and $x_1 = -1$, $y_1 = -4$.

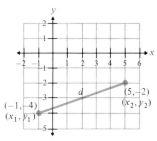

Figure 9.6

$$d = \sqrt{(x_2 - x_1)^2 + (y_2 - y_1)^2}$$
$$= \sqrt{[5 - (-1)]^2 + [-2 - (-4)]^2}$$
$$= \sqrt{(5 + 1)^2 + (-2 + 4)^2}$$
$$= \sqrt{6^2 + 2^2}$$
$$= \sqrt{36 + 4} = \sqrt{40} \quad \text{or} \quad \text{apprroximately 6.32}$$

Thus the distance between $(-1, -4)$ and $(5, -2)$ is approximately 6.32 units. ■

Other Applications

3 Radicals are often used in science and mathematics courses. Examples 5 through 7 illustrate some scientific applications of radicals.

EXAMPLE 5   During the sixteenth and seventeenth centuries Galileo Galilei did numerous experiments with objects falling freely under the influence of gravity. He showed, for example, that an object dropped from, say, 10 feet, hit the ground with a higher velocity than did an object dropped from 5 feet. A formula that can be used to determine the velocity of an object (neglecting air resistance) after it has fallen a certain height is

$$v = \sqrt{2gh}$$

where g is the acceleration of gravity and h is the height the object has fallen. On Earth the acceleration of gravity, g, is approximately 32 feet per second squared.

(a) Find the velocity of an object after it has fallen 5 feet.
(b) Find the velocity of an object after it has fallen 10 feet.
(c) Find the velocity of an object after it has fallen 100 feet.

Solution: (a) $v = \sqrt{2gh}$
$$= \sqrt{2(32)h}$$
$$= \sqrt{64h}$$

At $h = 5$ feet,

$$v = \sqrt{64(5)}$$
$$= \sqrt{320}$$
$$\approx 17.9 \text{ feet per second}$$

After an object has fallen 5 feet its velocity is approximately 17.9 feet per second. The symbol $\approx$ means "is approximately equal to."

(b) After falling 10 feet,

$$v = \sqrt{64(10)}$$
$$= \sqrt{640}$$
$$\approx 25.3 \text{ feet per second}$$

(c) After falling 100 feet,

$$v = \sqrt{64(100)}$$
$$= \sqrt{6400}$$
$$= 80 \text{ feet per second} \quad \blacksquare$$

EXAMPLE 6   The formula for the period in seconds, T, of a pendulum (the time required for the pendulum to make one complete swing both back and forth) is $T = 2\pi\sqrt{L/32}$, where L is the length of the pendulum in feet (see Fig. 9.7). Find the period of the pendulum if its length is 8 feet. Use 3.14 as an approximation for π.

Solution: $T = 2(3.14)\sqrt{\dfrac{8}{32}}$

$$= 6.28\sqrt{\dfrac{1}{4}}$$

$$= 6.28\left(\dfrac{1}{2}\right) = 3.14 \text{ seconds}$$

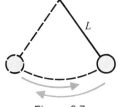

Figure 9.7

It takes a pendulum 8 feet long 3.14 seconds to make one complete swing. $\blacksquare$

EXAMPLE 7   For any planet, its "year" is the time it takes for the planet to revolve once around the sun. The number of "Earth days" in a given planet's year, N, is approximated by the formula

$$N = 0.2(\sqrt{R})^3$$

where R is the mean distance from the sun in millions of kilometers. Find the number of Earth days in the year of each of the planets illustrated in Fig. 9.8.

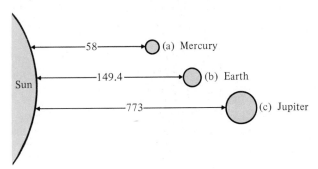

Figure 9.8

Solution: (a) Mercury:

$$N = 0.2(\sqrt{58})^3$$
$$\approx 0.2(7.6)^3$$
$$\approx 0.2(441.7)$$
$$\approx 88.3$$

It takes Mercury about 88 Earth days to revolve once around the sun.

(b) Earth:

$$N = 0.2(\sqrt{149.4})^3$$
$$\approx 0.2(12.2)^3$$
$$\approx 0.2(1826.1)$$
$$\approx 365.2$$

It takes Earth about 365 Earth days to revolve once around the sun (an answer that should not be surprising).

(c) Jupiter:

$$N \approx 0.2(\sqrt{773})^3$$
$$\approx 0.2(27.8)^3$$
$$\approx 0.2(21484.9)$$
$$\approx 4297$$

It takes Jupiter about 4297 Earth days (or about 11.8 of Earth's years) to revolve once about the sun.   ■

Exercise Set 9.6

Use the Pythagorean theorem to find the quantity indicated. You may leave your answer in square root form if a calculator with a square root key is not available. If a calculator is available round answers to the nearest hundredth.

1.

2.

3.

4.

5.

6.

7.

8.

9.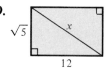

10.

11.

12.

13. A football field is 120 yards long from end zone to end zone. Find the length of the diagonal from one end zone to the other if the width of the field is 53.3 yards.

14. A boxing ring is a square 16 feet by 16 feet (actual ring size will vary with country and state). Find the distance from one boxer's corner to the other boxer's corner.

15. How long a length of wire is needed to reach from the top of a 4-meter telephone pole to a point 1.5 meters from the base of the pole?

16. An 8-meter extension ladder is placed against a house. The base of the ladder is 2 meters from the house. How high is the top of the ladder?

See Exercise 13

Use the distance formula to answer questions 17 through 20. You may leave your answer in square root form if a calculator with a square root keys is not available.

17. Find the length of the straight line between the points $(-4, 3)$ and $(-1, 4)$ (see Example 4).

18. Find the length of the straight line between the points $(4, -3)$ and $(6, 2)$.

19. Find the length of the straight line between the points $(-8, 4)$ and $(4, -8)$.

20. Find the length of the straight line between the points $(0, 5)$ and $(-6, -4)$.

Solve each problem. You may leave your answer in square root form if a calculator with a square root key is not available.

21. Find the side of a square that has an area of 144 square inches. Use $A = s^2$.

22. A formula for the area of a circle is $A = \pi r^2$, where π is approximately 3.14 and r is the radius of the circle. Find the radius of a circle of area 20 square inches.

23. Find the radius of a circle with an area of 80 square feet.

24. Find the length of the diagonal of a rectangle with a length of 12 inches and width of 5 inches.

25. Find the length of the diagonal of a rectangle with a length of 9 inches and width of 12 inches.

26. Find the length of the diagonal of a rectangle with a length of 25 inches and width of 10 inches.

27. Find the velocity of an object after it has fallen 80 feet (use $v = \sqrt{2gh}$, refer to Example 5).

28. Find the velocity of an object after it has fallen 1000 feet.

29. Find the velocity with which an object dropped from the Empire State Building, height 1250 feet, will strike the ground.

30. With what velocity will an object dropped from a plane 1 mile (5280 feet) high strike the ground?

31. Find the period of a 40-foot pendulum (use $T = 2\pi\sqrt{L/32}$, refer to Example 6.)

32. Find the period of a 60-foot pendulum.

33. Find the period of a 10-foot pendulum.

34. Find the number of Earth days in the year of the planet Mars, whose mean distance from the sun is 227 million kilometers (use $N = 0.2(\sqrt{R})^3$, refer to Example 7).

35. Find the number of Earth days in the year of the planet Saturn, whose mean distance from the sun is 1418 million kilometers.

36. When two forces, F_1 and F_2, pull at right angles to each other as illustrated, the resultant, or the effective force, R, can be found by the formula

$$R = \sqrt{F_1^2 + F_2^2}$$

Two cars at a 90° angle to each other are trying to pull a third out of the mud, as shown. If car A is exerting a force of 600 pounds and car B is exerting a force of 800 pounds, find the resulting force on the car stuck in the mud.

37. The escape velocity in meters per second, or the velocity needed for a spacecraft to escape a planet's gravitational field, is found by the formula

$$v_e = \sqrt{2gR}$$

where g is the force of gravity of the planet and R is the radius of the planet in meters. Find the escape velocity for Earth where $g = 9.75$ meters per second squared and $R = 6{,}370{,}000$ meters.

JUST FOR FUN

1. The length of a rectangle is 3 inches more than its width. If the length of the diagonal is 15 inches, find the dimensions of the rectangle.

2. The force of gravity on the moon is $\frac{1}{6}$ of that on Earth. If an object falls off a rocket 100 feet above the surface of the moon, with what velocity will it strike the moon? Use $v = \sqrt{2gh}$, see Example 5.

3. Find the length of a pendulum if the period is to be 2 seconds (use $T = 2\pi\sqrt{L/32}$, refer to Example 6.)

4. The length of the diagonal of a rectangular solid (see the accompanying figure) is given by

$$d = \sqrt{a^2 + b^2 + c^2}$$

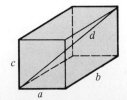

Find the distance of the diagonal of a suitcase of length 37 inches, width 15 inches, and depth 9 inches.

9.7

Higher Roots and Fractional Exponents (Optional)

■ *Evaluate cube and fourth roots.*

■ *Simplify cube and fourth roots.*

■ *Write radical expressions in exponential form.*

■ In this section we will use the same basic concepts used in Sections 9.1 through 9.4 to work with roots that contain indexes of 3 and 4. In this section we introduce cube and fourth roots.

$$\sqrt[3]{a} \text{ is read "the cube root of } a\text{"}$$

$$\sqrt[4]{a} \text{ is read "the fourth root of } a\text{"}$$

Note that,

$$\sqrt[3]{a} = b \quad \text{if} \quad b^3 = a$$

and

$$\sqrt[4]{a} = b \quad \text{if} \quad b^4 = a, \qquad b > 0$$

Examples

$$\sqrt[3]{8} = 2 \qquad \text{since } 2^3 = 2 \cdot 2 \cdot 2 = 8$$
$$\sqrt[3]{-8} = -2 \qquad \text{since } (-2)^3 = (-2)(-2)(-2) = -8$$
$$\sqrt[3]{27} = 3 \qquad \text{since } 3^3 = 3 \cdot 3 \cdot 3 = 27$$
$$\sqrt[4]{16} = 2 \qquad \text{since } 2^4 = 2 \cdot 2 \cdot 2 \cdot 2 = 16$$
$$\sqrt[4]{81} = 3 \qquad \text{since } 3^4 = 3 \cdot 3 \cdot 3 \cdot 3 = 81$$

EXAMPLE 1   Evaluate: (a) $\sqrt[3]{-27}$ (b) $\sqrt[3]{125}$

Solution: (a) To find $\sqrt[3]{-27}$ we must find the number that when cubed gives -27.

$$\sqrt[3]{-27} = -3 \qquad \text{since } (-3)^3 = -27$$

(b) To find $\sqrt[3]{125}$ we must find the number that when cubed gives 125.

$$\sqrt[3]{125} = 5 \qquad \text{since } 5^3 = 125 \qquad \blacksquare$$

Note that the cube root of a positive number is a positive number and the cube root of a negative number is a negative number. The radicand of a fourth root (or any even root) must be a nonnegative number for the expression to be a real number. For example, $\sqrt[4]{-16}$ is not a real number because no real number raised to the fourth power can be a negative number.

It will be helpful in the explanations that follow if we define perfect cube numbers. A **perfect cube number** is a number that is the cube of a natural number.

$$1, \ 2, \ 3, \ 4, \ 5, \ldots \qquad \text{Natural numbers}$$
$$1^3, 2^3, 3^3, 4^3, 5^3, \ldots$$
$$1, \ 8, \ 27, 64, 125, \ldots \qquad \text{Perfect cube numbers}$$

Note $\sqrt[3]{1} = 1$, $\sqrt[3]{8} = 2$, $\sqrt[3]{27} = 3$, $\sqrt[3]{64} = 4$, and so on.

Perfect fourth power numbers can be expressed in a similar manner.

$$1, \ 2, \ 3, \ 4, \ 5, \ldots \qquad \text{Natural numbers}$$
$$1^4, 2^4, 3^4, 4^4, \ 5^4, \ldots$$
$$1, 16, 81, 256, 625, \ldots \qquad \text{Perfect fourth power numbers}$$

Note $\sqrt[4]{1} = 1$, $\sqrt[4]{16} = 2$, $\sqrt[4]{81} = 3$, $\sqrt[4]{256} = 4$, and so on.

You may wish to refer back to these numbers when evaluating cube and fourth roots.

2 The product rule used in simplifying square roots can be expanded for indexes greater than 2. We will again use the product rule to simplify radicals.

Product Rule for Radicals

$$\sqrt[n]{a}\,\sqrt[n]{b} = \sqrt[n]{ab}, \qquad \text{for } a \geq 0, b \geq 0$$

To simplify a cube root whose radicand is a number, write the radicand as the product of a perfect cube number and another number. Then simplify, using the product rule.

EXAMPLE 2   Simplify: (a) $\sqrt[3]{32}$ (b) $\sqrt[3]{54}$ (c) $\sqrt[4]{32}$

Solution: (a) $\sqrt[3]{32} = \sqrt[3]{8 \cdot 4} = \sqrt[3]{8}\,\sqrt[3]{4} = 2\sqrt[3]{4}$

(b) $\sqrt[3]{54} = \sqrt[3]{27 \cdot 2} = \sqrt[3]{27}\,\sqrt[3]{2} = 3\sqrt[3]{2}$

(c) Write $\sqrt[4]{32}$ as a product of a perfect fourth power number and another number, then simplify.

$$\sqrt[4]{32} = \sqrt[4]{16 \cdot 2} = \sqrt[4]{16}\,\sqrt[4]{2} = 2\sqrt[4]{2} \qquad \blacksquare$$

3 A radical expression can be written in **exponential form** by using the following rule.

$$\sqrt[n]{a} = a^{1/n}, \qquad a \geq 0 \qquad \text{Rule 4}$$

Examples

$$\sqrt{8} = 8^{1/2} \qquad\qquad \sqrt{x} = x^{1/2}$$
$$\sqrt[3]{4} = 4^{1/3} \qquad\qquad \sqrt[4]{9} = 9^{1/4}$$
$$\sqrt[3]{x} = x^{1/3} \qquad\qquad \sqrt[4]{y} = y^{1/4}$$
$$\sqrt[3]{5x^2} = (5x^2)^{1/3} \qquad \sqrt[4]{3y^2} = (3y^2)^{1/4}$$

This concept just given can be expanded as follows.

Power   Index

$$\sqrt[n]{a^m} = (\sqrt[n]{a})^m = a^{m/n}, \text{ for } a \geq 0 \text{ and } m \text{ and } n \text{ integers} \qquad \text{Rule 5}$$

As long as the radicand is nonnegative we can change from one form to another.

Examples

$$\sqrt[3]{8^4} = (\sqrt[3]{8})^4 = 2^4 = 16$$
$$27^{2/3} = (\sqrt[3]{27})^2 = 3^2 = 9$$
$$\sqrt[3]{x^3} = x^{3/3} = x^1 = x$$
$$\sqrt[4]{y^{12}} = y^{12/4} = y^3$$

EXAMPLE 3   Simplify

(a) $\sqrt[3]{y^{15}}$ (b) $\sqrt[4]{x^{24}}$

Solution: (a) $\sqrt[3]{y^{15}} = y^{15/3} = y^5$

(b) $\sqrt[4]{x^{24}} = x^{24/4} = x^6$ ■

EXAMPLE 4   Evaluate

(a) $8^{5/3}$ (b) $16^{5/4}$ (c) $8^{-2/3}$

Solution: (a) $8^{5/3} = (\sqrt[3]{8})^5 = 2^5 = 32$

(b) $16^{5/4} = (\sqrt[4]{16})^5 = 2^5 = 32$

(c) Recall from Section 5.1 that $x^{-m} = \dfrac{1}{x^m}$. Thus

$$8^{-2/3} = \frac{1}{8^{2/3}} = \frac{1}{(\sqrt[3]{8})^2} = \frac{1}{2^2} = \frac{1}{4} \quad ■$$

EXAMPLE 5   Write each of the following radicals in exponential form.

(a) $\sqrt[3]{x^5}$ (b) $\sqrt[4]{y^7}$ (c) $\sqrt[4]{z^{15}}$

Solution: (a) $\sqrt[3]{x^5} = x^{5/3}$

(b) $\sqrt[4]{y^7} = y^{7/4}$

(c) $\sqrt[4]{z^{15}} = z^{15/4}$ ■

EXAMPLE 6   Simplify

(a) $\sqrt{x} \cdot \sqrt[4]{x}$ (b) $(\sqrt[4]{x^2})^8$

Solution: (a) $\sqrt{x} \cdot \sqrt[4]{x} = x^{1/2} \cdot x^{1/4}$ (b) $(\sqrt[4]{x^2})^8 = (x^{2/4})^8$

$= x^{1/2 + 1/4}$ $= (x^{1/2})^8$

$= x^{2/4 + 1/4}$ $= x^4$ ■

$= x^{3/4}$

$= \sqrt[4]{x^3}$

This section was meant to give you a brief introduction to roots other than square roots. If you take a course in intermediate algebra, you may study these concepts in more depth.

Exercise Set 9.7

Evaluate each of the following.

1. $\sqrt[3]{8}$ **2.** $\sqrt[3]{27}$ **3.** $\sqrt[3]{-8}$ **4.** $\sqrt[3]{-27}$

5. $\sqrt[4]{16}$ **6.** $\sqrt[3]{125}$ **7.** $\sqrt[4]{81}$ **8.** $\sqrt[4]{1}$

9. $\sqrt[3]{-1}$ **10.** $\sqrt[3]{-125}$ **11.** $\sqrt[3]{64}$ **12.** $\sqrt[3]{-64}$

Simplify each of the following.

13. $\sqrt[3]{54}$ **14.** $\sqrt[3]{32}$ **15.** $\sqrt[3]{16}$ **16.** $\sqrt[3]{24}$

17. $\sqrt[3]{81}$ **18.** $\sqrt[3]{128}$ **19.** $\sqrt[4]{32}$ **20.** $\sqrt[3]{250}$

21. $\sqrt[3]{40}$ **22.** $\sqrt[4]{48}$

Simplify each of the following.

23. $\sqrt[3]{x^3}$ **24.** $\sqrt[3]{y^6}$ **25.** $\sqrt[4]{y^{12}}$ **26.** $\sqrt[4]{y^{16}}$

27. $\sqrt[3]{x^{12}}$ **28.** $\sqrt[3]{x^9}$ **29.** $\sqrt[4]{y^4}$ **30.** $\sqrt[4]{y^{24}}$

31. $\sqrt[3]{x^{15}}$ **32.** $\sqrt[3]{x^{18}}$

Evaluate each of the following.

33. $8^{4/3}$ **34.** $27^{4/3}$ **35.** $16^{3/4}$ **36.** $81^{3/4}$

37. $1^{5/3}$ **38.** $16^{5/2}$ **39.** $9^{3/2}$ **40.** $64^{2/3}$

41. $16^{3/4}$ **42.** $25^{3/2}$ **43.** $125^{4/3}$ **44.** $8^{-1/3}$

45. $27^{-2/3}$ **46.** $16^{-3/4}$ **47.** $8^{-5/3}$ **48.** $64^{-2/3}$

Write each radical in exponential form.

49. $\sqrt[3]{x^7}$ **50.** $\sqrt[3]{x^6}$ **51.** $\sqrt[3]{x^4}$ **52.** $\sqrt[4]{x^7}$

53. $\sqrt[4]{y^{15}}$ **54.** $\sqrt[4]{x^9}$ **55.** $\sqrt[4]{y^{21}}$ **56.** $\sqrt[4]{x^5}$

Simplify each of the following and write the answer in exponential form.

57. $\sqrt[3]{x} \cdot \sqrt[3]{x}$ **58.** $\sqrt[3]{x} \cdot \sqrt[4]{x}$ **59.** $\sqrt[4]{x^2} \cdot \sqrt[4]{x^2}$ **60.** $\sqrt[3]{x} \cdot \sqrt[3]{x^5}$

61. $(\sqrt[3]{x^2})^6$ **62.** $(\sqrt[4]{x^3})^4$ **63.** $(\sqrt[4]{x^2})^4$ **64.** $(\sqrt[3]{x^6})^2$

JUST FOR FUN

1. Simplify $\sqrt[3]{xy} \cdot \sqrt[3]{x^2y^2}$

2. Simplify $\sqrt[4]{3x^2y} \cdot \sqrt[4]{27x^6y^3}$

3. Simplify $\sqrt[4]{32} - \sqrt[4]{2}$

4. Simplify $\sqrt[3]{3x^3y} + \sqrt[3]{24x^3y}$

5. Simplify $\dfrac{1}{\sqrt[3]{2}}$

6. Simplify $\dfrac{1}{\sqrt[3]{x}}$

Summary

Glossary

Conjugate: The conjugate of $a + b$ is $a - b$.

Hypotenuse: The side opposite the right angle in a right triangle.

Index of a radical: The root of a radical expression.

Irrational numbers: Real numbers that are not rational numbers.

Legs of a right triangle: The two smaller sides of the right triangle.

Like square roots: Square roots having the same radicand.

Perfect square factor: When a radicand contains a variable (or number) raised to an even exponent, that variable (or number) and exponent together form a perfect square factor.

Perfect square numbers: 1, 4, 9, 16, 25, 36, 49, 64, ...

Perfect cube numbers: 1, 8, 27, 64, 125, 216, 343, 512, ...

Principal or positive square root: The principal or positive square root of a positive real number x, written $\sqrt{x}$, is that positive number whose square equals x.

Radical equation: An equation that contains a variable in a radicand.

Radical expression: A mathematical expression containing a radical.

Radical sign: $\sqrt{}$.

Radicand: The expression within the radical sign.

Rationalize the denominator: Removing radical expressions from the denominator of a fraction.

Rational number: A number that can be written in the form $\frac{a}{b}$, where a and b are both integers, $b \neq 0$.

Right triangle: A triangle with a $90°$ angle.

Unlike square roots: Square roots having different radicands.

Important Facts

Product rule for radicals $\sqrt{a} \cdot \sqrt{b} = \sqrt{ab}$, $a \geq 0$, $b \geq 0$

$\sqrt{a^{2 \cdot n}} = a^n$, $a \geq 0$

$\sqrt{a^2} = a$, $a \geq 0$

Quotient rule for radicals $\dfrac{\sqrt{a}}{\sqrt{b}} = \sqrt{\dfrac{a}{b}}$, $a \geq 0$, $b > 0$

$\sqrt[n]{a} = a^{1/n}$, $a \geq 0$

$\sqrt[n]{a^m} = (\sqrt[n]{a})^m = a^{m/n}$, $a \geq 0$

Pythagorean theorem: $a^2 + b^2 = c^2$

Distance formula: $d = \sqrt{(x_2 - x_1)^2 + (y_2 - y_1)^2}$

Review Exercises

[9.1] Evaluate each expression.

1. $\sqrt{25}$

2. $\sqrt{36}$

3. $\sqrt{1}$

4. $\sqrt{100}$

5. $-\sqrt{81}$

6. $-\sqrt{144}$

[9.2] Simplify each expression.

7. $\sqrt{32}$

8. $\sqrt{44}$

9. $\sqrt{45x^5 y^4}$

10. $\sqrt{125x^4 y^6}$

11. $\sqrt{15x^5 yz^3}$

12. $\sqrt{48ab^4 c^5}$

Simplify each expression.

13. $\sqrt{8} \cdot \sqrt{12}$

14. $\sqrt{5x} \cdot \sqrt{5x}$

15. $\sqrt{18x} \cdot \sqrt{2xy}$

16. $\sqrt{25x^2 y} \cdot \sqrt{3y}$

17. $\sqrt{20xy^4} \cdot \sqrt{5xy^3}$

18. $\sqrt{8x^3 y} \cdot \sqrt{3y^4}$

[9.3] Simplify each expression.

19. $\dfrac{\sqrt{32}}{\sqrt{2}}$

20. $\sqrt{\dfrac{10}{250}}$

21. $\sqrt{\dfrac{7}{28}}$

22. $\dfrac{3}{\sqrt{5}}$

23. $\sqrt{\dfrac{5x}{12}}$

24. $\sqrt{\dfrac{x}{6}}$

25. $\sqrt{\dfrac{x^2}{2}}$

26. $\sqrt{\dfrac{x^5}{8}}$

27. $\sqrt{\dfrac{60xy^5}{4x^5y^3}}$

28. $\sqrt{\dfrac{30x^4y}{15x^2y^4}}$

29. $\dfrac{\sqrt{90}}{\sqrt{8x^3y^2}}$

30. $\dfrac{\sqrt{2x^4yz^4}}{\sqrt{7x^5yz^2}}$

31. $\dfrac{3}{1+\sqrt{2}}$

32. $\dfrac{5}{3-\sqrt{6}}$

33. $\dfrac{\sqrt{3}}{2+\sqrt{x}}$

34. $\dfrac{2}{\sqrt{x}-5}$

35. $\dfrac{\sqrt{5}}{\sqrt{x}+\sqrt{3}}$

[9.4] Simplify each expression.

36. $6\sqrt{3}-2\sqrt{3}$

37. $6\sqrt{2}-8\sqrt{2}+\sqrt{2}$

38. $3\sqrt{x}-5\sqrt{x}$

39. $\sqrt{x}+3\sqrt{x}-4\sqrt{x}$

40. $\sqrt{8}-\sqrt{2}$

41. $7\sqrt{40}-2\sqrt{10}$

42. $2\sqrt{98}-4\sqrt{72}$

43. $3\sqrt{18}+5\sqrt{50}-2\sqrt{32}$

44. $4\sqrt{27}+5\sqrt{80}+2\sqrt{12}$

[9.5] Solve each equation.

45. $\sqrt{x}=9$

46. $\sqrt{x}=-2$

47. $\sqrt{x-3}=6$

48. $\sqrt{3x+1}=5$

49. $\sqrt{2x+4}=\sqrt{3x-5}$

50. $4\sqrt{x}-x=4$

51. $\sqrt{x^2+4}=x+2$

52. $\sqrt{3x+5}-\sqrt{5x-9}=0$

53. $3\sqrt{2x+3}=9$

[9.6] Find the quantity indicated. You may leave your answer in square root form if the answer is not a perfect square.

54.

55.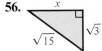

56.

57.

58. Jason leans a 12-foot ladder against a house. If the base of the ladder is 3 feet from the house, how high is the ladder on the house?

59. Find the diagonal of a rectangle with a length of 15 inches and a width of 6 inches.

60. Find the straight-line distance between the points $(-5, 4)$ and $(3, -2)$.

61. Find the straight-line distance between the points $(6, 5)$ and $(-6, 8)$.

62. Find the side of a square that has an area of 121 square feet.

[9.7] Evaluate.

63. $\sqrt[3]{8}$

64. $\sqrt[3]{-27}$

65. $\sqrt[4]{16}$

Simplify each of the following.

66. $\sqrt[3]{16}$ **67.** $\sqrt[3]{24}$ **68.** $\sqrt[4]{32}$
69. $\sqrt[3]{48}$ **70.** $\sqrt[3]{54}$ **71.** $\sqrt[4]{96}$

Simplify each of the following.

72. $\sqrt[3]{x^{15}}$ **73.** $\sqrt[3]{x^{12}}$ **74.** $\sqrt[4]{y^{16}}$ **75.** $\sqrt[4]{y^{20}}$

Evaluate each of the following.

76. $8^{2/3}$ **77.** $16^{1/2}$ **78.** $27^{-2/3}$
79. $64^{2/3}$ **80.** $16^{-3/4}$ **81.** $25^{3/2}$

Write each radical in exponential form.

82. $\sqrt[3]{x^5}$ **83.** $\sqrt[3]{x^{10}}$ **84.** $\sqrt[4]{y^9}$
85. $\sqrt{x^5}$ **86.** $\sqrt{y^{11}}$ **87.** $\sqrt[4]{x^7}$

Simplify each of the following.

88. $\sqrt{x} \cdot \sqrt{x}$ **89.** $\sqrt[3]{x} \cdot \sqrt[3]{x}$ **90.** $\sqrt[3]{x^2} \cdot \sqrt[3]{x^7}$
91. $\sqrt[4]{x^2} \cdot \sqrt[4]{x^6}$ **92.** $(\sqrt[3]{x^3})^2$ **93.** $(\sqrt[3]{x^2})^3$
94. $(\sqrt[4]{x^2})^6$ **95.** $(\sqrt[4]{x^3})^8$

Practice Test

Simplify each expression.

1. $\sqrt{(x+3)^2}$ **2.** $\sqrt{60}$ **3.** $\sqrt{96}$ **4.** $\sqrt{12x^2}$

5. $\sqrt{32x^4y^5}$ **6.** $\sqrt{8x^2y} \cdot \sqrt{6xy}$ **7.** $\sqrt{15xy^2} \cdot \sqrt{5x^3y^3}$ **8.** $\sqrt{\dfrac{5}{125}}$

9. $\dfrac{\sqrt{3xy^2}}{\sqrt{48x^3}}$ **10.** $\dfrac{1}{\sqrt{2}}$ **11.** $\sqrt{\dfrac{4x}{5}}$ **12.** $\sqrt{\dfrac{40x^2y^5}{6x^3y^7}}$

13. $\dfrac{3}{2+\sqrt{5}}$ **14.** $\dfrac{6}{\sqrt{x}-3}$ **15.** $\sqrt{48}+\sqrt{75}+2\sqrt{3}$ **16.** $4\sqrt{x}-6\sqrt{x}-\sqrt{x}$

Solve each equation.

17. $\sqrt{x+5} = 9$ **18.** $2\sqrt{x-4}+4 = x$

Solve each problems.

19. Find the value of x in the triangle shown.

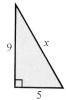

20. Find the length of the straight line between the points $(-2, -4)$ and $(5, 1)$.

*21. Evaluate $27^{-4/3}$

*22. Simplify $\sqrt[3]{x^4} \cdot \sqrt[3]{x^{11}}$

* From optional section

10

Quadratic Equations

See Section 10.1, Example 6.

■   *Know that every positive real number has two square roots.*

■   *Solve quadratic equations using the square root property.*

In Section 5.6 we solved quadratic equations by factoring. Recall that **quadratic equations** are equations of the form

$$ax^2 + bx + c = 0,$$

where a, b, and c are real numbers, $a \neq 0$. A quadratic equation in this form is said to be in **standard form.**

In this chapter we will give two techniques, completing the square and the quadratic formula, for solving quadratic equations that cannot be solved by factoring. In this section we introduce the square root property which will be used in the completing the square procedure in the next section.

■   In Section 9.1 we stated that every positive number has two square roots. Thus far we have been using only the positive or principal square root. In this section we use both the positive and negative square roots of a number.

The positive square root of 25 is 5.

$$\sqrt{25} = 5$$

The negative square root of 25 is -5.

$$-\sqrt{25} = -5$$

Notice that $5 \cdot 5 = 25$ and $(-5)(-5) = 25$. The two square roots of 25 are $+5$ and -5. A convenient way to indicate the two square roots of a number is to use the plus or minus symbol, $\pm$. For example, the square roots of 25 can be indicated ± 5, read "plus or minus 5."

Number	Both square roots
36	± 6
100	± 10
7	$\pm \sqrt{7}$

The value of a number like $-\sqrt{5}$ can be found by finding the value of $\sqrt{5}$ in Appendix B, or on your calculator, and then taking its opposite or negative value.

$$\sqrt{5} = 2.24$$
$$-\sqrt{5} = -2.24$$

Consider the equation

$$x^2 = 25$$

We can see by substitution that this equation has two solutions, 5 and -5.

check $x = 5$	check $x = -5$
$x^2 = 25$	$x^2 = 25$
$5^2 = 25$	$(-5)^2 = 25$
$25 = 25$ true	$25 = 25$ true

To solve the equation $x^2 = 25$, we can take the square root of both sides of the equation. We can solve the equation as follows:

$$x^2 = 25$$
$$\sqrt{x^2} = \pm\sqrt{25}$$
$$x = \pm 5$$

2   In general for any quadratic equation of the form $x^2 = a$, we can use the square root property to obtain the solution.

> **Square Root Property**
>
> $$\text{If} \qquad x^2 = a$$
> $$\text{then} \quad \sqrt{x^2} = \pm\sqrt{a}$$
> $$\text{and} \qquad x = \pm\sqrt{a}$$

EXAMPLE 1   Solve the equation $x^2 - 9 = 0$.

Solution: Add 9 to both sides of the equation to get the variable all by itself on one side of the equation.

$$x^2 = 9$$

Now take the square root of both sides of the equation.

$$\sqrt{x^2} = \pm\sqrt{9}$$
$$x = \pm 3$$

Check in the original equation.

$x = 3$	$x = -3$
$x^2 - 9 = 0$	$x^2 - 9 = 0$
$3^2 - 9 = 0$	$(-3)^2 - 9 = 0$
$9 - 9 = 0$	$9 - 9 = 0$
$0 = 0$ true	$0 = 0$ true ∎

EXAMPLE 2   Solve the equation $x^2 + 7 = 71$.

Solution: Begin by subtracting 7 from both sides of the equation.

$$x^2 + 7 = 71$$
$$x^2 = 64$$
$$\sqrt{x^2} = \pm\sqrt{64}$$
$$x = \pm 8 \quad ∎$$

When solving equations using the square root property the middle step is sometimes not shown.

EXAMPLE 3   Solve the equation $x^2 - 7 = 0$.

Solution: $x^2 - 7 = 0$

$$x^2 = 7$$

$$x = \pm\sqrt{7} \quad \blacksquare$$

EXAMPLE 4   Solve the equation $(x - 3)^2 = 4$.

Solution: Begin by taking the square root of both sides of the equation.

$$(x - 3)^2 = 4$$

$$x - 3 = \pm\sqrt{4}$$

$$x - 3 = \pm 2$$

$$x - 3 + 3 = 3 \pm 2 \qquad \text{add 3 to both sides of equation}$$

$$x = 3 \pm 2$$

$$x = 3 + 2 \quad \text{or} \quad x = 3 - 2$$

$$x = 5 \quad \text{or} \quad x = 1$$

The solutions are 1 and 5.   ■

EXAMPLE 5   Solve the equation $(3x + 4)^2 = 32$.

Solution: $(3x + 4)^2 = 32$

$$3x + 4 = \pm\sqrt{32}$$

$$3x + 4 = \pm 4\sqrt{2}$$

$$3x = -4 \pm 4\sqrt{2}$$

$$x = \frac{-4 \pm 4\sqrt{2}}{3}$$

Thus the solutions are $\dfrac{-4 + 4\sqrt{2}}{3}$ and $\dfrac{-4 - 4\sqrt{2}}{3}$.   ■

EXAMPLE 6   Mrs. Albert wants a rectangular garden. To make her garden look best, the length of the rectangle is to be 1.62 times its width (a rectangle with this length-to-width ratio is referred to as the *golden rectangle*); see Fig. 10.1. Find the dimensions of the rectangle if the rectangle is to have an area of 6000 square feet. Leave your answer in radical form if a calculator is not available.

Solution: Let x = width of rectangle, then $1.62x$ = length of rectangle.

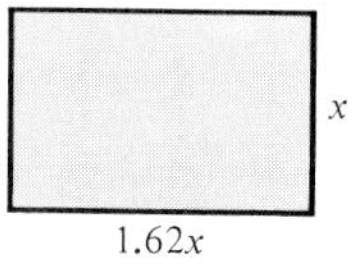

$$\text{area} = \text{length} \cdot \text{width}$$

$$6000 = (1.62x)x$$

$$6000 = 1.62x^2$$

$$x^2 = \frac{6000}{1.62} = 3703.7$$

$$x = \pm\sqrt{3703.7} = 60.86 \text{ feet}$$

1.62x

Figure 10.1

Since the distance cannot be negative the width, x, is 60.86 feet.

The length is $1.62(60.86) = 98.60$ feet.

Check: area = length · width

$$6000 = (98.60)(60.86)$$

$$6000 = 6000.79$$ true (slight round-off error due to rounding off decimal answers)   ■

Note that the answer did not come out to be a whole number. In many real-life situations this is the case. You should not feel uncomfortable when this occurs. If you do not have a calculator with a square root $\boxed{\sqrt{}}$ key, leave your answers in terms of square roots. To evaluate square roots on a calculator, enter the number and press the square root key.

Exercise Set 10.1

Solve each equation.

1. $x^2 = 16$ **2.** $x^2 = 25$ **3.** $x^2 = 100$ **4.** $x^2 = 49$
5. $y^2 = 36$ **6.** $z^2 = 9$ **7.** $x^2 = 10$ **8.** $a^2 = 15$
9. $x^2 = 8$ **10.** $w^2 = 24$ **11.** $3x^2 = 12$ **12.** $5y^2 = 45$
13. $2w^2 = 34$ **14.** $5x^2 = 90$ **15.** $2x^2 + 1 = 19$ **16.** $3x^2 - 4 = 8$
17. $4w^2 - 3 = 12$ **18.** $3y^2 + 8 = 36$ **19.** $5x^2 - 9 = 30$ **20.** $2x^2 + 3 = 51$

Solve each equation.

21. $(x + 1)^2 = 4$ **22.** $(x - 2)^2 = 9$ **23.** $(x - 3)^2 = 16$ **24.** $(x + 5)^2 = 25$
25. $(x + 4)^2 = 36$ **26.** $(x - 4)^2 = 100$ **27.** $(x - 1)^2 = 12$ **28.** $(x + 3)^2 = 18$
29. $(x + 6)^2 = 20$ **30.** $(x - 4)^2 = 32$ **31.** $(x + 2)^2 = 25$ **32.** $(x + 6)^2 = 75$
33. $(x - 9)^2 = 100$ **34.** $(x - 3)^2 = 15$ **35.** $(2x + 3)^2 = 18$ **36.** $(3x - 2)^2 = 30$
37. $(4x + 1)^2 = 20$ **38.** $(5x - 6)^2 = 100$ **39.** $(2x - 6)^2 = 18$ **40.** $(3x - 5)^2 = 90$

41. The length of a rectangle is twice its width. Find the length and width if the area is 80 square feet.

42. The length of a rectangle is 3 times its width. Find the length and width if its area is 96 square feet.

10.2
Solving Quadratic Equations by Completing the Square

1 *Write perfect square trinomials.*

2 *Solve quadratic equations by completing the square.*

Quadratic equations that cannot be solved using factoring can be solved by completing the square, or by the quadratic formula. In this section we focus on the technique of completing the square. In Section 10.3 we will use the quadratic formula.

1 A **perfect square trinomial** is a trinomial that can be expressed as the square of a binomial. Some examples follow.

Perfect square trinomials	*Factors*	*Square of a binomial*
$x^2 + 6x + 9$	$= (x + 3)(x + 3)$	$= (x + 3)^2$
$x^2 - 6x + 9$	$= (x - 3)(x - 3)$	$= (x - 3)^2$
$x^2 + 10x + 25$	$= (x + 5)(x + 5)$	$= (x + 5)^2$
$x^2 - 10x + 25$	$= (x - 5)(x - 5)$	$= (x - 5)^2$

Notice that each of the squared terms in the perfect square trinomials given on the previous page have a numerical coefficient of 1. When the coefficient of the squared term is 1 there is an important relationship between the coefficient of the x term and the constant. In every perfect square trinomial of this type *the constant term is the square of one-half the coefficient of the x term.*

Consider the perfect square trinomial $x^2 + 6x + 9$. The coefficient of the x term is 6 and the constant is 9. Note

$$\left[\frac{1}{2}(6)\right]^2 = 3^2 = 9.$$

Consider the perfect square trinomial $x^2 - 10x + 25$. The coefficient of the x term is -10 and the constant is 25. Note

$$\left[\frac{1}{2}(-10)\right]^2 = (-5)^2 = 25.$$

Consider the expression $x^2 + 8x +$ ▢ . Can you determine what number must be placed in the colored box to make the trinomial a perfect square trinomial? If you answered 16 you answered correctly. Note

$$\left[\frac{1}{2}(8)\right]^2 = 4^2 = 16$$
$$x^2 + 8x + 16 = (x + 4)^2$$

Let us examine perfect square trinomials a little further.

Perfect square trinomial   *Square of a binomial*

$$x^2 + 6x + 9 \quad = \quad (x + 3)^2$$
$$\frac{1}{2}(6) = 3$$

$$x^2 - 10x + 25 \quad = \quad (x - 5)^2$$
$$\frac{1}{2}(-10) = -5$$

Note that when a perfect square trinomial is written as the square of a binomial, *the constant in the binomial is one-half the value of the coefficient of the x term in the perfect square trinomial.*

2   The procedure for solving a quadratic equation by completing the square is illustrated in the following example.

EXAMPLE 1   Solve the equation $x^2 + 6x + 5 = 0$ by completing the square.

Solution: First we make sure that the squared term has a coefficient of 1. (We will explain what to do if the coefficient is not 1 in Example 5). Next we wish to get the squared and x terms by themselves on the left side of the equation. Therefore we subtract 5 from both sides of the equation.

$$x^2 + 6x + 5 = 0$$
$$x^2 + 6x = -5$$

Determine one-half the numerical coefficient of the x term. In this example the x term is $6x$.

$$\frac{1}{2}(6) = 3$$

Square the number determined above.

$$(3)^2 = (3)(3) = 9$$

and add this product to both sides of the equation.

$$x^2 + 6x + 9 = -5 + 9$$

or

$$x^2 + 6x + 9 = 4$$

By following this procedure, we produce a perfect square trinomial on the left side of the equation. The expression $x^2 + 6x + 9$ is a perfect square trinomial that can be expressed $(x + 3)^2$. Therefore

$$x^2 + 6x + 9 = 4$$

can be written $(x + 3)^2 = 4$

Now take the square root of both sides of the equation.

$$x + 3 = \pm\sqrt{4}$$
$$x + 3 = \pm 2$$

Finally, solve for x by subtracting 3 from both sides of the equation.

$$x + 3 - 3 = -3 \pm 2$$
$$x = -3 \pm 2$$

$$x = -3 + 2 \qquad \text{or} \qquad x = -3 - 2$$
$$x = -1 \qquad\qquad\qquad x = -5$$

Thus the solutions are -1 and -5. Check both solutions in the original equation.

$$x = -1 \qquad\qquad\qquad x = -5$$
$$x^2 + 6x + 5 = 0 \qquad\qquad x^2 + 6x + 5 = 0$$
$$(-1)^2 + 6(-1) + 5 = 0 \qquad (-5)^2 + 6(-5) + 5 = 0$$
$$1 - 6 + 5 = 0 \qquad\qquad 25 - 30 + 5 = 0$$
$$0 = 0 \quad \text{true} \qquad\qquad 0 = 0 \quad \text{true} \quad\blacksquare$$

EXAMPLE 2   Solve the equation $x^2 - 8x + 15 = 0$ by completing the square.

Solution: $x^2 - 8x + 15 = 0$
$$x^2 - 8x = -15$$

Take half the numerical coefficient of the x term, square it, and add this product to both sides of the equation.

$$\frac{1}{2}(-8) = -4 \qquad (-4)^2 = 16$$

Now add 16 to both sides of the equation

$$x^2 - 8x + 16 = -15 + 16$$
$$x^2 - 8x + 16 = 1$$

or $(x - 4)^2 = 1$

$$x - 4 = \pm\sqrt{1}$$
$$x - 4 = \pm 1$$
$$x = 4 \pm 1$$

$$x = 4 + 1 \qquad \text{or} \qquad x = 4 - 1$$
$$x = 5 \qquad\qquad\qquad x = 3$$

Check:

$x = 5$	$x = 3$
$x^2 - 8x + 15 = 0$	$x^2 - 8x + 15 = 0$
$(5)^2 - 8(5) + 15 = 0$	$(3)^2 - 8(3) + 15 = 0$
$25 - 40 + 15 = 0$	$9 - 24 + 15 = 0$
$0 = 0$ true	$0 = 0$ true

The solutions are 5 and 3. ■

EXAMPLE 3 Solve the equation $x^2 = 3x + 18$ by completing the square.

Solution: Place all terms except the constant on the left side of the equation.

$$x^2 = 3x + 18$$
$$x^2 - 3x = 18$$

Take half the numerical coefficient of the x term, square it and add this product to both sides of the equation.

$$\frac{1}{2}(-3) = -\frac{3}{2} \qquad \left(-\frac{3}{2}\right)^2 = \frac{9}{4}$$

$$x^2 - 3x + \frac{9}{4} = 18 + \frac{9}{4}$$

$$\left(x - \frac{3}{2}\right)^2 = 18 + \frac{9}{4}$$

$$\left(x - \frac{3}{2}\right)^2 = \frac{72}{4} + \frac{9}{4}$$

$$\left(x - \frac{3}{2}\right)^2 = \frac{81}{4}$$

$$x - \frac{3}{2} = \pm\sqrt{\frac{81}{4}}$$

$$x - \frac{3}{2} = \pm\frac{9}{2}$$

$$x = \frac{3}{2} \pm \frac{9}{2}$$

$$x = \frac{3}{2} + \frac{9}{2} \quad \text{or} \quad x = \frac{3}{2} - \frac{9}{2}$$

$$x = \frac{12}{2} = 6 \qquad\qquad x = -\frac{6}{2} = -3$$

The solutions are 6 and -3.   ∎

In the following examples we will not illustrate some of the intermediate steps.

EXAMPLE 4   Solve the equation $x^2 - 6x + 1 = 0$.

Solution: $x^2 - 6x + 1 = 0$

$$x^2 - 6x = -1$$
$$x^2 - 6x + 9 = -1 + 9$$
$$(x - 3)^2 = 8$$
$$x - 3 = \pm\sqrt{8}$$
$$x - 3 = \pm 2\sqrt{2}$$
$$x = 3 \pm 2\sqrt{2}$$

The solutions are $3 + 2\sqrt{2}$ and $3 - 2\sqrt{2}$.   ∎

EXAMPLE 5   Solve the equation $3m^2 - 9m + 6 = 0$ by completing the square.

Solution: To solve an equation by completing the square the numerical coefficient of the squared term must be 1. Since the numerical coefficient of the squared term is 3, we multiply both sides of the equation by $\frac{1}{3}$ to make the numerical coefficient of the squared term equal to 1.

$$3m^2 - 9m + 6 = 0$$
$$\frac{1}{3}(3m^2 - 9m + 6) = \frac{1}{3}(0)$$
$$m^2 - 3m + 2 = 0$$

Now proceed as before.

$$m^2 - 3m = -2$$
$$m^2 - 3m + \frac{9}{4} = -2 + \frac{9}{4}$$
$$\left(m - \frac{3}{2}\right)^2 = -\frac{8}{4} + \frac{9}{4}$$
$$\left(m - \frac{3}{2}\right)^2 = \frac{1}{4}$$
$$m - \frac{3}{2} = \pm\sqrt{\frac{1}{4}}$$
$$m - \frac{3}{2} = \pm\frac{1}{2}$$
$$m = \frac{3}{2} \pm \frac{1}{2}$$

$$m = \frac{3}{2} + \frac{1}{2} \quad \text{or} \quad m = \frac{3}{2} - \frac{1}{2}$$

$$m = \frac{4}{2} = 2 \qquad\qquad m = \frac{2}{2} = 1$$

The solutions are 2 and 1.   ∎

To Solve A Quadratic Equation by Completing the Square

1. Use the multiplication (or division) property if necessary to make the numerical coefficient of the squared term equal to 1.
2. Rewrite the equation with the constant by itself on the right side of the equation.
3. Take one-half the numerical coefficient of the first-powered term, square it, and add this quantity to both sides of the equation.
4. Replace the trinomial with its equivalent squared binomial.
5. Take the square root of both sides of the equation.
6. Solve for the variable.
7. Check your answers in the original equation.

Exercise Set 10.2

Solve each equation by completing the square.

1. $x^2 + 2x - 3 = 0$

2. $x^2 - 6x + 8 = 0$

3. $x^2 - 4x - 5 = 0$

4. $x^2 + 8x + 12 = 0$

5. $x^2 + 3x + 2 = 0$

6. $x^2 + 4x - 32 = 0$

7. $x^2 - 8x + 15 = 0$

8. $x^2 - 9x + 14 = 0$

9. $x^2 = -6x - 9$

10. $x^2 + 5x + 4 = 0$

11. $x^2 = -5x - 6$

12. $x^2 = 2x + 15$

13. $x^2 + 9x + 18 = 0$

14. $x^2 - 9x + 18 = 0$

15. $x^2 = 15x - 56$

16. $x^2 = 3x + 28$

17. $-4x = -x^2 + 12$

18. $-x^2 - 3x + 40 = 0$

19. $x^2 + 2x - 6 = 0$

20. $x^2 - 4x + 2 = 0$

21. $6x + 6 = -x^2$

22. $x^2 - x - 3 = 0$

23. $-x^2 + 5x = -8$

24. $x^2 + 3x - 6 = 0$

25. $2x^2 + 4x - 6 = 0$

26. $2x^2 + 2x - 24 = 0$

27. $2x^2 + 18x + 4 = 0$

28. $2x^2 = 8x + 90$

29. $3x^2 + 33x + 72 = 0$

30. $4x^2 = -28x + 32$

31. $2x^2 + 10x - 3 = 0$

32. $3x^2 - 8x + 4 = 0$

33. $3x^2 + 6x = 6$

34. $2x^2 - x = 5$

35. $x^2 + 4x = 0$

36. $2x^2 - 6x = 0$

37. $2x^2 - 4x = 0$

38. $3x^2 = 9x$

39. When three times a number is added to the square of a number the sum is 4. Find the number(s).

40. When five times a number is subtracted from two times the square of a number the difference is 12. Find the number(s).

41. If the square of three more than a number is 9, find the number(s).

42. If the square of two less than an integer is 16, find the number(s).

43. The product of two positive numbers is 21. Find the two numbers if the larger is 4 greater than the smaller.

Solve by completing the square.

1. $x^2 + \dfrac{3}{5}x - \dfrac{1}{2} = 0$

2. $x^2 - \dfrac{2}{3}x - \dfrac{1}{5} = 0$

3. $3x^2 + \dfrac{1}{2}x = 4$

10.3
Solving Quadratic Equations by the Quadratic Formula

1️⃣ *Solve quadratic equations by the quadratic formula.*

2️⃣ *Determine the number of solutions to a quadratic equation using the discriminant.*

1️⃣   A method that can be used to solve any quadratic equation is to use the quadratic formula. It is the most useful and most versatile method of solving quadratic equations.

The standard form of a quadratic equation is $ax^2 + bx + c = 0$, where a is the numerical coefficient of the squared term, b is the numerical coefficient of the first-powered term, and c is the constant.

Quadratic equation in standard form	*Values of a, b, and c*
$x^2 - 3x + 4 = 0$	$a = 1, b = -3, c = 4$
$-2x^2 + \dfrac{1}{2}x - 2 = 0$	$a = -2, b = \dfrac{1}{2}, c = -2$
$3x^2 - 4 = 0$	$a = 3, b = 0, c = -4$
$5x^2 + 3x = 0$	$a = 5, b = 3, c = 0$
$-\dfrac{1}{2}x^2 + 5 = 0$	$a = -\dfrac{1}{2}, b = 0, c = 5$
$-12x^2 + 8x = 0$	$a = -12, b = 8, c = 0$

To Solve A Quadratic Equation by the Quadratic Formula

1. Write the equation in standard form, $ax^2 + bx + c = 0$, and determine the numerical values for a, b, and c.
2. Substitute the values for a, b, and c in the quadratic formula and then evaluate to obtain the solution.

The quadratic formula

$$x = \frac{-b \pm \sqrt{b^2 - 4ac}}{2a}$$

EXAMPLE 1   Solve the equation $x^2 + 2x - 8 = 0$ using the quadratic formula.

Solution: $1x^2 + 2x - 8 = 0$

$$\overset{\uparrow}{a} \quad \overset{\uparrow}{b} \quad \overset{\uparrow}{c}$$

$$x = \frac{-b \pm \sqrt{b^2 - 4ac}}{2a}$$

$$= \frac{-(2) \pm \sqrt{(2)^2 - 4(1)(-8)}}{2(1)}$$

$$= \frac{-2 \pm \sqrt{4 + 32}}{2}$$

$$= \frac{-2 \pm \sqrt{36}}{2}$$

$$= \frac{-2 \pm 6}{2}$$

$$x = \frac{-2 + 6}{2} \qquad \text{or} \qquad x = \frac{-2 - 6}{2}$$

$$x = \frac{4}{2} = 2 \qquad\qquad x = \frac{-8}{2} = -4$$

Check:

$x = 2$	$x = -4$
$x^2 + 2x - 8 = 0$	$x^2 + 2x - 8 = 0$
$(2)^2 + 2(2) - 8 = 0$	$(-4)^2 + 2(-4) - 8 = 0$
$4 + 4 - 8 = 0$	$16 - 8 - 8 = 0$
$0 = 0$ true	$0 = 0$ true ∎

COMMON STUDENT ERROR

The **entire numerator** of the quadratic formula must be divided by $2a$.

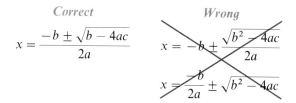

Correct

$$x = \frac{-b \pm \sqrt{b - 4ac}}{2a}$$

Wrong

$$x = -b \pm \frac{\sqrt{b^2 - 4ac}}{2a}$$

$$x = \frac{-b}{2a} \pm \sqrt{b^2 - 4ac}$$

EXAMPLE 2   Solve the equation $6x^2 - x - 2 = 0$ using the quadratic formula.

Solution:

$$6x^2 - x - 2 = 0$$

$$a = 6, \qquad b = -1, \qquad c = -2$$

$$x = \frac{-b \pm \sqrt{b^2 - 4ac}}{2a}$$

$$= \frac{-(-1) \pm \sqrt{(-1)^2 - 4(6)(-2)}}{2(6)}$$

$$= \frac{1 \pm \sqrt{1 + 48}}{12}$$

$$= \frac{1 \pm \sqrt{49}}{2}$$

$$= \frac{1 \pm 7}{12}$$

$$x = \frac{1 + 7}{12} = \frac{8}{12} = \frac{2}{3} \qquad \text{or} \qquad x = \frac{1 - 7}{12} = \frac{-6}{12} = \frac{-1}{2}$$

Check:

$$x = \frac{2}{3} \qquad\qquad\qquad\qquad x = -\frac{1}{2}$$

$$6x^2 - x - 2 = 0 \qquad\qquad\qquad 6x^2 - x - 2 = 0$$

$$6\left(\frac{2}{3}\right)^2 - \frac{2}{3} - 2 = 0 \qquad\qquad 6\left(-\frac{1}{2}\right)^2 - \left(-\frac{1}{2}\right) - 2 = 0$$

$$\overset{2}{\cancel{6}}\left(\frac{4}{\underset{3}{\cancel{9}}}\right) - \frac{2}{3} - 2 = 0 \qquad\qquad \overset{3}{\cancel{6}}\left(\frac{1}{\underset{2}{\cancel{4}}}\right) + \frac{1}{2} - 2 = 0$$

$$\frac{8}{3} - \frac{2}{3} - \frac{6}{3} = 0 \qquad\qquad\qquad \frac{3}{2} + \frac{1}{2} - \frac{4}{2} = 0$$

$$0 = 0 \quad \text{true} \qquad\qquad\qquad 0 = 0 \quad \text{true} \qquad \blacksquare$$

EXAMPLE 3   Solve the equation $2x^2 + 4x - 5 = 0$ using the quadratic formula.

Solution:

$$a = 2, \qquad b = 4, \qquad c = -5$$

$$x = \frac{-b \pm \sqrt{b^2 - 4ac}}{2a}$$

$$= \frac{-4 \pm \sqrt{(4)^2 - 4(2)(-5)}}{2(2)}$$

$$= \frac{-4 \pm \sqrt{16 + 40}}{4}$$

$$= \frac{-4 \pm \sqrt{56}}{4}$$

$$= \frac{-4 \pm 2\sqrt{14}}{4}$$

Now factor out a 2 from both terms in the numerator, then divide out common factors.

$$x = \frac{\overset{1}{\cancel{2}}(-2 \pm \sqrt{14})}{\underset{2}{\cancel{4}}}$$

$$x = \frac{-2 \pm \sqrt{14}}{2}$$

Thus the answers are

$$x = \frac{-2 + \sqrt{14}}{2} \quad \text{and} \quad x = \frac{-2 - \sqrt{14}}{2} \quad \blacksquare$$

COMMON STUDENT ERROR

Many students work the problems correctly until the last step, where they perform an error. Do not make the mistake of trying to simplify an answer which cannot be simplified any further. Below are listed answers which cannot be simplified, along with some common errors.

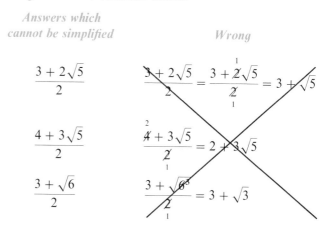

Answers which cannot be simplified

Wrong

$$\frac{3 + 2\sqrt{5}}{2}$$

$$\frac{\cancel{3} + 2\sqrt{5}}{2} = \frac{3 + \overset{1}{\cancel{2}}\sqrt{5}}{\underset{1}{\cancel{2}}} = 3 + \sqrt{5}$$

$$\frac{4 + 3\sqrt{5}}{2}$$

$$\frac{\overset{2}{\cancel{4}} + 3\sqrt{5}}{\underset{1}{\cancel{2}}} = 2 + 3\sqrt{5}$$

$$\frac{3 + \sqrt{6}}{2}$$

$$\frac{3 + \sqrt{\cancel{6}}^{3}}{\underset{1}{\cancel{2}}} = 3 + \sqrt{3}$$

When *both terms* in the numerator have a common factor, the common factor can sometimes be divided out as illustrated below.

Correct

$$\frac{2 + 4\sqrt{3}}{2} = \frac{\overset{1}{\cancel{2}}(1 + 2\sqrt{3})}{\underset{1}{\cancel{2}}} = 1 + 2\sqrt{3}$$

$$\frac{6 + 3\sqrt{3}}{6} = \frac{\overset{1}{\cancel{3}}(2 + \sqrt{3})}{\underset{2}{\cancel{6}}} = \frac{2 + \sqrt{3}}{2}$$

EXAMPLE 4   Solve the equation $x^2 = -4x + 6$ using the quadratic formula.

Solution: Write the equation in standard form.

$$x^2 + 4x - 6 = 0$$

$$a = 1, \qquad b = 4, \qquad c = -6$$

$$x = \frac{-b \pm \sqrt{b^2 - 4ac}}{2a}$$

$$= \frac{-4 \pm \sqrt{(4)^2 - 4(1)(-6)}}{2(1)}$$

$$= \frac{-4 \pm \sqrt{16 + 24}}{2}$$

$$= \frac{-4 \pm \sqrt{40}}{2}$$

$$= \frac{-4 \pm 2\sqrt{10}}{2}$$

$$= \frac{\overset{1}{\cancel{2}}(-2 \pm \sqrt{10})}{\underset{1}{\cancel{2}}}$$

$$= -2 \pm \sqrt{10}$$

$$x = -2 + \sqrt{10} \qquad \text{or} \qquad x = -2 - \sqrt{10} \qquad \blacksquare$$

EXAMPLE 5   Solve the equation $x^2 = 4$ using the quadratic formula.

Solution: Write in standard form.

$$x^2 - 4 = 0$$

$$a = 1, \qquad b = 0, \qquad c = -4$$

$$x = \frac{-b \pm \sqrt{b^2 - 4ac}}{2a}$$

$$= \frac{0 \pm \sqrt{0^2 - 4(1)(-4)}}{2(1)}$$

$$= \frac{\pm\sqrt{16}}{2} = \frac{\pm 4}{2} = \pm 2$$

Thus the solutions are 2 and -2.   ∎

EXAMPLE 6   Solve the quadratic equation $2x^2 + 5x = -6$.

Solution:

$$2x^2 + 5x + 6 = 0$$

$$a = 2, \qquad b = 5, \qquad c = 6$$

$$x = \frac{-b \pm \sqrt{b^2 - 4ac}}{2a}$$

$$= \frac{-5 \pm \sqrt{(5)^2 - 4(2)(6)}}{2(2)}$$

$$= \frac{-5 \pm \sqrt{25 - 48}}{4}$$

$$= \frac{-5 \pm \sqrt{-23}}{4}$$

Since $\sqrt{-23}$ is not a real number, we can go no further. This equation has no real solution. *When given a problem of this type, your answer should be "no real solution." Do not leave the answer blank, and do not write 0 for the answer.* ■

2   The expression under the square root sign in the quadratic formula is called the **discriminant.**

$$\underbrace{b^2 - 4ac}$$

discriminant

The discriminant can be used to determine the number of solutions to a quadratic equation.

When the discriminant is:
1. **Greater than zero,** $b^2 - 4ac > 0$, the quadratic equation has **two distinct solutions.**
2. **Equal to zero,** $b^2 - 4ac = 0$, the quadratic equation has a **single unique solution.** This single solution is often referred to as a **double root.**
3. **Less than zero,** $b^2 - 4ac < 0$, the quadratic equation has **no real solution.**

$b^2 - 4ac$	Number of Solutions
Positive	Two distinct solutions
0	A single unique solution
Negative	No real solution

EXAMPLE 7   (a) Find the discriminant of the equation $x^2 - 8x + 16 = 0$.
(b) Use the quadratic formula to find the solution.

Solution: (a) $a = 1, \qquad b = -8, \qquad c = 16$

$$b^2 - 4ac = (-8)^2 - 4(1)(16) = 64 - 64 = 0$$

Since the discriminant equals zero, there should be a single unique solution.

(b) $x = \dfrac{-b \pm \sqrt{b^2 - 4ac}}{2a}$

$ = \dfrac{-(-8) \pm \sqrt{0}}{2(1)}$

$ = \dfrac{8 \pm 0}{2} = \dfrac{8}{2} = 4$

The only solution is 4.   ∎

EXAMPLE 8   Without actually finding the solutions, determine if the following equations have two distinct solutions, a single unique solution, or no real solution.

(a) $2x^2 - 4x + 6 = 0$ (b) $x^2 - 5x - 8 = 0$ (c) $4x^2 - 12x = -9$

Solution: We use the discriminant of the quadratic formula to answer these equations.

(a) $b^2 - 4ac = (-4)^2 - 4(2)(6) = 16 - 48 = -32$

Since the discriminant is negative, this equation has no real solution.

(b) $b^2 - 4ac = (-5)^2 - 4(1)(-8) = 25 + 32 = 57$

Since the discriminant is positive, this equation has two distinct solutions.

(c) First rewrite $4x^2 - 12x = -9$ as $4x^2 - 12x + 9 = 0$.

$$b^2 - 4ac = (-12)^2 - 4(4)(9) = 144 - 144 = 0$$

(b) Since the discriminant is zero, this equation has a single unique solution.   ∎

Before we leave this section let us look at one of many application problems that may be solved using the quadratic formula.

EXAMPLE 9   Mr. Jackson is planning to plant a grass walk of uniform width around his rectangular swimming pool, which measures 18 feet by 24 feet. How far will the walkway extend from the pool if Mr. Jackson has only enough seed to plant 2000 square feet of grass?

Solution: Let us make a diagram of the situation (see Fig. 10.2). Let x = the uniform width of the grass area. Then the total length of the larger rectangular area becomes $24 + 2x$ or $2x + 24$. The total width of the larger rectangular area becomes $18 + 2x$ or $2x + 18$.

The grassy area can be found by subtracting the area of the pool from the larger rectangular area.

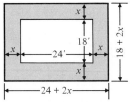

Figure 10.2

$$\text{area of pool} = l \cdot w = (24)(18) = 432 \text{ square feet}$$

$$\text{area of large rectangle} = l \cdot w = (2x + 24)(2x + 18)$$
$$= 4x^2 + 84x + 432 \text{ (pool plus grassy area)}$$

$$\text{grassy area} = \text{area of large rectangle} - \text{area of pool}$$
$$= 4x^2 + 84x + 432 - (432)$$
$$= 4x^2 + 84x$$

The total grassy area must be 2000 square feet.

$$\text{grassy area} = 4x^2 + 84x$$
$$2000 = 4x^2 + 84x$$
$$4x^2 + 84x - 2000 = 0$$
$$4(x^2 + 21x - 500) = 0$$

By the quadratic formula,

$$x = \frac{-b \pm \sqrt{b^2 - 4ac}}{2a}$$

$$a = 1, \qquad b = 21, \qquad c = -500$$

$$x = \frac{-21 \pm \sqrt{(21)^2 - 4(1)(-500)}}{2(1)}$$

$$= \frac{-21 \pm \sqrt{441 + 2000}}{2}$$

$$= \frac{-21 \pm \sqrt{2441}}{2}$$

$$= \frac{-21 \pm 49.41}{2}$$

$$x = \frac{-21 - 49.41}{2} \qquad \text{or} \qquad x = \frac{-21 + 49.41}{2}$$

$$x = \frac{-70.41}{2} \qquad \text{or} \qquad x = \frac{28.41}{2}$$

$$x = -35.205 \qquad \text{or} \qquad x = 14.21$$

The only possible answer is $x = 14.21$. Thus there will be a walkway about 14.2 feet wide all around the pool. ■

Exercise Set 10.3 _____

Determine whether each equation has two distinct solutions, a single unique solution, or no real solution.

1. $x^2 + 3x - 5 = 0$
2. $2x^2 + 6x + 3 = 0$
3. $3x^2 - 4x + 7 = 0$
4. $-4x^2 + x - 8 = 0$
5. $5x^2 + 3x - 7 = 0$
6. $2x^2 = 16x - 32$
7. $4x^2 - 24x = -36$
8. $5x^2 - 4x = 7$
9. $x^2 - 8x + 5 = 0$
10. $x^2 - 5x - 9 = 0$
11. $-3x^2 + 5x - 8 = 0$
12. $x^2 + 4x - 8 = 0$
13. $x^2 + 7x - 3 = 0$
14. $2x^2 - 6x + 9 = 0$
15. $4x^2 - 9 = 0$
16. $6x^2 - 5x = 0$

Use the quadratic formula to solve each equation. If the equation has no real solution, so state.

17. $x^2 - 3x + 2 = 0$
18. $x^2 + 6x + 8 = 0$
19. $x^2 - 9x + 20 = 0$
20. $x^2 - 3x - 10 = 0$
21. $x^2 + 5x - 24 = 0$
22. $x^2 - 6x = -5,$
23. $x^2 = 13x - 36$
24. $x^2 - 36 = 0$
25. $x^2 - 25 = 0 \quad 5,$
26. $x^2 - 6x = 0$
27. $x^2 - 3x = 0$
28. $z^2 - 17z + 72 = 0$
29. $p^2 - 7p + 12 = 0$
30. $2x^2 - 3x + 2 = 0$
31. $2y^2 - 7y + 4 = 0$

32. $2x^2 - 7x = -5$

33. $6x^2 = -x + 1$

34. $4r^2 + r - 3 = 0$

35. $2x^2 - 4x - 1 = 0$

36. $3w^2 - 4w + 5 = 0$

37. $2s^2 - 4s + 3 = 0$

38. $x^2 - 7x + 3 = 0$

39. $4x^2 = x + 5$

40. $x^2 - 2x - 1 = 0$

41. $2x^2 - 7x = 9$

42. $-x^2 + 2x + 15 = 0$

43. $-2x^2 + 11x - 15 = 0$

44. $6x^2 + 5x + 9 = 0$

45. How many real solutions does a quadratic equation have if the discriminant equals
(a) -4 (b) 0 (c) $\frac{1}{2}$?

46. The product of two consecutive positive integers is 20. Find the two consecutive numbers.

47. The length of a rectangle is 3 feet longer than its width. Find the dimensions of the rectangle if its area is 28 square feet.

48. The length of a rectangle is 3 feet smaller than twice its width. Find the length and width of the rectangle if its area is 20 square feet.

49. Lisa wishes to plant a uniform strip of grass around her pool. If her pool measures 20 feet by 30 feet and she has only enough seed to cover 336 square feet, what will be the width of the uniform strip? See Example 9.

50. The McDonald's garden is 30 feet by 40 feet. They wish to lay a uniform border of pine bark around their garden. How large a strip should they lay if they only have enough bark to cover 296 square feet?

JUST FOR FUN

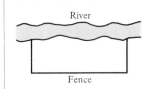
River

Fence

1. A farmer wishes to form a rectangular region along a river bank by constructing fencing as illustrated in the diagram. If she only has 400 feet of fencing and wishes to enclose an area of 15,000 square feet, find the dimensions of the rectangular region.

10.4

Graphing Quadratic Equations

 **1** *Graph quadratic equations in two variables.*

2 *Find the coordinates of the vertex of a parabola.*

3 *Use symmetry when graphing quadratic equations.*

4 *Find the roots of a quadratic equation.*

---

1   In Section 7.2 we learned how to graph linear equations in two variables. In this section we graph quadratic equations in two variables of the form

$$y = ax^2 + bx + c, \qquad a \neq 0$$

Every quadratic equation of the given form when graphed will be a **parabola.** The graph of $y = ax^2 + bx + c$ will have the general shapes indicated in Fig. 10.3.

Parabola opens upward   Parabola opens downward

(a)   (b)

Figure 10.3

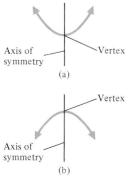

Axis of
symmetry

Vertex

(a)

Vertex

Axis of
symmetry

(b)

Figure 10.4

When a quadratic equation is in the form $y = ax^2 + bx + c$, the sign of the numerical coefficient of the squared term, a, will determine whether the parabola will open upward (Fig. 10.3a) or downward (Fig. 10.3b). When a is positive, the parabola will open upward, and when a is negative, the parabola will open downward. The **vertex** is the lowest point on a parabola that opens upward and the highest point on a parabola that opens downward (see Fig. 10.4).

Graphs of quadratic equations of the form $y = ax^2 + bx + c$ will have **symmetry** about a line through the vertex. This means that if we fold the paper along this imaginary line, called the **axis of symmetry,** the right and left sides of the graph would coincide.

One method that can be used to graph quadratic equations is to plot it point by point. When determining points to plot, select values for x and determine the corresponding values for y.

EXAMPLE 1   Graph the equation $y = x^2$.

Solution: Since $a = 1$ which is greater than 0, this parabola will open upward.

$$y = x^2$$

		x	y
Let $x = 3$,	$y = (3)^2 = 9$	3	9
Let $x = 2$,	$y = (2)^2 = 4$	2	4
Let $x = 1$,	$y = (1)^2 = 1$	1	1
Let $x = 0$,	$y = (0)^2 = 0$	0	0
Let $x = -1$,	$y = (-1)^2 = 1$	-1	1
Let $x = -2$,	$y = (-2)^2 = 4$	-2	4
Let $x = -3$,	$y = (-3)^2 = 9$	-3	9

Connect the points with a smooth curve (Fig. 10.5). Note how the graph is symmetric about the line $x = 0$ (or the y axis).

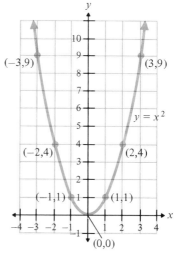

Figure 10.5

EXAMPLE 2   Graph the equation $y = -2x^2 + 16x - 24$.

Solution: Since $a = -2$, which is less than 0, this parabola opens downwards.

$$y = -2x^2 + 16x - 24$$

Let			x	y
Let $x = 0$,	$y = -2(0)^2 + 16(0) - 24 = -24$		0	-24
Let $x = 1$,	$y = -2(1)^2 + 16(1) - 24 = -10$		1	-10
Let $x = 2$,	$y = -2(2)^2 + 16(2) - 24 = 0$		2	0
Let $x = 3$,	$y = -2(3)^2 + 16(3) - 24 = 6$		3	6
Let $x = 4$,	$y = -2(4)^2 + 16(4) - 24 = 8$		4	8
Let $x = 5$,	$y = -2(5)^2 + 16(5) - 24 = 6$		5	6
Let $x = 6$,	$y = -2(6)^2 + 16(6) - 24 = 0$		6	0
Let $x = 7$,	$y = -2(7)^2 + 16(7) - 24 = -10$		7	-10
Let $x = 8$,	$y = -2(8)^2 + 16(8) - 24 = -24$		8	-24

Note how the graph (Fig. 10.6) is symmetric about the line $x = 4$, which has been dashed. The vertex of this parabola is the point (4, 8). Since the y values are large, the y axis has been marked with 4-unit intervals to allow us to graph the points $(0, -24)$ and $(8, -24)$. The arrows on the ends of the graph indicate that the parabola continues indefinitely.

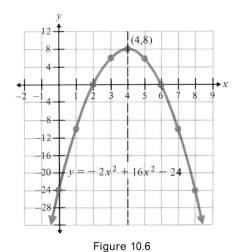

Figure 10.6

2   When graphing quadratic equations, how do we decide what values to use for x? When the location of the vertex is unknown, this is a difficult question to answer. When the location of the vertex is known, it becomes more obvious which values to use.

In Example 2, the axis of symmetry is $x = 4$ and the x coordinate of the vertex is also 4. When given an equation in the form $y = ax^2 + bx + c$, both the axis of symmetry and the x coordinate of the vertex can be found by the following formula

> **x-coordinate of vertex**
>
> $$x = \frac{-b}{2a}$$

In Example 2, $a = -2$, $b = 16$, and $c = -24$. Substituting these values in the formula above gives

$$x = \frac{-b}{2a}$$

$$= \frac{-(16)}{2(-2)}$$

$$= \frac{-16}{-4}$$

$$= 4$$

Thus the graph is symmetric about the line $x = 4$, so the x coordinate of the vertex must also be 4.

The y coordinate of the vertex can be found by substituting the value of the x coordinate of the vertex into the original equation and solving for y.

$$y = -2x^2 + 16x - 24$$

$$= -2(4)^2 + 16(4) - 24$$

$$= -2(16) + 64 - 24$$

$$= -32 + 64 - 24$$

$$= 8$$

Thus the vertex is at the point (4, 8).

When given an equation of the form $y = ax^2 + bx + c$, the y coordinate of the vertex can also be found by the following formula.

> **y coordinate of vertex**
>
> $$y = \frac{4ac - b^2}{4a}$$

For Example 2,

$$y = \frac{4ac - b^2}{4a}$$

$$= \frac{4(-2)(-24) - (16)^2}{4(-2)}$$

$$= \frac{192 - 256}{-8}$$

$$= \frac{-64}{-8}$$

$$= 8$$

You may use the method of your choice to find the y coordinate of the vertex.

3   One method to use in selecting points to plot when graphing parabolas is to determine the axis of symmetry or the vertex of the graph. Then select nearby values of x on any one side of the axis of symmetry. When graphing the equation, make use of your knowledge of the symmetry of the graph.

EXAMPLE 3   (a) Find the axis of symmetry of the equation $y = x^2 + 8x + 15$.
(b) Find the vertex of the graph.
(c) Graph the equation.

Solution: (a) $a = 1, \qquad b = 8, \qquad c = 15$

$$x = \frac{-b}{2a} = \frac{-8}{2(1)} = -4$$

The parabola is symmetric about the line $x = -4$

(b) $y = x^2 + 8x + 15$

$y = (-4)^2 + 8(-4) + 15 = 16 - 32 + 15 = -1$

The vertex is $(-4, -1)$.

(c) Since the axis of symmetry is $x = -4$, we will select values for x that are greater than or equal to -4. It is often helpful to plot each point as it is determined. If a point does not appear to be part of the curve of the parabola, check it.

$$y = x^2 + 8x + 15$$

x	y
-3	0
-2	3
-1	8

Let $x = -3, \quad y = (-3)^2 + 8(-3) + 15 = 0$
Let $x = -2, \quad y = (-2)^2 + 8(-2) + 15 = 3$
Let $x = -1, \quad y = (-1)^2 + 8(-1) + 15 = 8$

The points are plotted in Fig. 10.7a. The graph of the equation is illustrated in Fig. 10.7b. Note how we make use of symmetry to complete the graph. The points $(-3, 0)$ and $(-5, 0)$ are each 1 horizontal unit from the axis of symmetry, $x = -4$. The points $(-2, 3)$ and $(-6, 3)$ are each 2 horizontal units from the axis of symmetry, and the points $(-1, 8)$ and $(-7, 8)$ are each 3 horizontal units from the axis of symmetry.

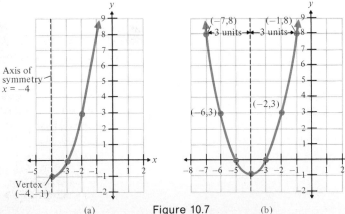

(a)   Figure 10.7   (b)

EXAMPLE 4   Graph the equation $y = -2x^2 + 3x - 4$.

Solution: $a = -2$, $b = 3$, $c = -4$

Since $a < 0$, this parabola will open downward.

$$\text{Axis of symmetry:}\quad x = \frac{-b}{2a}$$

$$x = \frac{-3}{2(-2)} = \frac{-3}{-4} = \frac{3}{4}$$

Since the x value of the vertex is a fraction, we will use the formula to find the y value of the vertex.

$$y = \frac{4ac - b^2}{4a}$$

$$y = \frac{4(-2)(-4) - 3^2}{4(-2)} = \frac{32 - 9}{-8}$$

$$= \frac{23}{-8} = -\frac{23}{8}$$

The vertex of this graph is at $(\frac{3}{4}, -\frac{23}{8})$. Since the axis of symmetry is $x = \frac{3}{4}$, we will begin by selecting values of x that are greater than $\frac{3}{4}$.

$$y = -2x^2 + 3x - 4$$

		x	y
Let $x = 1$,	$y = -2(1)^2 + 3(1) - 4 = -3$	1	-3
Let $x = 2$,	$y = -2(2)^2 + 3(2) - 4 = -6$	2	-6
Let $x = 3$,	$y = -2(3)^2 + 3(3) - 4 = -13$	3	-13

When the axis of symmetry is a fractional value, be very careful when constructing the graph. You should plot as many additional points as needed. In this example, when $x = 0$, $y = -4$, and when $x = -1$, $y = -9$. Figure 10.8a shows the points found on the right side of the axes of symmetry. Figure 10.8b shows the completed graph.

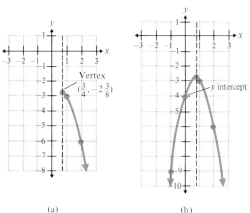

(a)   (b)

Figure 10.8

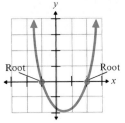

Figure 10.9

4   In Example 4 the graph crossed the y axis at $y = -4$. Recall from earlier sections that to find the y intercept, we let $x = 0$ and solve for y. The location of the y intercept is often helpful when graphing quadratic equations.

The value or values of x where the graph crosses the x axis (the x intercepts) are called the **roots** of the equation. Such points must have a y value of zero (see Fig. 10.9).

A quadratic equation of the form $y = ax^2 + bx + c$ will have either two distinct real roots (Fig. 10.10a), a double root (Fig. 10.10b), or no real roots (Fig. 10.10c). In Section 10.3 we mentioned that when the discriminant, $b^2 - 4ac$, is greater than zero, there are two distinct solutions; equal to zero, there is a single unique solution (also called a *double root*); and less than zero, there is no real solution. This concept is illustrated in Fig. 10.10.

The roots of an equation can be found graphically, or algebraically by one of the methods presented earlier in this chapter. Sometimes the roots can also be found by factoring.

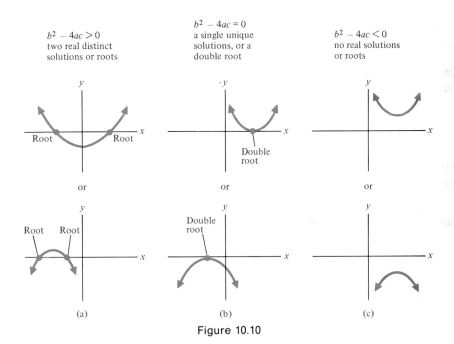

Figure 10.10

EXAMPLE 5   (a) Find the roots of the equation $y = x^2 - 2x - 24$ by factoring, by completing the square, and by the quadratic formula.

(b) Graph the equation.

Solution: (a) We will find the roots using all three algebraic methods. Each root must have a y value of zero; thus to find the roots, we set $y = 0$ and solve the equation for x.

$$0 = x^2 - 2x - 24$$

or

$$x^2 - 2x - 24 = 0$$

Method 1: Factoring

$$x^2 - 2x - 24 = 0$$
$$(x - 6)(x + 4) = 0$$

$$x - 6 = 0 \quad \text{or} \quad x + 4 = 0$$
$$x = 6 \qquad\qquad x = -4$$

Method 2: Completing the square

$$x^2 - 2x - 24 = 0$$
$$x^2 - 2x = 24$$
$$x^2 - 2x + 1 = 24 + 1$$
$$(x - 1)^2 = 25$$
$$x - 1 = \pm 5$$
$$x = 1 \pm 5$$

$$x = 1 + 5 = 6 \quad \text{or} \quad x = 1 - 5 = -4$$

Method 3: Quadratic formula

$$x^2 - 2x - 24 = 0$$
$$a = 1, \qquad b = -2, \qquad c = -24$$

$$x = \frac{-b \pm \sqrt{b^2 - 4ac}}{2a}$$

$$= \frac{-(-2) \pm \sqrt{(-2)^2 - 4(1)(-24)}}{2(1)}$$

$$= \frac{2 \pm \sqrt{4 + 96}}{2}$$

$$= \frac{2 \pm \sqrt{100}}{2}$$

$$= \frac{2 \pm 10}{2}$$

$$x = \frac{2 + 10}{2} = \frac{12}{2} = 6 \quad \text{or} \quad x = \frac{2 - 10}{2} = \frac{-8}{2} = -4$$

Note that the same values were obtained by all three methods.

(b) $y = x^2 - 2x - 24$

Since $a > 0$, this parabola opens upward.

Axis of symmetry: $x = \dfrac{-b}{2a} = \dfrac{-(-2)}{2(1)} = \dfrac{2}{2} = 1$

$$y = x^2 - 2x - 24$$

	x	y
Let $x = 1$, $\quad y = 1^2 - 2(1) - 24 = -25$	1	−25
Let $x = 2$, $\quad y = 2^2 - 2(2) - 24 = -24$	2	−24
Let $x = 3$, $\quad y = 3^2 - 2(3) - 24 = -21$	3	−21
Let $x = 4$, $\quad y = 4^2 - 2(4) - 24 = -16$	4	−16
Let $x = 5$, $\quad y = 5^2 - 2(5) - 24 = -9$	5	−9
Let $x = 6$, $\quad y = 6^2 - 2(6) - 24 = 0$	6	0

Again we make use of symmetry in completing the graph, Fig. 10.11. The roots 6 and −4 agree with the answer obtained in part (a).

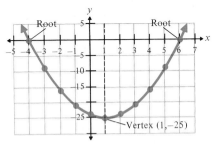

Figure 10.11

COMMON STUDENT ERROR

When solving an equation like $-x^2 + 2x + 8 = 0$, we can multiply both sides of the equation by −1 to obtain the solution.

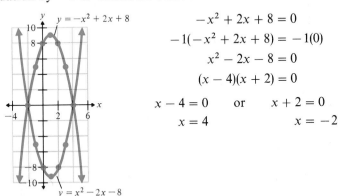

$$-x^2 + 2x + 8 = 0$$
$$-1(-x^2 + 2x + 8) = -1(0)$$
$$x^2 - 2x - 8 = 0$$
$$(x - 4)(x + 2) = 0$$

$$x - 4 = 0 \quad \text{or} \quad x + 2 = 0$$
$$x = 4 \qquad\qquad x = -2$$

When graphing an equation like $y = -x^2 + 2x + 8$, many students incorrectly multiply just the right side of the equation by −1 to get $y = x^2 - 2x - 8$, then graph $y = x^2 - 2x - 8$. This is **wrong** because $y = -x^2 + 2x + 8$ and $y = x^2 - 2x - 8$ are two different graphs as illustrated above.

Note that although the graphs are different the roots −2 and 4 are the same. If you just wish to find the *roots* of an equation you can set $y = 0$, and then if desired multiply both sides of the equation by −1.

Exercise Set 10.4 _____

Indicate the axis of symmetry, the coordinates of the vertex, and whether the parabola opens up or down.

1. $y = x^2 + 2x - 7$ **2.** $y = x^2 + 4x - 9$ **3.** $y = -x^2 + 5x - 6$

4. $y = 3x^2 + 6x - 9$ **5.** $y = -3x^2 + 5x + 8$ **6.** $y = x^2 + 3x - 6$

7. $y = -4x^2 - 8x - 12$ **8.** $y = 2x^2 + 3x + 8$ **9.** $y = 3x^2 - 2x + 2$

10. $y = -x^2 + x + 8$ **11.** $y = 4x^2 + 12x - 5$ **12.** $y = -2x^2 - 6x - 5$

Graph each quadratic equation, and determine the roots, if they exist.

13. $y = x^2 - 1$ **14.** $y = x^2 + 4$ **15.** $y = -x^2 + 3$ **16.** $y = -x^2 - 2$

17. $y = x^2 + 2x + 3$ **18.** $y = x^2 + 4x + 3$ **19.** $y = x^2 + 2x - 15$ **20.** $y = -x^2 + 10x - 21$

21. $y = -x^2 + 4x - 5$ **22.** $y = x^2 + 8x + 15$ **23.** $y = x^2 - x - 12$ **24.** $y = x^2 - 5x + 4$

25. $y = x^2 - 6x + 9$ **26.** $y = x^2 - 6x$ **27.** $y = -x^2 + 5x$ **28.** $y = x^2 - 4x + 4$

29. $y = 2x^2 - 6x + 4$ **30.** $y = x^2 - 2x + 1$ **31.** $y = -x^2 + 11x - 28$ **32.** $y = 4x^2 + 12x + 9$

33. $y = x^2 - 2x - 15$ **34.** $y = -x^2 + 5x - 4$ **35.** $y = -2x^2 + 7x - 3$ **36.** $y = 2x^2 + 3x - 2$

37. $y = -2x^2 + 3x - 2$ **38.** $y = -4x^2 - 6x + 4$ **39.** $y = 2x^2 - x - 15$ **40.** $y = 6x^2 + 10x - 4$

41. What are the coordinates of the vertex of a parabola?

42. What determines whether the graph of a quadratic equation of the form $y = ax^2 + bx + c$, $a \neq 0$ is a parabola that opens upwards or downwards?

43. What are the roots of a graph?

44. How can you find the roots of a graph algebraically?

45. How many real roots will the graph of a quadratic equation have if the discriminant has a value of **(a)** 5   **(b)** −2   **(c)** 0?

Summary

Glossary

Axis of symmetry: The imaginary line in a graph about which symmetry occurs.

Parabola: The graph of a quadratic equation is a parabola.

Perfect square trinomial: A trinomial that can be expressed as the square of a binomial.

Root: The value of x where a graph crosses the x axis.

Standard form of a quadratic equation: $ax^2 + bx + c = 0$, $a \neq 0$.

Vertex of a parabola: The lowest point on a parabola that opens upward and the highest point on a parabola that opens downward.

Important Facts

Square root property: If $x^2 = a$, then $\sqrt{x^2} = \pm\sqrt{a}$, and $x = \pm\sqrt{a}$.

Quadratic formula: $x = \dfrac{-b \pm \sqrt{b^2 - 4ac}}{2a}$

Discriminant: $b^2 - 4ac$

If $b^2 - 4ac > 0$ the quadratic equation has two distinct solutions.

If $b^2 - 4ac = 0$ the quadratic equation has a single unique solution.

If $b^2 - 4ac < 0$ the quadratic equation has no real solution.

Coordinates of the vertex of a parabola $\left(\dfrac{-b}{2a}, \dfrac{4ac - b^2}{4a} \right)$.

Review Exercises _____

[10.1] Solve each equation.

1. $x^2 = 25$ **2.** $x^2 = 8$ **3.** $2x^2 = 12$ **4.** $x^2 + 3 = 9$
5. $x^2 - 4 = 16$ **7.** $3x^2 + 8 = 32$ **8.** $(x - 3)^2 = 12$
9. $(2x + 4)^2 = 30$ **10.** $(3x - 5)^2 = 50$

[10.2] Solve each equation by completing the square.

11. $x^2 - 10x + 16 = 0$ **12.** $x^2 - 8x + 15 = 0$ **13.** $x^2 - 14x + 13 = 0$ **14.** $x^2 + x - 6 = 0$
15. $x^2 - 3x - 54 = 0$ **16.** $x^2 = -5x + 6$ **17.** $x^2 + 2x - 5 = 0$ **18.** $x^2 - 3x - 8 = 0$
19. $2x^2 - 8x = 64$ **20.** $2x^2 - 4x = 30$ **21.** $4x^2 + 2x - 12 = 0$ **22.** $6x^2 - 19x + 15 = 0$

[10.3] Determine whether each equation has two distinct real solutions, a single unique solution, or no real solution.

23. $3x^2 - 4x - 20 = 0$ **24.** $-3x^2 + 4x = 9$ **25.** $2x^2 + 6x + 7 = 0$
26. $x^2 - x + 8 = 0$ **27.** $x^2 - 12x = -36$ **28.** $3x^2 - 4x + 5 = 0$
29. $-3x^2 - 4x + 8 = 0$ **30.** $x^2 - 9x + 6 = 0$

Solve each equation using the quadratic formula. If an equation has no real solution, so state.

31. $x^2 - 9x + 14 = 0$ **32.** $x^2 + 7x - 30 = 0$ **33.** $x^2 = 7x - 10$
34. $5x^2 - 7x = 6$ **35.** $x^2 - 18 = 7x$ **36.** $x^2 - x - 30 = 0$
37. $6x^2 + x - 15 = 0$ **38.** $2x^2 + 4x - 3 = 0$ **39.** $-2x^2 + 3x + 6 = 0$
40. $x^2 - 6x + 7 = 0$ **41.** $3x^2 - 4x + 6 = 0$ **42.** $3x^2 - 6x - 8 = 0$
43. $2x^2 + 3x = 0$ **44.** $2x^2 - 5x = 0$

[10.1–10.3] Find the solution to each quadratic equation using the method of your choice.

45. $x^2 - 11x + 24 = 0$ **46.** $x^2 - 16x + 63 = 0$ **47.** $x^2 = -3x + 40$
48. $x^2 + 6x = 27$ **49.** $x^2 - 4x - 60 = 0$ **50.** $x^2 - x - 42 = 0$
51. $x^2 + 11x - 12 = 0$ **52.** $x^2 = 25$ **53.** $x^2 + 6x = 0$
54. $2x^2 + 5x = 3$ **55.** $2x^2 = 9x - 10$ **56.** $6x^2 + 5x = 6$
57. $x^2 + 3x - 6 = 0$ **58.** $3x^2 - 11x + 10 = 0$ **59.** $-3x^2 - 5x + 8 = 0$
60. $-2x^2 + 6x = -9$ **61.** $2x^2 - 5x = 0$ **62.** $3x^2 + 5x = 0,$

[10.4] Indicate the axis of symmetry, the coordinates of the vertex, and whether the parabola opens upward or downward.

63. $y = x^2 - 2x - 3 \ x =$ **64.** $y = x^2 - 10x + 24$ **65.** $y = x^2 + 7x + 12$
66. $y = -x^2 - 2x + 15$ **67.** $y = x^2 - 3x$ **68.** $y = 2x^2 + 7x + 3$
69. $y = -x^2 - 8$ **70.** $y = -4x^2 + 8x + 5$
71. $y = -x^2 - x + 20$ **72.** $y = 3x^2 + 5x - 8$

Graph each quadratic equation, and determine the real roots if they exist. If they do not exist, so state.

73. $y = x^2 + 6x$ **74.** $y = -2x^2 + 8$ **75.** $y = x^2 + 2x - 8$

76. $y = x^2 - x - 2$

77. $y = x^2 + 5x + 4$

78. $y = x^2 + 4x + 3$

79. $y = -2x^2 + 3x - 2$

80. $y = 3x^2 - 4x + 1$

81. $y = 4x^2 - 8x + 6$

82. $y = -3x^2 - 14x + 5$

83. $y = -x^2 - 6x - 4$

84. $y = 2x^2 + 5x - 12$

[10.2–10.3] Solve each problem.

85. The product of two positive integers is 88. Find the two numbers if the larger one is 3 greater than the smaller.

86. The length of a rectangle is 5 feet less than twice its width. Find the length and width of the rectangle if its area is 63 square feet.

Practice Test

1. Solve the equation $x^2 + 1 = 21$.
2. Solve the equation $(2x - 3)^2 = 35$.
3. Solve by completing the square: $x^2 - 4x = 60$.
4. Solve by completing the square: $x^2 = -x + 12$.
5. Solve by the quadratic formula: $x^2 - 5x - 6 = 0$.
6. Solve by the quadratic formula: $2x^2 + 5 = -8x$.
7. Solve by the method of your choice: $3x^2 - 5x = 0$.
8. Solve by the method of your choice: $2x^2 + 9x = 5$.
9. Determine whether the following equation has two distinct real solutions, a single unique solution, or no real solution: $3x^2 - 4x + 2 = 0$.

10. Indicate the axis of symmetry, the coordinates of the vertex, and whether the graph opens upward or downward: $y = -x^2 + 3x + 8$.
11. Graph the following equation, and determine the roots, if they exist: $y = x^2 + 2x - 8$.
12. Graph the following equation, and determine the roots, if they exist: $y = -x^2 + 6x - 9$.
13. The length of a rectangle is 1 foot greater than three times its width. Find the length and width of the rectangle if its area is 30 square feet.

Appendix

A. Review of Decimals and Percent
B. Squares and Square Roots

Review of Decimals and Percent

> **To Add or Subtract Numbers Containing Decimal Points**
>
> 1. Align the numbers by the decimal points.
> 2. Add or subtract the numbers in the corresponding columns.
> 3. Place the decimal point in the answer directly below the decimal points in the numbers being added or subtracted.

EXAMPLE 1 Add $4.6 + 13.813 + 9.02$.

Solution:

$$
\begin{array}{r}
4.600 \\
13.813 \\
+\ 9.020 \\
\hline
27.433 \quad \blacksquare
\end{array}
$$

EXAMPLE 2 Subtract 3.062 from 25.9.

Solution:

$$
\begin{array}{r}
25.900 \\
-\ 3.062 \\
\hline
22.838 \quad \blacksquare
\end{array}
$$

> **To Multiply Numbers Containing Decimal Points**
>
> 1. Multiply as if the factors were whole numbers.
> 2. Determine the total number of digits to the right of the decimal points in the factors.
> 3. Place the decimal point in the product so that the product contains the same number of digits to the right of the decimal as the total found in Step 2. For example, if there are a total of three digits to the right of the decimal points in the factors, there must be three digits to the right of the decimal point in the product.

EXAMPLE 3   Multiply 2.34 × 1.9.

Solution: 2.34 ⟵ two digits to the right of decimal point
× 1.9 ⟵ one digit to the right of decimal point
2106
234
4.446 ⟵ three digits to the right of decimal in answer   ■

EXAMPLE 4   Multiply 2.13 × 0.02.

Solution: 2.13 ⟵ two digits to the right of decimal point
× 0.02 ⟵ two digits to the right of decimal point
0.0426 ⟵ four digits to the right of decimal point in answer

Note that it was necessary to add a zero preceding the digit 4 in the answer in order to have four digits to the right of the decimal point.   ■

To Divide Numbers Containing Decimal Points

1. Multiply both the dividend and divisor by a number that will result in the divisor being a whole number.
2. Divide as if working with whole numbers.
3. Place the decimal point in the quotient directly above the decimal point in the dividend.

To make the divisor a whole number, multiply *both* the dividend and divisor by 10 if the divisor is given in tenths, by 100 if the divisor is given in hundredths, by 1000 if the divisor is given in thousandths, and so on. Multiplying both the numerator and denominator by the same nonzero number is the same as multiplying the fraction by 1. Therefore, the value of the fraction is unchanged.

EXAMPLE 5   Divide 1.956/0.12.

Solution: Since the divisor, 0.12, is twelve hundredths, we multiply both the divisor and dividend by 100.

$$\frac{1.956}{0.12} \times \frac{100}{100} = \frac{195.6}{12.}$$

Now divide.

$$
\begin{array}{r}
16.3 \\
12.\overline{)195.6} \\
\underline{12} \\
75 \\
\underline{72} \\
36 \\
\underline{36} \\
0
\end{array}
$$

The decimal point in the answer is placed directly above the decimal point in the dividend. Thus $1.956/0.12 = 16.3$. ■

EXAMPLE 6 Divide 0.26 by 10.4.

Solution: $\dfrac{0.26}{10.4} \times \dfrac{10}{10} = \dfrac{2.6}{104.}$

Now divide.

$$
\begin{array}{r}
0.025 \\
104\overline{)2.600} \\
\underline{2\,08} \\
520 \\
\underline{520} \\
0
\end{array}
$$

Note that a zero had to be placed before the digit 2 in the quotient.

$$\frac{0.26}{10.4} = 0.025 \quad ■$$

Percent

One of the main uses of decimals comes from percent problems. The word percent means "per one hundred." The symbol % means percent. One percent means "one per hundred," or

$$1\% = \frac{1}{100} \quad \text{or} \quad 1\% = 0.01$$

EXAMPLE 7 Convert 16% to a decimal.

Solution: Since

$$1\% = 0.01$$
$$16\% = 16(0.01) = 0.16 \quad ■$$

EXAMPLE 8   Convert 2.3% to a decimal.

Solution: $2.3\% = 2.3(0.01) = 0.023$ ■

Often you will need to find an amount that is a certain percent of a number. For example, when you purchase an item in a state or county that has a sales tax you must often pay a percent of the item's price as the sales tax. Examples 9 and 10 show how to find a certain percent of a number.

EXAMPLE 9   Find 12% of 200.

Solution: The word "of" indicates multiplication. Change 12% to a decimal number, then multiply by 200.

$$(0.12)(200) = 24$$

Thus 12% of 200 is 24.   ■

EXAMPLE 10   In Monroe County in New York State there is a 7% sales tax.
(a) Find the sales tax on a stereo system that cost $580.
(b) Find the total cost of the system, including tax.

Solution: (a) The sales tax is 7% of 580.

$$(0.07)(580) = 40.60$$

The sales tax is $40.60.
(b) The toal cost is the purchase price plus the sales tax:

$$\text{total cost} = \$580 + \$40.60 = \$620.60$$ ■

Appendix B
Squares and Square Roots

Number	Square	Square Root	Number	Square	Square Root
1	1	1.000	51	2,601	7.141
2	4	1.414	52	2,704	7.211
3	9	1.732	53	2,809	7.280
4	16	2.000	54	2,916	7.348
5	25	2.236	55	3,025	7.416
6	36	2.449	56	3,136	7.483
7	49	2.646	57	3,249	7.550
8	64	2.828	58	3,364	7.616
9	81	3.000	59	3,481	7.681
10	100	3.162	60	3,600	7.746
11	121	3.317	61	3,721	7.810
12	144	3.464	62	3,844	7.874
13	169	3.606	63	3,969	7.937
14	196	3.742	64	4,096	8.000
15	225	3.873	65	4,225	8.062
16	256	4.000	66	4,356	8.124
17	289	4.123	67	4,489	8.185
18	324	4.243	68	4,624	8.246
19	361	4.359	69	4,761	8.307
20	400	4.472	70	4,900	8.367
21	441	4.583	71	5,041	8.426
22	484	4.690	72	5,184	8.485
23	529	4.796	73	5,329	8.544
24	576	4.899	74	5,476	8.602
25	625	5.000	75	5,625	8.660
26	676	5.099	76	5,776	8.718
27	729	5.196	77	5,929	8.775
28	784	5.292	78	6,084	8.832
29	841	5.385	79	6,241	8.888
30	900	5.477	80	6,400	8.944
31	961	5.568	81	6,561	9.000
32	1,024	5.657	82	6,724	9.055
33	1,089	5.745	83	6,889	9.110
34	1,156	5.831	84	7,056	9.165
35	1,225	5.916	85	7,225	9.220
36	1,296	6.000	86	7,396	9.274
37	1,369	6.083	87	7,569	9.327
38	1,444	6.164	88	7,744	9.381
39	1,521	6.245	89	7,921	9.434
40	1,600	6.325	90	8,100	9.487
41	1,681	6.403	91	8,281	9.539
42	1,764	6.481	92	8,464	9.592
43	1,849	6.557	93	8,649	9.644
44	1,936	6.633	94	8,836	9.695
45	2,025	6.708	95	9,025	9.747
46	2,116	6.782	96	9,216	9.798
47	2,209	6.856	97	9,409	9.849
48	2,304	6.928	98	9,604	9.899
49	2,401	7.000	99	9,801	9.950
50	2,500	7.071	100	10,000	10.000

Answers

Exercise Set 1.1

1. $\frac{2}{5}$　**3.** $\frac{3}{4}$　**5.** $\frac{1}{2}$　**7.** $\frac{3}{7}$　**9.** $\frac{5}{8}$　**11.** Lowest terms　**13.** $\frac{3}{4}$　**15.** $\frac{3}{10}$　**17.** $\frac{5}{28}$　**19.** $\frac{1}{12}$　**21.** $\frac{5}{3}$　**23.** $\frac{5}{16}$
25. 6　**27.** $\frac{5}{12}$　**29.** $\frac{13}{45}$　**31.** $\frac{19}{14}$　**33.** 12　**35.** $\frac{3}{5}$　**37.** $\frac{1}{4}$　**39.** 1　**41.** $\frac{3}{29}$　**43.** $\frac{37}{30}$　**45.** $\frac{1}{5}$　**47.** $\frac{4}{15}$
49. $\frac{2}{9}$　**51.** $\frac{29}{24}$ or $1\frac{5}{24}$　**53.** $\frac{1}{6}$　**55.** $\frac{47}{12}$ or $3\frac{11}{12}$　**57.** $\frac{23}{6}$ or $3\frac{5}{6}$　**59.** $\frac{52}{15}$ or $3\frac{7}{15}$　**61.** $\frac{21}{20}$ or $1\frac{1}{20}$　**63.** $13\frac{3}{4}$ yd
65. $13\frac{11}{16}$ in.　**67.** $8\frac{7}{16}$ ft　**69.** $11\frac{7}{8}$ ft

Just for Fun　**1.** Rice and water 1 cup, salt $\frac{3}{8}$ tsp, butter $1\frac{1}{2}$ tsp

Exercise Set 1.2

1. $\{\ldots, -3, -2, -1, 0, 1, 2, 3, \ldots\}$　**3.** $\{1, 2, 3, 4, \ldots\}$　**5.** $\{\ldots, -3, -2, -1\}$　**7.** T　**9.** T　**11.** F　**13.** F
15. T　**17.** F　**19.** T　**21.** F　**23.** T　**25.** T　**27.** F　**29.** T　**31.** T　**33.** T　**35.** F　**37.** F
39. (a) 7, 9　(b) 7, 0, 9　(c) $-6, 7, 0, 9$　(d) $-6, 7, 12.4, -\frac{9}{5}, -2\frac{1}{4}, 0, 9, 0.35$　(e) $\sqrt{3}, \sqrt{7}$
(f) $-6, 7, 12.4, -\frac{9}{5}, -2\frac{1}{4}, \sqrt{3}, 0, 9, \sqrt{7}, 0.35$　**41.** (a) 5　(b) 5　(c) -300　(d) $5, -300$
(e) $\frac{1}{2}, 4\frac{1}{2}, \frac{5}{12}, -1.67, 5, -300, -9\frac{1}{2}$　(f) $\sqrt{2}, -\sqrt{2}$,　(g) $\frac{1}{2}, \sqrt{2}, -\sqrt{2}, 4\frac{1}{2}, \frac{5}{12}, -1.67, 5, -300, -9\frac{1}{2}$
43. $-\frac{2}{3}, \frac{1}{2}, 6.3$　**45.** $-\sqrt{7}, \sqrt{3}, \sqrt{6}$　**47.** $-5, 0, 4$　**49.** $-13, -5, -1$　**51.** $1.5, 3, 6\frac{1}{4}$　**53.** $-7, 1, 5$

Exercise Set 1.3

1. $2 < 3$　**3.** $-3 < 0$　**5.** $\frac{1}{2} > -\frac{2}{3}$　**7.** $0.2 < 0.4$　**9.** $\frac{2}{5} > -1$　**11.** $4 > -4$　**13.** $-2.1 < -2$　**15.** $\frac{5}{9} > -\frac{5}{9}$
17. $-\frac{3}{2} < \frac{3}{2}$　**19.** $0.49 > 0.43$　**21.** $5 > -7$　**23.** $-0.006 > -0.007$　**25.** $\frac{3}{5} < 1$　**27.** $-\frac{2}{3} > -3$　**29.** $8 > |-7|$
31. $|0| < \frac{2}{3}$　**33.** $|-3| < |-4|$　**35.** $4 < |-\frac{9}{2}|$　**37.** $|-\frac{6}{2}| > |-\frac{2}{6}|$　**39.** $4, -4$　**41.** $2, -2$

Exercise Set 1.4

1. -18　**3.** 32　**5.** 0　**7.** $-\frac{5}{3}$　**9.** $-\frac{3}{5}$　**11.** -0.63　**13.** $-2\frac{1}{2}$　**15.** 3.1　**17.** 7　**19.** 1　**21.** -6
23. 0　**25.** 0　**27.** -10　**29.** 0　**31.** -10　**33.** 0　**35.** -6　**37.** 3　**39.** -9　**41.** 9　**43.** -27
45. -44　**47.** -26　**49.** 5　**51.** -20　**53.** -31　**55.** 91　**57.** -140　**59.** 0　**61.** \$81　**63.** 1407 m
65. 2600 ft

Exercise Set 1.5

1. 3　**3.** -1　**5.** 0　**7.** -3　**9.** -6　**11.** 6　**13.** -6　**15.** 6　**17.** -8　**19.** 2　**21.** 2　**23.** 9
25. 0　**27.** -18　**29.** -2　**31.** 0　**33.** 0　**35.** 0　**37.** -1　**39.** -5　**41.** -41　**43.** -3　**45.** -180

47. -110 **49.** 140   **51.** 0   **53.** -82 **55.** 5   **57.** -18 **59.** -16 **61.** 10   **63.** -2 **65.** 13   **67.** 0
69. -4 **71.** 11   **73.** 81   **75.** 7   **77.** -2 **79.** -15 **81.** -2 **83.** 0   **85.** 43   **87.** -6 **89.** -9
91. 35   **93.** -21 **95.** -12 **97.** -3 **99.** 18   **101.** -9 **103.** 3500 ft   **105.** 65,226 ft

Just for Fun   **1.** -5 **2.** -50 **3.** 50

Exercise Set 1.6
1. 12   **3.** -9 **5.** -32 **7.** -9 **9.** 12   **11.** 36   **13.** -30 **15.** -60 **17.** 0   **19.** 16   **21.** 24
23. -15 **25.** 96   **27.** 81   **29.** -10 **31.** 36   **33.** 0   **35.** -1 **37.** -120 **39.** -12 **41.** 84
43. $-\frac{3}{10}$ **45.** $\frac{14}{27}$ **47.** 4   **49.** $-\frac{15}{28}$ **51.** 3   **53.** 4   **55.** 4   **57.** -4 **59.** -18 **61.** 5   **63.** 6   **65.** 5
67. -1 **69.** -4 **71.** 9   **73.** 0   **75.** 1   **77.** 0   **79.** -4 **81.** 3   **83.** -12 **85.** -4 **87.** 45
89. $-\frac{3}{4}$ **91.** $\frac{3}{80}$ **93.** 1   **95.** $-\frac{144}{5}$ **97.** $-\frac{36}{5}$ **99.** 0   **101.** Indeterminate   **103.** 0   **105.** Undefined
107. 0   **109.** True   **111.** False   **113.** False   **115.** True

Just for Fun   **1.** 1   **2.** -1

Exercise Set 1.7
1. 9   **3.** 8   **5.** 27   **7.** 216   **9.** -8 **11.** -1 **13.** 27   **15.** -36 **17.** 36   **19.** 16   **21.** 5   **23.** 16
25. -16 **27.** -64 **29.** 225   **31.** 80   **33.** 32   **35.** -75 **37.** 144   **39.** xyz^2 **41.** x^4z **43.** a^3b^3
45. x^2yz^2 **47.** x^3y^2 **49.** xy^4 **51.** 5^2y^2z **53.** xxy **55.** $xyyy$ **57.** $xyyzzz$ **59.** $3\cdot3\cdot yz$
61. $2\cdot2\cdot2\cdot xxxy$ **63.** $(-2)(-2)yyyz$ **65.** 9, -9 **67.** 16, -16 **69.** 4, -4 **71.** 49, -49 **73.** 1, -1
75. $\frac{1}{4}$, $-\frac{1}{4}$ **77.** False   **79.** False   **81.** True   **83.** True   **85.** Any nonzero number will be positive when squared.

Just for Fun   **1. (a)** Nenno gets $30,000; Kelly gets $10,737,418.23, Nenno loses $10,707,418.23
(b) $2^0 + 2^1 + 2^2 + \cdots + 2^{29} = 2^{30} - 1$ cents or $(2^{30} - 1)/100$ dollars   **2. (a)** 2^5 **(b)** 3^5 **(c)** x^{m+n} **(d)** $2^1 = 2$ **(e)** 3^2
(f) x^{m-n} **(g)** 2^6 **(h)** 3^6 **(i)** x^{mn} **(j)** 2^2x^2 **(k)** 3^2x^2 **(l)** a^2x^2

Exercise Set 1.8
1. 23   **3.** 5   **5.** 13   **7.** -4 **9.** 16   **11.** 29   **13.** -13 **15.** -2 **17.** 18   **19.** 10   **21.** 7   **23.** 36
25. -441 **27.** $\frac{1}{2}$ **29.** 10   **31.** 12   **33.** 9   **35.** 129.81   **37.** 26.04   **39.** $\frac{71}{112}$ **41.** $\frac{1}{4}$ **43.** $\frac{170}{9}$
45. $[(6\cdot3)-4]-2$, 12   **47.** $9[[(20\div5)+12]-8]$, 72   **49.** $(\frac{4}{5}+\frac{3}{7})\cdot\frac{2}{3}$, $\frac{86}{105}$ **51.** 2   **53.** 10   **55.** 3
57. -7 **59.** -25 **61.** 75   **63.** -20 **65.** 0   **67.** -5 **69.** 21   **71.** 33   **73.** -18

Just for Fun   **1.** 160   **2.** 177   **3.** -312

Exercise Set 1.9
1. Distributive property   **3.** Commutative property of multiplication   **5.** Distributive property
7. Associative property of multiplication   **9.** Distributive property   **11.** $4 + 3$ **13.** $(-6\cdot4)\cdot2$ **15.** $(y)(6)$
17. $1\cdot x + 1\cdot y$ or $x + y$ **19.** $3y + 4x$ **21.** $5(x + y)$ **23.** $3(x + 2)$ **25.** $3x + (4 + 6)$ **27.** $(x + y)3$
29. $4x + 4y + 12$ **31.** Commutative property of addition   **33.** Distributive property
35. Commutative property of addition   **37.** Distributive property

Chapter 1 Review Exercises
1. $\frac{1}{2}$ **2.** $\frac{9}{25}$ **3.** $\frac{25}{36}$ **4.** $\frac{7}{6}$ or $1\frac{1}{6}$ **5.** $\frac{19}{72}$ **6.** $\frac{17}{15}$ or $1\frac{2}{15}$ **7.** $\{1, 2, 3, \ldots\}$ **8.** $\{0, 1, 2, 3, \ldots\}$
9. $\{\ldots, -3, -2, -1, 0, 1, 2, 3, \ldots\}$ **10.** {quotient of two integers, denominator not 0}
11. {all numbers that can be represented on the real number line}   **12. (a)** $3, 426$ **(b)** $3, 0, 426$ **(c)** $3, -5, -12, 0, 426$
(d) $3, -5, -12, 0, \frac{1}{2}, -0.62, 426, -3\frac{1}{4}$ **(e)** $\sqrt{7}$ **(f)** $3, -5, -12, 0, \frac{1}{2}, -0.62, \sqrt{7}, 426, -3\frac{1}{4}$ **13. (a)** 1   **(b)** 1
(c) $-8, -9$,   **(d)** $-8, -9, 1$ **(e)** $-2.3, -8, -9, 1\frac{1}{2}, 1, -\frac{3}{17}$ **(f)** $-2.3, -8, -9, 1\frac{1}{2}, \sqrt{2}, -\sqrt{2}, 1, -\frac{3}{17}$ **14.** $>$ **15.** $<$
16. $>$ **17.** $>$ **18.** $<$ **19.** $>$ **20.** $<$ **21.** $>$ **22.** $<$ **23.** $=$ **24.** 3   **25.** -9 **26.** 0   **27.** -5
28. -3 **29.** -6 **30.** -6 **31.** -5 **32.** 8   **33.** -2 **34.** -9 **35.** -10 **36.** 5   **37.** -5 **38.** 4
39. -12 **40.** 5   **41.** -4 **42.** -12 **43.** -7 **44.** 6   **45.** 6   **46.** -1 **47.** 9   **48.** -28 **49.** 27
50. -36 **51.** -6 **52.** $-\frac{6}{35}$ **53.** $-\frac{6}{11}$ **54.** $\frac{15}{56}$ **55.** 0   **56.** 48   **57.** 12   **58.** -70 **59.** -60
60. -24 **61.** 144   **62.** -5 **63.** -3 **64.** -4 **65.** 18   **66.** 0   **67.** 0   **68.** -8 **69.** 5   **70.** 9
71. $-\frac{3}{32}$ **72.** $-\frac{3}{4}$ **73.** $\frac{56}{27}$ **74.** $-\frac{35}{9}$ **75.** 1   **76.** 0   **77.** 0   **78.** Undefined   **79.** Undefined
80. Indeterminate   **81.** 0   **82.** 24   **83.** -8 **84.** 1   **85.** 3   **86.** -8 **87.** 18   **88.** -2 **89.** -4

90. 10　**91.** 1　**92.** 15　**93.** -4　**94.** 16　**95.** 36　**96.** 729　**97.** 1　**98.** 81　**99.** 16　**100.** -27
101. -1　**102.** -32　**103.** $\frac{4}{49}$　**104.** $\frac{9}{25}$　**105.** $\frac{8}{125}$　**106.** x^2y　**107.** xy^2　**108.** x^3y^2　**109.** y^2z^2
110. $2^2 \cdot 3^3 xy^2$　**111.** $5 \cdot 7^2 x^2 y$　**112.** $x^2 y^2 z$　**113.** xxy　**114.** $xzzz$　**115.** $yyyz$　**116.** $2xxxyy$　**117.** -9
118. -16　**119.** -27　**120.** -16　**121.** 23　**122.** -2　**123.** 23　**124.** 22　**125.** 26　**126.** -19
127. -39　**128.** -3　**129.** -4　**130.** -60　**131.** 10　**132.** 20　**133.** 20　**134.** 114　**135.** 9　**136.** -35
137. 14　**138.** 2　**139.** 26　**140.** 9　**141.** 0　**142.** -3　**143.** -11　**144.** -3　**145.** 21
146. Associative property of addition　**147.** Commutative property of multiplication　**148.** Distributive property
149. Commutative property of multiplication　**150.** Commutative property of addition
151. Associative property of addition　**152.** Commutative property of addition

Chapter 1 Practice Test

1. (a) 42　**(b)** 42, 0　**(c)** $-6, 42, 0, -7, -1$　**(d)** $-6, 42, -3\frac{1}{2}, 0, 6.52, \frac{5}{9}$　**(e)** $\sqrt{5}$
(f) $-6, 42, -3\frac{1}{2}, 0, 6.52, \sqrt{5}, \frac{5}{9}, -7, -1$　**2.** $<$　**3.** $>$　**4.** -12　**5.** -11　**6.** 16　**7.** -14　**8.** 8
9. -24　**10.** $\frac{16}{63}$　**11.** -2　**12.** -69　**13.** -2　**14.** 12　**15.** 81　**16.** $\frac{27}{125}$　**17.** $2^2 5^2 y^2 z^3$
18. $2 \cdot 2 \cdot 3 \cdot 3 \cdot 3 xxxxyy$　**19.** 26　**20.** 10　**21.** 11　**22.** Commutative property of addition
23. Distributive property　**24.** Associative property of addition　**25.** Commutative property of multiplication

CHAPTER 2

Exercise Set 2.1

1. $8x$　**3.** $-x$　**5.** $x + 9$　**7.** $3x$　**9.** $4x - 7$　**11.** $2x + 3$　**13.** $5x + 5$　**15.** $5x + 3y + 3$　**17.** $2x - 4$
19. $x + 7$　**21.** $-8x + 2$　**23.** $-2x + 11$　**25.** $3x - 6$　**27.** $-5x + 3$　**29.** $6y + 6$　**31.** $4x - 10$　**33.** $3x - 4$
35. $x + 5$　**37.** $x + 7$　**39.** $x - y$　**41.** $-4x - 8$　**43.** $x + 7$　**45.** $7x - 16$　**47.** $7x - 1$　**49.** $x - 8$
51. $-x - 1$　**53.** $-2x - 5$　**55.** $2x + 8$　**57.** $4x + 20$　**59.** $-2x + 8$　**61.** $-x + 2$　**63.** $x - 4$　**65.** $\frac{1}{4}x - 3$
67. $-18x + 30$　**69.** $-14x + 42$　**71.** $x + 6$　**73.** $x - y$　**75.** $-2x + 6y - 8$　**77.** $12 - 6x + 3y$
79. $x - 4 + y$　**81.** $x + 3y - 9$　**83.** $x - 4 - 2y$　**85.** $3x - 8$　**87.** $2x - 5$　**89.** $14x + 18$　**91.** $4x - 2y + 3$
93. $-x + y + 3$　**95.** $7x + 3$　**97.** $x - 9$　**99.** $6x - 12$　**101.** $-x - 2$　**103.** $-x + 6$　**105.** $8x - 19$
107. $x + 2$　**109.** $3x$　**111.** $x + 15$　**113.** $2x + 4y + 12$　**115.** $-6x + 3y - 3$　**117.** $x - 5$

Exercise Set 2.2

1. 5　**3.** -10　**5.** -7　**7.** -61　**9.** 18　**11.** 43　**13.** -12　**15.** 72　**17.** -26　**19.** -58　**21.** 10
23. -12　**25.** 3　**27.** -9　**29.** 49　**31.** 3　**33.** 5　**35.** 1　**37.** 11　**39.** 15　**41.** -36　**43.** -47
45. 46　**47.** -21　**49.** -3　**51.** 72　**53.** Two or more equations with the same solution　**55.** Subtract 3

Exercise Set 2.3

1. 3　**3.** 8　**5.** -2　**7.** -12　**9.** 5　**11.** 3　**13.** $-\frac{3}{2}$　**15.** 4　**17.** 2　**19.** 49　**21.** $-\frac{1}{2}$　**23.** 6
25. $\frac{35}{19}$　**27.** $-\frac{3}{14}$　**29.** -1　**31.** $-\frac{3}{40}$　**33.** -75　**35.** 125　**37.** -35　**39.** 24　**41.** -5　**43.** 12
45. -16　**47.** -36　**49.** 6　**51.** $-\frac{9}{4}$　**53.** $-\frac{5}{4}$　**55.** 9　**57.** Divide by 3　**59.** Multiply by $\frac{3}{2}$, 6
61. Multiply by $\frac{7}{3}$, $\frac{28}{15}$

Exercise Set 2.4

1. 2　**3.** -6　**5.** 5　**7.** $\frac{12}{5}$　**9.** -12　**11.** 3　**13.** $\frac{15}{2}$　**15.** $-\frac{21}{2}$　**17.** $-\frac{56}{9}$　**19.** 5　**21.** $-\frac{7}{6}$　**23.** 3
25. -4　**27.** 1　**29.** $\frac{15}{2}$　**31.** 2　**33.** 0　**35.** -1　**37.** 0　**39.** $-\frac{7}{6}$　**41.** 1　**43.** -4　**45.** 4　**47.** $-\frac{3}{2}$
49. -5　**51.** -1　**53.** 6　**55.** 3　**57.** Addition property

Exercise Set 2.5

1. 4　**3.** 1　**5.** $\frac{3}{5}$　**7.** 3　**9.** -21　**11.** 1　**13.** $\frac{3}{10}$　**15.** 4　**17.** $-\frac{17}{7}$　**19.** No solution　**21.** $\frac{34}{5}$
23. -4　**25.** $-\frac{13}{2}$　**27.** All real numbers　**29.** All real numbers　**31.** 0　**33.** All real numbers　**35.** $-\frac{112}{15}$
37. 9　**39.** $-\frac{5}{3}$　**41.** 12　**43.** 16　**45.** $-\frac{10}{3}$　**47.** Same expression will appear on both sides of equation.

Just for Fun　**1.** $\frac{1}{4}$　**2.** $-\frac{17}{13}$

Exercise Set 2.6

1. (a) 5:8 **(b)** 1:5 **(c)** 2:1 **(d)** 11:25 **(e)** 25:4 **3.** $\frac{1}{4}$ gal oil, $3\frac{3}{4}$ gal gas **5.** 19.8 tons copper, 13.2 tons zinc
7. 80°, 60°, 40° **9.** Client $32,000, lawyer $16,000 **11.** 1 **13.** -40 **15.** -15 **17.** 1 **19.** -96
21. -400 **23.** 6831 mi **25.** $1114.27 **27.** 1.27 ft **29.** 260 in or 21.67 ft **31.** 24 teaspoons

Just for Fun **1.** $\frac{1}{3}$ cup flour, $\frac{1}{6}$ tsp salt, $1\frac{1}{3}$ tbsp butter, $\frac{2}{3}$ tsp nutmeg, $\frac{2}{3}$ tsp cinnamon, 1 cup sugar **2.** 0.625 cc

Exercise Set 2.7

1. $x > 4$ **3.** $x \geq -2$ **5.** $x > -5$ **7.** $x < 10$

9. $x \leq -4$ **11.** $x > -\frac{3}{2}$ **13.** $x \leq 1$ **15.** $x < -3$

17. $x < \frac{3}{2}$ **19.** $x > \frac{35}{9}$ **21.** $x > -\frac{8}{3}$ **23.** $x \leq -\frac{11}{3}$

25. $x \geq -6$ **27.** $x < 1$ **29.** $x < 2$ **31.** All real numbers

33. $x > \frac{3}{4}$ **35.** $x > \frac{23}{10}$ **37.** No solution **39.** $x \geq -\frac{7}{11}$

41. When multiplying or dividing by a negative number **43.** All real numbers

Just for Fun **1.** $x \geq -2$ **2.** $x > \frac{5}{6}$

Chapter 2 Review Exercises

1. $2x + 8$ **2.** $3x - 6$ **3.** $8x - 6$ **4.** $-2x - 8$ **5.** $-x - 2$ **6.** $-x + 2$ **7.** $-16 + 4x$ **8.** $18 - 6x$
9. $20x - 24$ **10.** $-6x + 15$ **11.** $36x - 36$ **12.** $-4x + 12$ **13.** $-3x - 3y$ **14.** $-6x + 4$ **15.** $-3 - 2y$
16. $-x - 2y + z$ **17.** $3x + 9y - 6z$ **18.** $-4x + 6y - 14$ **19.** $5x$ **20.** $7y + 2$ **21.** $-2y + 7$ **22.** $5x + 1$
23. $6x + 3y$ **24.** $-3x + 3y$ **25.** $6x + 8y$ **26.** $6x + 3y + 2$ **27.** $-x - 1$ **28.** $3x + 3y + 6$ **29.** 3
30. $-12x + 3$ **31.** $5x + 6$ **32.** $-2x$ **33.** $5x + 7$ **34.** $-10x + 12$ **35.** $5x + 3$ **36.** $4x - 4$ **37.** $22x - 42$
38. $3x - 3y + 6$ **39.** $-x + 5y$ **40.** $3x + 2y + 16$ **41.** 3 **42.** $-x - 2y + 4$ **43.** 2 **44.** -8 **45.** 11
46. -27 **47.** 2 **48.** $\frac{11}{2}$ **49.** -2 **50.** -3 **51.** 12 **52.** 1 **53.** 6 **54.** -3 **55.** $-\frac{21}{5}$ **56.** -5
57. -19 **58.** -1 **59.** $\frac{2}{3}$ **60.** $\frac{1}{5}$ **61.** $\frac{9}{2}$ **62.** $\frac{10}{7}$ **63.** -3 **64.** -1 **65.** -8 **66.** $-\frac{23}{5}$ **67.** -10
68. $-\frac{4}{3}$ **69.** No solution **70.** All real numbers **71.** $\frac{17}{3}$ **72.** $-\frac{20}{7}$ **73.** $\frac{22}{5}$ **74.** 3 **75.** 3 **76.** 9
77. $\frac{135}{4}$ **78.** -4 **79.** -24 **80.** -5 **81.** 600 **82.** 36 **83.** 90 **84.** $x \geq 2$

85. $x < 3$ **86.** $x \geq -\frac{12}{5}$ **87.** No solution **88.** All real numbers

89. $x < -3$ **90.** $x \leq \frac{9}{5}$ **91.** $x > \frac{8}{5}$ **92.** $x < -\frac{5}{3}$

93. $x \leq \frac{5}{11}$ **94.** No solution **95.** All real numbers **96.** $x \leq \frac{11}{2}$

97. $33,000, $22,000 **98.** $36,000, $27,000 **99.** 64, 16, 16 face cords **100.** 67.5 mi **101.** 15.75 min
102. $6\frac{1}{3}$ in. **103.** 9.45 ft

Chapter 2 Practice Test

1. $4x - 8$ **2.** $-x - 3y + 4$ **3.** $2x + 4$ **4.** $-x + 10$ **5.** $-6x + y - 6$ **6.** $7x - 5y + 3$ **7.** $8x - 1$
8. 4 **9.** -2 **10.** 2 **11.** -1 **12.** $-\frac{1}{7}$ **13.** No solution **14.** All real numbers **15.** -45

16. $x > -7$ **17.** $x \le 12$ **18.** No solution ———   **19.** $64,000, $32,000

20. 150 gal

CHAPTER 3

Exercise Set 3.1
1. 16   **3.** 48   **5.** 12   **7.** 12.56   **9.** 10   **11.** 1000   **13.** 10   **15.** 2   **17.** 1080   **19.** 50.24   **21.** 54
23. 15   **25.** 68   **27.** 60   **29.** $y = -2x + 8, 4$ **31.** $y = (2x + 4)/6, 4$ **33.** $y = (-3x + 6)/2, 0$
35. $y = (4x - 20)/5, -\frac{4}{5}$ **37.** $y = (3x + 18)/6, 3$ **39.** $y = (-x + 8)/2, 6$ **41.** $t = d/r$ **43.** $p = i/rt$ **45.** $d = C/\pi$
47. $b = 2A/h$ **49.** $w = (P - 2l)/2$ **51.** $n = (m - 3)/4$ **53.** $b = y - mx$ **55.** $r = (I - P)/Pt$ **57.** $d = (3A - m)/2$
59. $b = d - a - c$ **61.** $E = 2I - L$ **63.** $y = (-ax + c)/b$ **65.** $F = \frac{9}{5}C + 32$ **67.** $1440   **69.** 35   **71.** 88
73. 10°C   **75.** 95°F   **77.** $P = 10$ **79.** $K = 4$ **81.** 12   **83.** 110

Just for Fun   **1.** (a) $A = d^2 - \pi(d/2)^2$ (b) 3.44 sq ft (c) 7.74 sq ft   **2.** (a) $V = 18x^3 - 3x^2$ (b) 6027 cm³
(c) $S = 54x^2 - 8x$ (d) 2590 cm²   **3.** (a) 342.56 ft (b) 192 ft

Exercise Set 3.2
1. $x + 5$ **3.** $4x$ **5.** $6x - 3$ **7.** $\frac{3}{4}x + 7$ **9.** $2(x + 8)$ **11.** $25x$ **13.** $12x$ **15.** $16a + b$ **17.** $x, x + 12$
19. $x, x + 1$ **21.** $x, 100 - x$ **23.** $x, 4x - 5$ **25.** $x, 1.7x$ **27.** $x, 80 - x$ **29.** $4x$ **31.** $0.23x$ **33.** $15x$
35. $10x$ **37.** $300n$ **39.** $0.075x$ **41.** $10a$ **43.** $5p$ **45.** $x + 5x = 18$ **47.** $x + (x + 1) = 47$ **49.** $2x - 8 = 12$
51. $\frac{1}{5}(x + 10) = 150$ **53.** $x + (2x - 8) = 1000$ **55.** $x + 0.08x = 92$ **57.** $x + (2x - 3) = 21$ **59.** $40t = 180$
61. $15y = 215$ **63.** $25q = 150$

Exercise Set 3.3
1. 48, 49   **3.** 76, 78   **5.** 8, 19   **7.** 12, 13, 14   **9.** 5, 10, 14   **11.** 25, 42   **13.** 20 years   **15.** 136 weeks
17. 200 mi   **19.** 170 mi   **21.** $0.95   **23.** 10   **25.** Younger workers 12, third 24, fourth 36   **27.** $12.30
29. 5 hrs (more than 4.83)

Just for Fun   **1.** 37,500 mi   **2.** n
$4n$
$4n + 6$
$(4n + 6)/2 = 2n + 3$
$2n + 3 - 3 = 2n$

Exercise Set 3.4
1. 6 mi   **3.** 4.62 hr   **5.** 6136.36 min (102.27 hr)   **7.** 33.75 ft   **9.** 2.4 cm   **11.** 250 cm³/hr   **13.** 3.5 hr
15. 52 mph, 82 mph   **17.** 6 PM   **19.** 4 hr

Just for Fun   **1.** 46.2 sec   **2.** 0.35 ft/sec

Exercise Set 3.5
1. (a) 13 shares Barber Oil, 52 shares Teledyne   (b) $434   **3.** $3800 at 9%, $2200 at 10%
5. $7500 at 8%, $5000 at 12%   **7.** 49 nickels, 13 dimes   **9.** 14 quarters, 11 half-dollars
11. 11.25 lb almonds, 18.75 lb walnuts   **13.** 250 adults, 350 children   **15.** $0.21l$, 60%, $0.29l$, 25%
17. 40% pure juice   **19.** 4 m, 4 m, 2 m   **21.** 40°, 60°, 80°   **23.** width = $2\frac{2}{3}$ ft, height = $4\frac{2}{3}$ ft

Just for Fun   **1.** 3.5 qt punch   **2.** 6 qt

Chapter 3 Review Exercises
1. 12.56   **2.** 48   **3.** 20   **4.** 300   **5.** 240   **6.** 28.26   **7.** 4   **8.** 20   **9.** 21   **10.** -11 **11.** -8
12. 15   **13.** 4.5   **14.** $y = 2x - 12, 8$ **15.** $y = (3x + 4)/2, 5$ **16.** $y = (3x - 5)/2, -7$ **17.** $y = -3x - 10, -10$
18. $y = -\frac{3}{2}x - 3, 6$ **19.** $m = F/a$ **20.** $h = 2A/b$ **21.** $K = PV/T$ **22.** $t = i/pr$ **23.** $w = (P - 2l)/2$

24. $y = (2x - 6)/3$ **25.** $y = x^2 + 4x - 3$ **26.** $y = 2x^2 - 3x - 6$ **27.** $h = 2A/(b + d)$ **28.** $C = \frac{5}{9}(F - 32)$
29. $F = \frac{9}{5}C + 32$ **30.** $C = -2A + B$ **31.** \$180   **32. (a)** \$3,000   **(b)** \$8,000   **33.** 29, 33   **34.** 127, 128
35. 103, 105   **36.** 38, 7   **37.** 8.44 years   **38.** 426 mi   **39.** \$8,000   **40.** \$12,000   **41.** 4 hr
42. 538.46 mph   **43.** 57.14 hr   **44.** 4 hr   **45.** 2 hr   **46.** \$4000 at 13%, \$8000 at $7\frac{1}{4}$%
47. 600 gal of 87% octane, 600 gal of 91% octane   **48.** 1.2 liter of 10% acid, 0.8 liter of 5% acid
49. 60 lb of \$2.50, 20 lb of \$3.10   **50.** twenty 22¢ stamps, ten 15¢ stamps   **51.** $w = 15.5$ ft, $l = 19.5$ ft

Chapter 3 Practice Test
1. 18 ft   **2.** 145   **3.** $R = P/I$ **4.** $y = (3x - 6)/2$ **5.** $a = 3A - b$ **6.** $c = (D - Ra)/R$ **7.** 56, 102
8. 15, 30, 30 in.   **9.** \$16.39   **10.** 13, 14, 15 in.   **11.** 80 mph   **12.** 20 liters

CHAPTER 4

Exercise Set 4.1
1. x^7 **3.** y^3 **5.** 243   **7.** y^5 **9.** y^5 **11.** x^8 **13.** 25   **15.** x^4 **17.** y **19.** 1   **21.** 1   **23.** 3
25. 1   **27.** 1   **29.** x^{10} **31.** x^{25} **33.** x^3 **35.** x^{12} **37.** x^8 **39.** $9x^2$ **41.** x^2 **43.** $64x^6$ **45.** $-27x^9$
47. $8x^6y^3$ **49.** $25x^4y^{10}$ **51.** x^2/y^2 **53.** $x^3/125$ **55.** y^5/x^5 **57.** $216/x^3$ **59.** $27x^3/y^3$ **61.** $4x^2/25$
63. $64y^9/x^3$ **65.** $-27x^9/64$ **67.** $1/x^4$ **69.** $\frac{1}{243}$ **71.** x **73.** $1/x^8$ **75.** 16   **77.** $1/x^8$ **79.** x^3 **81.** $1/x^5$
83. 128   **85.** $1/x^8$ **87.** $1/4y^4$ **89.** $1/2y^2$ **91.** y **93.** $x^{12}/27$ **95.** y^4/x^7 **97.** $5x^2/y^2$ **99.** $1/4x^2y$
101. $7/2x^5y^5$ **103.** $2y^3/x^4$ **105.** $1/2x$ **107.** $-6x^6y^7$ **109.** $3x^6y^8$ **111.** $10x^5y$ **113.** 2   **115.** $-6x^2y^4$
117. $1/xy^2$ **119.** $x^2y^2/36$ **121.** y^9/x^3 **123.** $y^{10}/9x^2$ **125.** $72x^5$ **127.** $16/x^3$ **129.** $128x^8$ **131.** $16/x^4$
133. $4x^6y^4/z^2$ **135.** $4x^6y^8/25$ **137.** $9x^4/16y^6$ **139.** $64y^6/27x^9z^3$ **143.** $2/x$

Just for Fun   **1.** $-9x^{17}/16$ **2.** $-8x^3/y^3$ **3. (a)** and **(b)** $z^2/9x^4y^6$

Exercise Set 4.2
1. 4.2×10^4 **3.** 9×10^2 **5.** 5.3×10^{-2} **7.** 1.9×10^4 **9.** 1.86×10^{-6} **11.** 9.14×10^{-6} **13.** 1.07×10^2
15. 1.53×10^{-1} **17.** 4200   **19.** 40,000,000   **21.** 0.0000213   **23.** 0.312   **25.** 9,000,000   **27.** 535   **29.** 35,000
31. 10,000   **33.** 120,000,000   **35.** 0.0153   **37.** 320   **39.** 0.021   **41.** 20   **43.** 4.2×10^{12} **45.** 4.5×10^{-7}
47. 2×10^3 **49.** 2×10^{-7} **51.** 3×10^8 **53.** $9.2 \times 10^{-5}, 1.3 \times 10^{-1}, 8.4 \times 10^3, 6.2 \times 10^4$ **55.** 3.2×10^7 seconds
57. 8,640,000,000 cu ft

Just for Fun   **1. (a)** About 5.87×10^{12} miles   **(b)** About 500 seconds or $8\frac{1}{3}$ minutes

Exercise Set 4.3
1. Monomial   **3.** Monomial   **5.** Binomial   **7.** Trinomial   **9.** Not polynomial   **11.** Binomial   **13.** Monomial
15. Polynomial   **17.** Trinomial   **19.** Not polynomial   **21.** First degree   **23.** $2x^2 + x - 6$, second
25. $-x^2 - 4x - 8$, second   **27.** Third   **29.** Second   **31.** $-6x^3 + x^2 - 3x + 4$, third   **33.** $5x^2 - 2x - 4$, second
35. $-2x^3 + 3x^2 + 5x - 6$, third   **37.** $6x + 1$ **39.** $-2x + 11$ **41.** $-x - 2$ **43.** $21x - 21$ **45.** $x^2 + 6x + 3$
47. $2x^2 + 8x + 5$ **49.** $5x^2 + x + 20$ **51.** $-x^2 + x + 20$ **53.** $6x^2 - 5x + 4$ **55.** $-2x^3 - 3x^2 + 4x - 3$
57. $3x^2 - 2xy$ **59.** $7x^2y - 3x + 2$ **61.** $x + 7$ **63.** $2x + 1$ **65.** $7x - 1$ **67.** $-3x + 16$ **69.** $x^2 + x + 16$
71. $-3x^2 + 2x - 12$ **73.** $7x^2 + 7x - 13$ **75.** $-2x^2 - 4x - 6$ **77.** $2x^3 - x^2 + 6x - 2$ **79.** $5x^3 - 7x^2 - 2$
81. $3xy + 3x + 3$ **83.** $x - 6$ **85.** $3x + 4$ **87.** -5 **89.** $-8x + 7$ **91.** $-7x + 3$ **93.** $6x^2 + 7x - 10$
95. $-7x^2 + x - 12$ **97.** $-x^2 - 6x - 4$ **99.** $5x^2 - 2x + 7$ **101.** $2x^2 + 5x + 2$ **103.** $4x^3 - 8x^2 - x - 4$
105. $9x^3 - x^2 - 5x - 4$ **107.** $-x + 11$ **109.** $2x^2 - 9x + 14$ **111.** $-x^3 + 11x^2 + 9x - 7$ **113.** $3x + 17$
115. $4x + 7$ **117.** $5x^2 + 7x - 2$ **119.** $4x^2 - 5x - 6$ **121.** $4x^3 - 7x^2 + x - 2$

Exercise Set 4.4
1. $3x + 12$ **3.** $2x^2 - 6x$ **5.** $8x^2 - 24x$ **7.** $2x^3 + 6x^2 - 2x$ **9.** $-2x^3 + 4x^2 - 10x$ **11.** $-20x^3 + 30x^2 - 20x$
13. $24x^3 + 32x^2 - 40x$ **15.** $6x^2y + 15x^2 - 18xy$ **17.** $xy - y^2 - 3y$ **19.** $x^2 + 7x + 12$ **21.** $6x^2 + 3x - 30$
23. $4x^2 - 16$ **25.** $-6x^2 - 8x + 30$ **27.** $-2x^2 + x + 15$ **29.** $x^2 + 7x + 12$ **31.** $x^2 + 2x - 8$
33. $6x^2 + 23x + 20$ **35.** $6x^2 - x - 12$ **37.** $x^2 - 1$ **39.** $4x^2 - 12x + 9$ **41.** $-2x^2 + 5x + 12$
43. $-4x^2 + 2x + 12$ **45.** $x^2 - y^2$ **47.** $6x^2 - 5xy - 6y^2$ **49.** $8xy - 12x - 6y^2 + 9y$ **51.** $3x^2 + 17x + 20$

53. $10x^2 + 23x + 12$ **55.** $x^2 - 16$ **57.** $4x^2 - 1$ **59.** $x^2 + 2xy + y^2$ **61.** $x^2 - 4x + 4$ **63.** $9x^2 - 25$
65. $2x^3 + 10x^2 + 11x - 3$ **67.** $5x^3 - x^2 + 16x + 16$ **69.** $-14x^3 - 22x^2 + 19x - 3$ **71.** $18x^3 - 69x^2 + 54x - 27$
73. $6x^4 + 5x^3 + 5x^2 + 10x + 4$ **75.** $x^4 - 3x^3 + 5x^2 - 6x$ **77.** $3x^4 - 7x^3 - 7x^2 + 3x$ **79.** $a^3 + b^3$

Just for Fun **1.** $2x^3\sqrt{5} + 5x^2 - (x\sqrt{5})/2$ **2.** $(1/3)x^2 + (11/45)x - 4/15$
3. $6x^6 - 18x^5 + 3x^4 + 35x^3 - 54x^2 + 38x - 12$ **4.** $x^2 - y^2 + 2y - 1$ **5.** $x^2 + 2xy - 2x + y^2 - 2y + 1$

Exercise Set 4.5
1. $x + 2$ **3.** $x + 3$ **5.** $\frac{3}{2}x + 4$ **7.** $-3x + 2$ **9.** $3x + 1$ **11.** $3 + (6/x)$ **13.** $(-3/x) + 1$ **15.** $x^2 + 2x - 3$
17. $-2x^2 + 3x + 4$ **19.** $x + 4 - (3/x)$ **21.** $3x - 2 + (6/x)$ **23.** $-x^2 - (3/2)x + (2/x)$ **25.** $3x + 1 - (4/x^2)$
27. $x + 3$ **29.** $2x + 3$ **31.** $2x + 4$ **33.** $x - 2$ **35.** $x + 5 - [3/(2x - 3)]$ **37.** $2x + 3$
39. $2x - 3 + [2/(4x + 9)]$ **41.** $4x - 3 - [3/(2x + 3)]$ **43.** $x^2 + 2x + 3$ **45.** $2x^2 - x - 4 + [2/(x - 1)]$
47. $2x^2 - 8x + 38 - [156/(x + 4)]$ **49.** $x^2 - 2x + 4$ **51.** $x^2 - 3x + 9$ **53.** $3x^2 + 2x + 1 + [5/(3x - 2)]$

Just for Fun **1.** $2x^2 - 3x + \frac{5}{2} - [3/2(2x + 3)]$ **2.** $x^2 + \frac{2}{3}x + \frac{4}{9} - [37/9(3x - 2)]$

Chapter 4 Review Exercises
1. x^6 **2.** x^8 **3.** 243 **4.** 32 **5.** $1/x^2$ **6.** $1/x^5$ **7.** $1/x^3$ **8.** x^8 **9.** x^3 **10.** 1 **11.** 9 **12.** 16
13. $1/x^2$ **14.** x^5 **15.** x^7 **16.** $1/x^6$ **17.** x^2 **18.** x^6 **19.** 1 **20.** 3 **21.** 1 **22.** 1 **23.** $4x^2$
24. $27x^3$ **25.** $4x^2$ **26.** $-27x^3$ **27.** $16x^8$ **28.** $1/9x^8$ **29.** $-x^{12}$ **30.** x^{12} **31.** $4x^6/y^2$ **32.** $x^9/64y^3$
33. $-125/x^6$ **34.** $27x^{12}/8y^3$ **35.** $24x^5$ **36.** $-8x^8$ **37.** x^6 **38.** $4x/y$ **39.** $12x^5y^2$ **40.** $27y^3/x^6$
41. $x^4/16y^6$ **42.** $9x^2$ **43.** 3.64×10^5 **44.** 1.64×10^6 **45.** 7.63×10^{-3} **46.** 1.76×10^{-1} **47.** 2.08×10^3
48. 3.14×10^{-4} **49.** 0.0042 **50.** 16,500 **51.** 970,000 **52.** 0.00000438 **53.** 0.914 **54.** 536 **55.** 4,600,000
56. 1260 **57.** 19.84 **58.** 340,000 **59.** 0.09 **60.** 0.00003 **61.** 1.2×10^9 **62.** 2.4×10^1 **63.** 9.2
64. 2×10^8 **65.** 3.4×10^{-3} **66.** 5×10^8 **67.** Binomial, 1 **68.** Monomial, 0 **69.** $x^2 + 3x - 4$, trinomial, second
70. $4x^2 - x - 3$, trinomial, second **71.** Binomial, second **72.** Not polynomial **73.** $-4x^2 + x$, binomial, second
74. Not polynomial **75.** $x^3 + 4x^2 - 2x - 6$, third **76.** $3x + 7$ **77.** $9x + 1$ **78.** $2x - 5$ **79.** $4x^2 + 14$
80. $-3x^2 + 10x - 15$ **81.** $11x^2 - 2x - 3$ **82.** $x - 7$ **83.** $-2x + 2$ **84.** $9x^2 - 8x + 4$ **85.** $6x^2 - 18x - 4$
86. $-5x^2 + 8x - 19$ **87.** $4x + 2$ **88.** $2x + 10$ **89.** $2x^2 - 4x$ **90.** $2x^3 - 6x^2$ **91.** $6x^3 - 12x^2 + 21x$
92. $-3x^3 + 6x^2 + x$ **93.** $24x^3 - 16x^2 + 8x$ **94.** $x^2 + 9x + 20$ **95.** $2x^2 - 2x - 12$ **96.** $16x^2 + 48x + 36$
97. $-6x^2 + 14x + 12$ **98.** $x^2 - 16$ **99.** $3x^3 + 7x^2 + 14x + 4$ **100.** $3x^3 + x^2 - 10x + 6$
101. $10x^3 - 19x^2 + 36x - 12$ **102.** $x + 2$ **103.** $x - 2$ **104.** $8x + 4$ **105.** $2x^2 + 3x - \frac{4}{3}$ **106.** $8x + 6 - (4/x)$
107. $4x - 2$ **108.** $-8x + 2$ **109.** $-2x - 4$ **110.** $(5/2)x + 5 + (1/x)$ **111.** $x + 4$ **112.** $2x - 3$
113. $5x - 2 + [2/(x + 6)]$ **114.** $2x^2 + 3x - 4$ **115.** $2x^2 + x - 2 + [2/(2x - 1)]$

Chapter 4 Practice Test
1. $6x^6$ **2.** $27x^6$ **3.** $4x^3$ **4.** $y^4/4x^6$ **5.** $x^3/8y^6$ **6.** $1/5x^3y^6$ **7.** Trinomial **8.** Monomial
9. Not polynomial **10.** $6x^3 - 2x^2 + 5x - 5$, third degree **11.** $3x^2 - 3x + 1$ **12.** $-2x^2 + 4x$ **13.** $3x^2 - x + 3$
14. $12x^3 - 6x^2 + 15x$ **15.** $8x^2 + 2x - 21$ **16.** $-12x^2 - 2x + 30$ **17.** $6x^3 - 4x^2 - 28x + 24$ **18.** $4x^2 + 2x - 1$
19. $-x + 2 - (5/3x)$ **20.** $4x + 5$ **21.** 1.26×10^9 **22.** 2×10^{-7}

CHAPTER 5 _____

Exercise Set 5.1
1. $2^3 \cdot 5$ **3.** $2 \cdot 3^2 \cdot 5$ **5.** $2^3 \cdot 5^2$ **7.** 4 **9.** 12 **11.** 18 **13.** x **15.** $3x$ **17.** 1 **19.** xy **21.** x^3y^5
23. 8 **25.** $3x^3y^7$ **27.** x^2 **29.** $x + 3$ **31.** $2x - 3$ **33.** $3x - 4$ **35.** $2(x + 2)$ **37.** $5(3x - 1)$
39. Cannot be factored **41.** $4x(4x - 3)$ **43.** $2p(10 - 9p)$ **45.** $2x(3x^2 - 4)$ **47.** $12x^8(3x^4 - 2)$ **49.** $3y^3(8y^{12} - 3)$
51. $x(1 + 3y^2)$ **53.** Cannot be factored **55.** $4xy(4yz + x^2)$ **57.** $2xy^2(17x + 8y^2)$ **59.** $36xy^2z(z^2 + x^2)$
61. $y^3z^5(14 - 9x)$ **63.** $3(x^2 + 2x + 3)$ **65.** $3(3x^2 + 6x + 1)$ **67.** $3x(x^2 - 2x + 4)$ **69.** Cannot be factored
71. $3(5p^2 - 2p + 3)$ **73.** $4x^3(6x^3 + 2x - 1)$ **75.** $xy(48x + 16y + 33)$ **77.** $(x + 3)(x + 2)$ **79.** $(7x - 4)(4x - 3)$
81. $(4x + 1)(2x + 1)$ **83.** $(4x + 1)(2x + 1)$

Just for fun   **1.** $2(x - 3)[2x^2(x - 3)^2 - 3x(x - 3) + 2]$ **2.** $2x^2(2x + 7)(3x^3 + 2x - 1)$
3. (a) $3(0) + 3(1) + 3(2) + 3(3) + 3(4)$ **(b)** $3(0 + 1 + 2 + 3 + 4)$ **(c)** $3(10) = 30$ **(d)** $3(55) = 165$

Exercise Set 5.2
1. $(x + 3)(x + 4)$ **3.** $(x + 5)(x + 2)$ **5.** $(x + 2)(x + 3)$ **7.** $(x - 5)(x + 3)$ **9.** $(2x - 3)(2x + 3)$
11. $(x + 4)(2x + 3)$ **13.** $(3x + 1)(x + 3)$ **15.** $(2x - 1)(2x - 1) = (2x - 1)^2$ **17.** $(8x + 1)(x + 4)$
19. $2(2x - 1)(2x - 5)$ **21.** $(x + 1)(3x - 2)$ **23.** $(x - 1)(3x - 2)$ **25.** $(3x - 4)(5x - 6)$ **27.** $(x - 3y)(x + 2y)$
29. $(3x + y)(2x - 3y)$ **31.** $(2x - 5y)(5x - 6y)$ **33.** $2(x + 4)(x - 6)$ **35.** $4(x + 2)(x + 2) = 4(x + 2)^2$
37. $x(3x - 1)(2x + 3)$ **39.** $2(x + 4y)(x - 2y)$ **41.** Determine if all terms have a common factor, if so factor out the GCF.

Just for Fun   **1.** $x(3x^2 + 2)(x^2 - 5)$ **2.** $(x - y)(x^2 + y)$ **3.** $(3a - x)(6a + x^2)$ **4.** $y(2x - 3y)(x^2y^2 + 3z^2)$
5. $(2a - 3c)(a^3b - c^2)$

Exercise Set 5.3
1. $(x + 2)(x + 5)$ **3.** $(x + 3)(x + 2)$ **5.** $(x + 4)(x + 3)$ **7.** Cannot be factored   **9.** $(y - 15)(y - 1)$
11. $(x + 3)(x - 2)$ **13.** $(k - 5)(k + 3)$ **15.** $(b - 9)(b - 2)$ **17.** Cannot be factored   **19.** $(a + 11)(a + 1)$
21. $(x + 15)(x - 2)$ **23.** $(x + 2)^2$ **25.** $(x + 3)^2$ **27.** $(x + 5)^2$ **29.** $(w - 15)(w - 3)$ **31.** $(x + 24)(x - 2)$
33. $(x - 5)(x + 4)$ **35.** $(y - 7)(y - 2)$ **37.** $(x + 3y)(x + y)$ **39.** $(x + 5y)(x + 3y)$ **41.** $2(x - 6)(x - 1)$
43. $5(x + 3)(x + 1)$ **45.** $2(x - 4)(x - 3)$ **47.** $x(x - 6)(x + 3)$ **49.** $2x(x + 7)(x - 4)$ **51.** $x(x + 2)^2$
53. Multiplying factors using the FOIL method.

Just for Fun   **1.** $(x + 0.4)(x + 0.2)$ **2.** $(x - 0.6)(x + 0.1)$ **3.** $(x + \frac{1}{5})(x + \frac{1}{5})$ **4.** $(x - \frac{1}{3})(x - \frac{1}{3})$

Exercise Set 5.4
1. $(3x + 2)(x + 1)$ **3.** $(2x + 3)(3x + 2)$ **5.** $(2x + 3)(x + 1)$ **7.** $(2x + 5)(x + 3)$ **9.** $(3x + 2)(x - 4)$
11. $(5y - 3)(y - 1)$ **13.** Cannot be factored   **15.** $(4x + 1)(x + 3)$ **17.** Cannot be factored   **19.** $(5y - 1)(y - 3)$
21. $(3x - 1)(x + 5)$ **23.** $(7x - 2)(x - 2)$ **25.** $(x - 1)(3x - 7)$ **27.** $(5z + 2)(z - 7)$ **29.** $3(2x + 1)(x + 5)$
31. $2(3x + 5)(x - 1)$ **33.** $x(2x - 1)(3x + 4)$ **35.** $2x(2x + 3)(x - 1)$ **37.** $2x(3x + 5)(x - 1)$ **39.** $5(6x + 1)(2x + 1)$
41. $(2x + y)(x + 2y)$ **43.** $(2x - y)(x - 3y)$ **45.** $2(3x - y)(3x + 4y)$ **47.** Factor out the GCF if there is one.

Exercise Set 5.5
1. $(x + 2)(x - 2)$ **3.** $(y + 5)(y - 5)$ **5.** $(x + 7)(x - 7)$ **7.** $(x + y)(x - y)$ **9.** $(3y + 4)(3y - 4)$
11. $4(4a + 3b)(4a - 3b)$ **13.** $(5x + 4)(5x - 4)$ **15.** $(z^2 + 9x)(z^2 - 9x)$ **17.** $9(x^2 + 3y)(x^2 - 3y)$
19. $(7m^2 - 4n)(7m^2 + 4n)$ **21.** $20(x + 3)(x - 3)$ **23.** $(x + y)(x^2 - xy + y^2)$ **25.** $(a - b)(a^2 + ab + b^2)$
27. $(x + 2)(x^2 - 2x + 4)$ **29.** $(x - 3)(x^2 + 3x + 9)$ **31.** $(a + 1)(a^2 - a + 1)$ **33.** $(2x + 3)(4x^2 - 6x + 9)$
35. $(3a - 4)(9a^2 + 12a + 16)$ **37.** $(3 - 2y)(9 + 6y + 4y^2)$ **39.** $(2x - 3y)(4x^2 + 6xy + 9y^2)$ **41.** $2(x - 3)(x + 2)$
43. $y(x + 4)(x - 4)$ **45.** $3(x + 1)^2$ **47.** $5(x + 3)(x - 1)$ **49.** $3(x + 4)(x - 6)$ **51.** $2(x + 6)(x - 6)$
53. $3y(x + 3)(x - 3)$ **55.** $3y^2(x + 1)(x^2 - x + 1)$ **57.** $2(x - 2)(x^2 + 2x + 4)$ **59.** $2(x + 4)(3x - 2)$
61. $x(3x + 2)(x - 4)$ **63.** $(x + 2)(4x - 3)$ **65.** $25(b + 2)(b - 2)$ **67.** $a^3b^2(a + 2b)(a - 2b)$ **69.** $3x^2(x - 3)^2$
71. $x(x^2 + 25)$ **73.** $(y^2 + 4)(y + 2)(y - 2)$ **75.** $5(2a - 3b)(a + 4b)$ **77.** $(3x + 5)(3x - 1)$ **79.** $x(x + 5)(x - 5)$

Just for Fun   **1.** $(x^2 + 1)(x^4 - x^2 + 1)$ **2.** $(x^2 - 3y^3)(x^4 + 3x^2y^3 + 9y^6)$
3. Cannot divide both sides of equation by $a - b$ since it equals 0.

Exercise Set 5.6
1. $0, -3$ **3.** $0, 9$ **5.** $-\frac{5}{2}, 3$ **7.** $4, -4$ **9.** $0, 12$ **11.** $0, -2$ **13.** $-4, 3$ **15.** $2, 10$ **17.** $3, -6$
19. $-4, 6$ **21.** $2, -21$ **23.** $-5, -6$ **25.** $4, -2$ **27.** $32, -2$ **29.** $6, -3$ **31.** $\frac{1}{3}, 7$ **33.** $\frac{2}{3}, -1$
35. $-4, 3$ **37.** $\frac{4}{3}, -\frac{1}{2}$ **39.** $2, 3$ **41.** $0, 16$ **43.** $6, -6$ **45.** $3, -3$ **47.** $8, 10$ **49.** $7, 9$ **51.** $5, 7$
53. $w = 3$ ft, $l = 12$ ft   **55.** 2 **57.** 3

Just for Fun   **1. (a)** 192 ft   **(b)** 8 sec

Chapter 5 Review Exercises

1. x^2 **2.** $3p$ **3.** 6   **4.** $4x^2y^2$ **5.** 1   **6.** $8x^2$ **7.** $2x - 5$ **8.** $x + 5$ **9.** $5(x - 4)$ **10.** $3(3x + 11)$
11. $4y(4y - 3)$ **12.** $5p^2(11p - 4)$ **13.** $6x^2y(4 + 3xy)$ **14.** $9xy(2x - 1)$ **15.** $2(x^2 + 2x - 4)$
16. $6x^4y^2(10y^2 + x^5y - 3x)$ **17.** Cannot be factored   **18.** $(x - 2)(5x + 3)$ **19.** $(3x - 2)(x - 1)$ **20.** $(2x + 1)(4x - 3)$
21. $(x + 2)(x + 3)$ **22.** $(x + 3)(x - 5)$ **23.** $(x + 7)(x - 7)$ **24.** $(x + 3)(x - 3)$ **25.** $(3x + 1)(x + 3)$
26. $(x + 3)(3x + 1)$ **27.** $(5x - 1)(x + 4)$ **28.** $(x + 4y)(5x - y)$ **29.** $(4x + 5y)(3x - 2y)$ **30.** $(3x - 2y)(4x + 5y)$
31. $(4x - 1)(x + 6)$ **32.** $(3x - 1)(4x - 3)$ **33.** $(4x + 3)(5x - 3)$ **34.** $(3x - 1)(2x + 3)$ **35.** $(x + 2)(x + 4)$
36. $(x - 5)(x - 3)$ **37.** $(x - 5)(x + 4)$ **38.** $(x + 5)(x - 4)$ **39.** $(x - 6)(x + 3)$ **40.** $(x - 7)(x - 2)$
41. $(x - 6)(x - 4)$ **42.** $(x - 9)(x + 3)$ **43.** $(x - 15)(x + 3)$ **44.** $(x + 8)(x + 3)$ **45.** $x(x + 4)(x + 1)$
46. $x(x - 8)(x + 5)$ **47.** $(x + 3y)(x + 2y)$ **48.** $(x + 3y)(x - 5y)$ **49.** $2x(x + 4y)(x + 2y)$ **50.** $4x(x + 5y)(x + 3y)$
51. $(2x - 1)(x + 4)$ **52.** $(3x + 1)(x + 4)$ **53.** $(2x + 3)(2x - 5)$ **54.** $(4x - 5)(x - 1)$ **55.** $(3x + 4)(x - 3)$
56. $(2x - 3)(2x + 5)$ **57.** $(5x - 2)(x - 6)$ **58.** $(3x + 4)(x + 3)$ **59.** $(6x + 1)(x + 5)$ **60.** $3(2x - 3)(x - 4)$
61. $(2x - 5)(x + 7)$ **62.** $(3x - 2)(2x + 5)$ **63.** $(4x + 5)(2x - 7)$ **64.** $(2x + 5)^2$ **65.** $x(3x - 2)^2$
66. $2x(3x + 1)(3x - 5)$ **67.** $(2x - 5y)(x + 2y)$ **68.** $(2x - 3y)(2x - 5y)$ **69.** $(2x - 3y)(3x + 7y)$
70. $(8x + y)(2x - 3y)$ **71.** $(x + 5)(x - 5)$ **72.** $(x + 8)(x - 8)$ **73.** $4(x + 2)(x - 2)$ **74.** $9(3x + y)(3x - y)$
75. $(8x^2 + 9y^2)(8x^2 - 9y^2)$ **76.** $(4 + 5y)(4 - 5y)$ **77.** $(2x^2 + 3y^2)(2x^2 - 3y^2)$ **78.** $(10x^2 + 11y^2)(10x^2 - 11y^2)$
79. $(x - y)(x^2 + xy + y^2)$ **80.** $(x + y)(x^2 - xy + y^2)$ **81.** $(a + 2)(a^2 - 2a + 4)$ **82.** $(a - 1)(a^2 + a + 1)$
83. $(a + 3)(a^2 - 3a + 9)$ **84.** $(x - 2)(x^2 + 2x + 4)$ **85.** $(2x - y)(4x^2 + 2xy + y^2)$ **86.** $(3 - 2y)(9 + 6y + 4y^2)$
87. $8(x + 3)(x - 1)$ **88.** $2(x - 4)^2$ **89.** $4(x + 3)(x - 3)$ **90.** $3(y + 3)(y - 3)$ **91.** $8(x - 1)(x^2 + x + 1)$
92. $y(x - 3)(x^2 + 3x + 9)$ **93.** $y(x + 4)(x - 1)$ **94.** $3x(2x + 3)(x + 5)$ **95.** $(2x - 5y)^2$ **96.** $(4y + 7z)(4y - 7z)$
97. $(a + 6)(b + 7)$ **98.** $y^5(4 + 5y)(4 - 5y)$ **99.** $2x(4x + 1)(4x + 3)$ **100.** $(y^2 + 1)(y + 1)(y - 1)$ **101.** 0, 4
102. $-3, -4$ **103.** $5, -\frac{2}{3}$ **104.** 0, 3   **105.** $0, -4$ **106.** $6, -4$ **107.** $-3, -5$ **108.** $-4, 2$ **109.** $-4, 3$
110. $-2, -5$ **111.** 2, 4   **112.** $1, -2$ **113.** $\frac{1}{4}, -\frac{3}{2}$ **114.** $\frac{1}{2}, -8$ **115.** $2, -2$ **116.** $\frac{7}{6}, -\frac{7}{6}$ **117.** 10, 11
118. 6, 8   **119.** 5, 8   **120.** $w = 7$ ft, $l = 9$ ft   **121.** 5 in., 9 in.

Chapter 5 Practice Test

1. $2x^2$ **2.** $3xy^2$ **3.** $4xy(x - 2)$ **4.** $3x(8xy - 2y + 3)$ **5.** $(x + 2)(x - 3)$ **6.** $(3x + 1)(x - 4)$
7. $(5x - 3y)(x - 3y)$ **8.** $(x + 4)(x + 8)$ **9.** $(x + 8)(x - 3)$ **10.** $(x - 5y)(x - 4y)$ **11.** $2(x - 5)(x - 6)$
12. $x(2x - 1)(x - 1)$ **13.** $(3x + 2y)(4x - 3y)$ **14.** $(x + 3y)(x - 3y)$ **15.** $(x + 3)(x^2 - 3x + 9)$ **16.** $2, \frac{5}{2}$
17. $-2, -3$ **18.** $-5, 1$ **19.** 4, 9   **20.** $w = 4$ in., $l = 6$ in.

CHAPTER 6 _____

Exercise Set 6.1

1. $1/(1 + y)$ **3.** 4   **5.** $(x^2 + 6x + 3)/2$ **7.** $x + 1$ **9.** $x/(x - 2)$ **11.** $(x - 3)/(x - 2)$ **13.** $2(x + 1)$ **15.** -1
17. $-(x + 2)$ **19.** $-(x + 6)/2x$ **21.** $-(x + 3)$ **23.** $-(x - 4)$ **25.** $(x + 5)/(x - 5)$ **27.** $(x - 2)/(2x + 3)$
29. $x + 1$ **31.** The variable does not make the denominator equal to 0.   **33.** $x \neq -5$ **35.** $x \neq 5, x \neq -5$

Exercise Set 6.2

1. $xy/4$ **3.** $80x^4/y^6$ **5.** $9x^2/8$ **7.** $32x^7/35my^2$ **9.** $36x^9y^2/25z^7$ **11.** $\frac{1}{2}$ **13.** x **15.** $(-3x + 2)/(3x + 2)$
17. 1   **19.** 1   **21.** $(x + 3)/(x - 4)$ **23.** $(a + b)(a - b)$ or $a^2 - b^2$ **25.** $(a + 3)/(a + 2)$ **27.** $(x - 6)(x - 3)$
29. $(x + 3)/2(x + 2)$ **31.** 1   **33.** $\frac{1}{2}$ **35.** $3x^2y$ **37.** $10z/x$ **39.** $2x^3y^5/9z^2$ **41.** $-wy/3x$ **43.** $6a^2b$
45. $9/5xy^4$ **47.** $a/2$ **49.** $3x^2/2$ **51.** $-1/x$ **53.** 1   **55.** $x/(x + 6)$ **57.** $(x - 8)/(x + 2)$ **59.** $(x - 8)(x + 4)$
61. $-3x(x + 2)$ **63.** $(x + 3)/(x - 1)$ **65.** 1

Exercise Set 6.3

1. $(2x - 1)/6$ **3.** $(x - 11)/3$ **5.** $(x - 3)/x$ **7.** $(x + 3)/x$ **9.** $(x + 7)/(x + 2)$ **11.** $-8/x$ **13.** $(2x - 11)/(x - 7)$
15. $(2x + 1)/2x^2$ **17.** $1/(x + 1)$ **19.** $1/(x - 3)$ **21.** 0   **23.** $-(4x + 1)/(x - 7)$ **25.** $x - 3$ **27.** 3
29. $(x^2 + 9)/(x + 3)$ **31.** -1 **33.** $(x - 3)/(x - 1)$ **35.** $x - 4$ **37.** $(x + 2)/(x - 8)$ **39.** $\frac{3}{4}$
41. $(x - 8)/(x + 8)$ **43.** $(3x + 2)/(x - 4)$ **45.** The signs change.   **47.** Should be $[6x - 2 - (3x^2 - 4x + 5)]/(x^2 - 4x + 3)$

Exercise Set 6.4

1. 3 **3.** $6x$ **5.** $10x$ **7.** x^2 **9.** $2x+3$ **11.** $x^2(x+1)$ **13.** $144x^3y$ **15.** $36x(x+5)$ **17.** $x(x+1)$
19. $180x^2y^3$ **21.** $6(x+4)(x-3)$ **23.** $(x+6)(x+5)$ **25.** $(x-8)(x+3)(x+8)$ **27.** $(x+3)(x-3)(x-1)$
29. $(x-2)(x+1)(x+3)$ **31.** $(x+2)^2$ **33.** $2x(x+3)$

Exercise Set 6.5

1. $11/2x$ **3.** $(3x+12)/2x^2$ **5.** $(2x^2-1)/x^2$ **7.** $(3x+5)/5x^2$ **9.** $(28x+15y)/20x^2y^2$ **11.** $x(y+1)/y$
13. $(9x-1)/3x$ **15.** $(5x^2+y^2)/xy$ **17.** $(4y-30x^2)/5x^2y$ **19.** $(8x-10)/x(x-2)$ **21.** $(11a+6)/a(a+3)$
23. $(-2x^2+4x+8)/3x(x+2)$ **25.** $2/(x-2)$ **27.** $9/(x+3)$ **29.** $(7x+1)/(x+1)(x-1)$ **31.** $20x/(x-5)(x+5)$
33. $(5x-12)/(x+3)(x-3)$ **35.** $(-x+6)/(x+2)(x-2)$ **37.** $5(x+1)/(x+4)(x-2)(x-1)$
39. $(2x+13)/(x-8)(x-1)(x+2)$ **41.** $(4x-3)/(x+3)(x+2)/(x-2)$ **43.** $11/(x-3)(x-4)$
45. $-(x-3)/(x+4)$ **47.** $(x+3)(x-1)/(x+2)(x+2)$

Just for Fun **1.** $(2x-3)/(2-x)$ **2.** $(3x^3+8x^2-10x+4)/(x-2)(x+3)$
3. $(8x^3-8x^2-14x+33)/(x+2)(x-3)(2x+3)$

Exercise Set 6.6

1. $\frac{8}{11}$ **3.** $\frac{57}{32}$ **5.** $\frac{25}{1224}$ **7.** $x^3y/8$ **9.** $6xz^2$ **11.** $(xy+1)/x$ **13.** $3/x$ **15.** $(3y-1)/(2y-1)$
17. $(x-y)/y$ **19.** $-a/b$ **21.** -1 **23.** $2(x+2)/x^3$ **25.** $a+b$ **27.** $(a^2+b)/b(b+1)$ **29.** $(y-x)/(x+y)$

Just for Fun **1. (a)** $\frac{2}{7}$ **(b)** $\frac{4}{13}$

Exercise Set 6.7

1. 4 **3.** 48 **5.** 30 **7.** -1 **9.** 1 **11.** 4 **13.** 3 **15.** $-\frac{1}{5}$ **17.** $\frac{1}{4}$ **19.** $\frac{14}{3}$ **21.** No solution
23. 2 **25.** $-\frac{12}{7}$ **27.** 8 **29.** -14 **31.** $-2, -3$ **33.** 4 **35.** $-\frac{5}{2}$ **37.** 5 **39.** No solution

Just for Fun **1.** 15 cm **2. (a)** 120 ohms **(b)** 900 ohms

Exercise Set 6.8

1. $b=16$ cm, $h=10$ cm **3.** 1, 3 **5.** $\frac{15}{8}$ **7.** 1 mph **9.** 7.5 mi **11.** 50 km/hr, 80 km/hr
13. walks 3 mph, jogs 6 mph **15.** $2\frac{2}{9}$ hr **17.** $3\frac{3}{7}$ days **19.** $2\frac{2}{5}$ hr **21.** $15\frac{3}{11}$ min

Just for Fun **1.** All real numbers except 3 **2.** 1 or $\frac{2}{3}$

Chapter 6 Review Exercises

1. $1/(1-y)$ **2.** $x/(2x+4)$ **3.** $x^2+4x+12$ **4.** $3x+2y$ **5.** $x+4$ **6.** $x+2$ **7.** $-(2x-1)$ **8.** $-(x+1)$
9. $(x-6)/(x+2)$ **10.** $(3x+4)/(x-4)$ **11.** $8xy^2/3$ **12.** $6xz^2$ **13.** $16b^3c^2/a^2$ **14.** $-\frac{1}{2}$ **15.** $-2x$
16. $(x+y)y^2/2x^3$ **17.** 1 **18.** $(x+y)/x$ **19.** 36 **20.** $32z/x^3$ **21.** $3/(x-y)$ **22.** $1/3(a+3)$
23. $(a-2)/2x(a+2)$ **24.** 1 **25.** $2x(x-5y)$ **26.** $16(x-2)/3(x+2y)$ **27.** 1 **28.** $(x-2)/(x+2)$
29. 4 **30.** $2(3x-4)/3y$ **31.** 9 **32.** $4/(x+10)$ **33.** $3x+2$ **34.** $3x+4$ **35.** 24 **36.** $15x^2$ **37.** $x(x+1)$
38. $(x+2)(x-3)$ **39.** $x(x+1)$ **40.** $(x+y)(x-y)$ **41.** $x-7$ **42.** $(x+7)(x-5)(x+2)$ **43.** $3/x$
44. $(24x+y)/4xy$ **45.** $(5x^2-12y)/3x^2y$ **46.** $(7x+12)/(x+2)$ **47.** $(5x+12)/(x+3)$ **48.** $(a^2+c^2)/ac$
49. $(7x+12)/x(x+3)$ **50.** $-(x+4)/3x(x-2)$ **51.** $(8x+25)/(x+3)(x+4)$ **52.** $(4x+26)/(x+5)^2$
53. $3(x-1)/(x+3)(x-3)$ **54.** $4/(x+2)(x-3)(x-2)$ **55.** $2(x-4)/(x-3)(x-5)$
56. $(22x+5)/(x-5)(x-10)(x+5)$ **57.** $\frac{34}{9}$ **58.** $\frac{55}{26}$ **59.** $5yz/6$ **60.** $16x^3z^2/y^3$ **61.** $(xy+1)/y^3$
62. $(xy-x)/(x+1)$ **63.** $(4x+2)/x(6x-1)$ **64.** $2/x$ **65.** a **66.** $(2a+1)/2$ **67.** $(x+1)/(-x+1)$
68. $x^2(3-y)/y(y-x)$ **69.** 9 **70.** 1 **71.** 6 **72.** 6 **73.** 52 **74.** -20 **75.** 18 **76.** $\frac{1}{2}$ **77.** -6
78. -18 **79.** $2\frac{2}{9}$ hr **80.** $4\frac{8}{19}$ hr **81.** $\frac{5}{2}, 10$ **82.** Bus 80 km/hr, train 120 km/hr

Chapter 6 Practice Test

1. $2x^3z/3y^3$ **2.** $a+3$ **3.** $(x-3y)/3$ **4.** $4/y(y+5)$ **5.** $(7x-2)/2y$ **6.** $(7x^2-6x-11)/(x+3)$
7. $(10x+3)/2x^2$ **8.** $(-x+10)/(x+2)$ **9.** $-1/(x+4)(x-4)$ **10.** $\frac{29}{10}$ **11.** $x^2(1+y)/y$ **12.** 60 **13.** 12
14. $3\frac{1}{13}$ hr

Exercise Set 7.1

1. $A(3, 1)$, $B(-3, 0)$, $C(1, -3)$, $D(-2, -3)$, $E(0, 3)$, $F(3/2, -1)$ **3.**

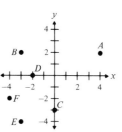

5. Yes   **7.** The x coordinate

Exercise Set 7.2

1. a, d, e   **3.** b, c, d   **5.** b, c, d   **7.**

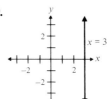

9.

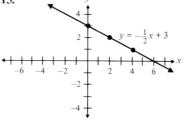

11.

13.

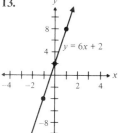

15.

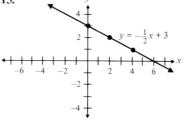

17.

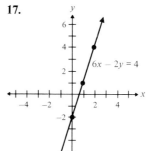

19.

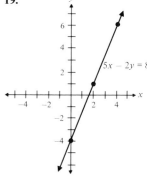

21.

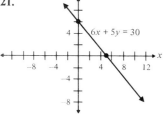

23.

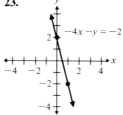

25.

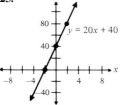

27.

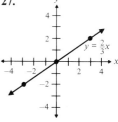

29.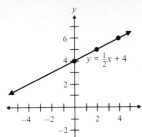

31.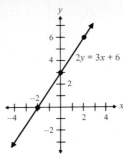

33.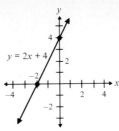

35.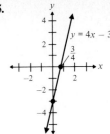

37.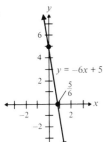

39.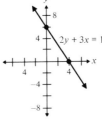

41.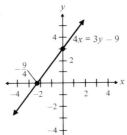

43.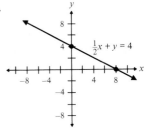

45.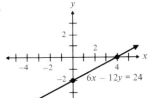

47.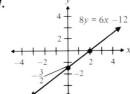

49.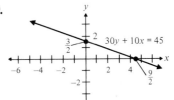

51.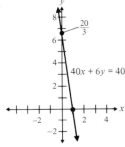

53.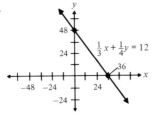

55.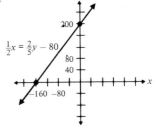

57. The set of points whose coordinates satisfy the equation.   **59.** 2, 3   **61.** A vertical line

Exercise Set 7.3

1. $m = 2, b = -1$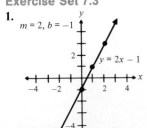

3. $m = -1, b = 5$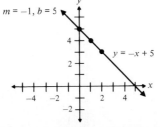

5. $m = -4, b = 0$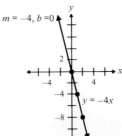

395

7. $m = 2, b = -3$

$-2x + y = -3$

9. $m = -1, b = 3$

$3x + 3y = 9$

11. $m = \frac{1}{2}, b = 4$

$-x + 2y = 8$

13. $m = \frac{2}{3}, b = -\frac{3}{2}$

$4x = 6y + 9$

15. $m = 3, b = 4$

$-6x = -2y + 8$

17. $m = \frac{3}{8}, b = -1$

$-3x + 8y = -8$

19. $m = \frac{3}{2}, b = 2$

$3x = 2y - 4$

21. $m = \frac{1}{4}, b = -\frac{1}{2}$

$20x = 80y + 40$

23. $x = 3$ **25.** $y = -\frac{9}{2}$ **27.** $y = x + 2$

29. $y = -\frac{1}{3}x + 2$ **31.** $y = -\frac{3}{2}x + 15$ **33.** $y = -\frac{1}{2}x - 2$ **35.** Yes   **37.** No   **39.** No   **41.** Yes
43. $y = -2x - 3$ **45.** $y = \frac{1}{2}x - \frac{9}{2}$ **47.** $y = \frac{3}{5}x - \frac{22}{5}$ **49.** $y = 3x + 10$ **51.** $y = -\frac{3}{2}x$ **53.** $y = \frac{1}{2}x - 2$
57. The values of y decrease as the values of x increase (graph falls as it moves to the right).
59. Standard form: $6x + 2y = 4$, slope-intercept form $y = -3x + 2$, point-slope form $y - 2 = -3(x - 0)$
61. $5x - 3y = 8$, $y = \frac{5}{3}x - \frac{8}{3}$, $y + \frac{8}{3} = \frac{5}{3}(x - 0)$

Just for Fun   **1.** $y = \frac{3}{4}x + 2$ **2.** $y = -\frac{2}{5}x + \frac{13}{10}$
3. (a) The angle (less than or equal to 90°) that the hill makes with the horizontal is measured to determine the slope.
(b) The slope of a line has no specific units. The slope of a hill is measured in degrees.

Exercise Set 7.4

1.

3.

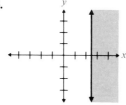

5.

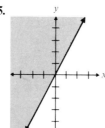

7.

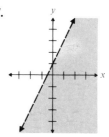

396

9.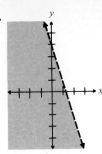

11.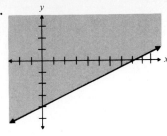

13.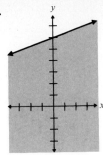

15.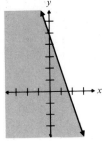

17.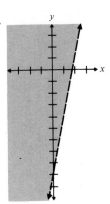

19.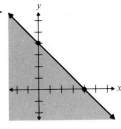

21.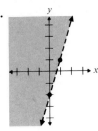

23.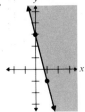

25.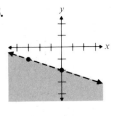

27. Points on line satisfy equation (=) but not an inequality that is strictly greater than or less than.

Chapter 7 Review Exercises

1. b, d   **2.** b, c, d   **3.**

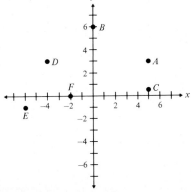

4. No   **5.**

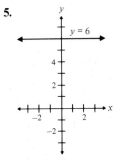

6.

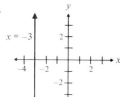

$x = -3$

7.

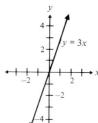

$y = 3x$

8.

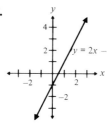

$y = 2x - 1$

9.

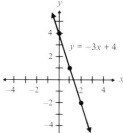

$y = -3x + 4$

10.

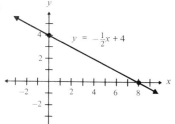

$y = -\frac{1}{2}x + 4$

11.

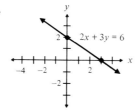

$2x + 3y = 6$

12.

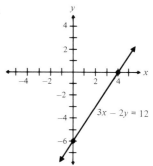

$3x - 2y = 12$

13.

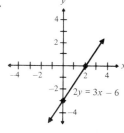

$2y = 3x - 6$

14.

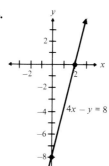

$4x - y = 8$

15.

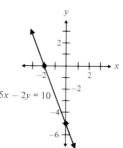

$-5x - 2y = 10$

16.

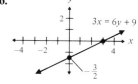

$3x = 6y + 9$

$-\frac{3}{2}$

17.

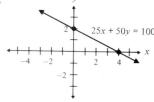

$25x + 50y = 100$

18.

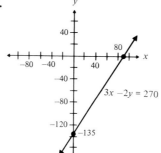

$3x - 2y = 270$

19.

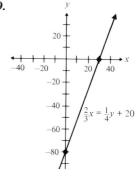

$\frac{2}{3}x = \frac{1}{4}y + 20$

398

20. $m = -1, b = 4$ **21.** $m = 3, b = 5$ **22.** $m = -4, b = \frac{1}{2}$ **23.** $m = -\frac{2}{3}, b = \frac{8}{3}$ **24.** $m = -\frac{1}{2}, b = \frac{3}{2}$

25. $m = \frac{3}{2}, b = 3$ **26.** $m = -\frac{3}{5}, b = \frac{12}{5}$ **27.** $m = -\frac{9}{7}, b = \frac{15}{7}$ **28.** $m = \frac{1}{2}, b = -2$

29. Slope is undefined, no y intercept   **30.** $m = 0, b = -3$ **31.** $y = 2x + 2$ **32.** $y = x - \frac{5}{2}$

33. $y = 2x + 1$ **34.** $y = -(1/2)x + 2$ **35.** Yes   **36.** No   **37.** Yes   **38.** Yes   **39.** $y = 2x - 2$

40. $y = -3x + 2$ **41.** $y = -(2/3)x + 4$ **42.** $y = 2$ **43.** $x = 3$ **44.** $y = x - 1$ **45.** $y = -(7/2)x - 4$

46. $x = -4$ **47.**

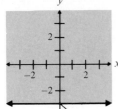

48.

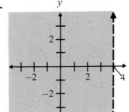

49.

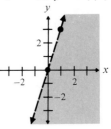

50.

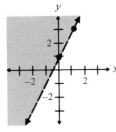

51.

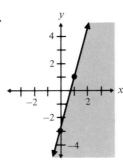

52.

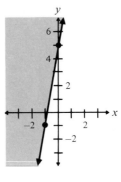

53.

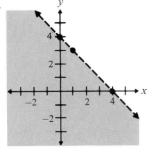

54.

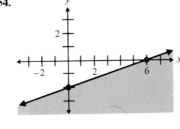

Chapter 7 Practice Test

1. a, b   **2.** $m = \frac{4}{9}, b = -\frac{5}{3}$ **3.** $y = -x - 1$ **4.** $y = 3x$ **5.** $y = -\frac{3}{7}x + \frac{2}{7}$

6.

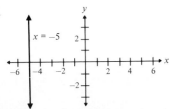

7.

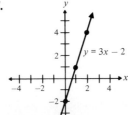

8.

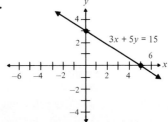

399

9.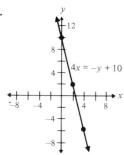

$4x = -y + 10$

10.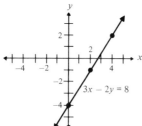

$3x - 2y = 8$

11.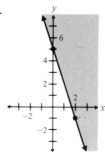

12.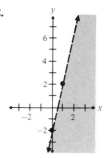

CHAPTER 8 _____

Exercise Set 8.1
1. c   **3.** a   **5.** a   **7.** a, c   **9.** a, c   **11.** b

Exercise Set 8.2
1. Consistent—one   **3.** Dependent—infinite number   **5.** Consistent—one   **7.** Dependent—infinite number
9. One solution   **11.** No solution   **13.** One solution   **15.** Infinite number of solutions   **17.** No solution
19. 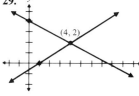(0,2)   **21.** 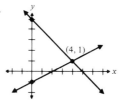(3,3)   **23.** 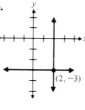(2,−3)   **25.** 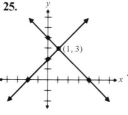(1, 3)   **27.** 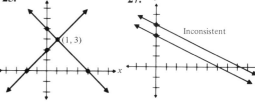Inconsistent

29. 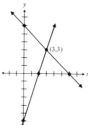(4, 2)   **31.** (4, 1)   **33.** (3, 3)   **35.** 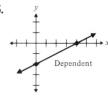Dependent

37. 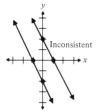Inconsistent   **39.** 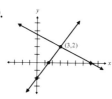(3,2)   **41.** 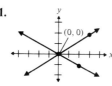(0, 0)   **43.** 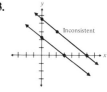Inconsistent

Exercise Set 8.3
1. (2, 1)   **3.** (−1, −1)   **5.** Inconsistent—no solution   **7.** (4, −9)   **9.** Dependent   **11.** (−1, 2)   **13.** (5, −3)
15. (2, 1)   **17.** Inconsistent   **19.** (−4, −2)   **21.** (6, −2)
23. The x in the first equation   **25.** You will obtain a true statement, such as $2 = 2$.

Exercise Set 8.4

1. $(6, 2)$ **3.** $(-2, 3)$ **5.** $(4, \frac{11}{2})$ **7.** No solution   **9.** $(-8, -26)$ **11.** $(4, -2)$ **13.** $(5, 4)$ **15.** $(\frac{3}{2}, -\frac{1}{2})$

17. Dependent   **19.** $(\frac{2}{5}, \frac{8}{5})$ **21.** $(10, \frac{15}{2})$ **23.** $(\frac{31}{28}, -\frac{23}{28})$ **25.** $(\frac{20}{39}, -\frac{16}{39})$ **27.** $(\frac{14}{5}, -\frac{12}{5})$

29. You will obtain a false statement like $0 = 6$.

Just for Fun   1. $(8, -1)$ **2.** $(-\frac{105}{41}, \frac{447}{82})$

Exercise Set 8.5

1. $x + y = 37$; 12, 25   **3.** $x + y = 76$; 37, 39   **5.** $x - y = 28$; 14, 42
   $x = 2y + 1$ $y = x + 2$ $x = 3y$

7. $x + y = 4.5$; 3.85 mph—canoe, 0.65 mph—current   **9.** $y = 36 + 0.12x$; 200 mi   **11.** $y = 1600 + 6x$; 200 books
   $x - y = 3.2$ $y = 30 + 0.15x$ $y = 1200 + 8x$

13. $x + y = 30$; 25 lb at \$3, 5 lb at \$5   **15.** $x + y = 100$; 70 gal at 5%, 30 gal of skim
   $3x + 5y = 100$ $0.05x + 0.00y = 0.035(100)$

17. $x + y = 8,000$; \$2500 at 8%, \$5500 at 10%   **19.** $c = 60,000 + 1500n$; $26\frac{2}{3}$—use mini if 27 or more terminals used
   $0.1x + 0.08y = 750$ $c = 20,000 + 3000n$

21. $d = 154$; 5.5 hr
   $d = 70t - 42t$ or $d = 28t$

Just for Fun   1. 1.125 mi   **2.** 200 g of first alloy, 100 g of second alloy

Exercise Set 8.6

1.   **3.**   **5.**   **7.**   **9.**

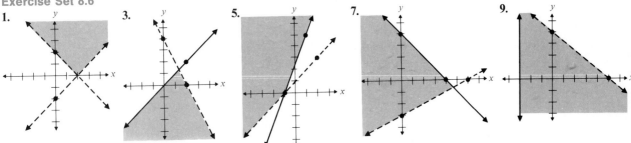

11.   **13.**   **15.**

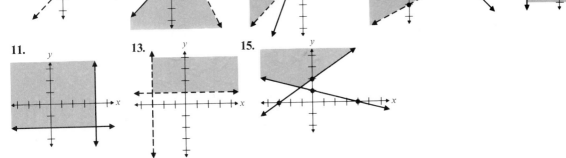

Chapter 8 Review Exercises

1. c   **2.** a   **3.** Consistent, one   **4.** Inconsistent, none   **5.** Dependent, infinite number   **6.** Consistent, one

7. No solution   **8.** One solution   **9.** One solution   **10.** One solution

11.   **12.**   **13.**   **14.**

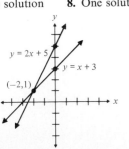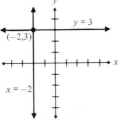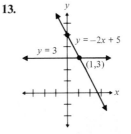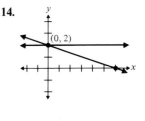

15.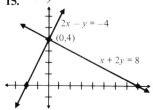

$2x - y = -4$
$(0,4)$
$x + 2y = 8$

16.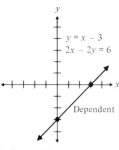

$y = x - 3$
$2x - 2y = 6$

Dependent

17.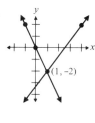

$(1, -2)$

18.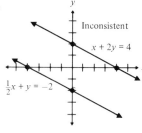

Inconsistent
$x + 2y = 4$
$\frac{1}{2}x + y = -2$

19. $(5, 2)$ **20.** $(-3, 2)$ **21.** $(2, 1)$ **22.** $(5, 4)$ **23.** $(5, 2)$ **24.** $(-18, 6)$ **25.** $(2, 2)$
26. Dependent equations—infinite number of solutions   **27.** $(\frac{5}{2}, -1)$ **28.** $(\frac{20}{9}, \frac{26}{9})$ **29.** $(8, -2)$ **30.** $(1, -2)$
31. $(26, -16)$ **32.** $(-7, 19)$ **33.** $(\frac{32}{13}, \frac{8}{13})$ **34.** $(-1, \frac{13}{3})$ **35.** $(1, 2)$ **36.** $(\frac{7}{5}, \frac{13}{5})$ **37.** $(6, -2)$ **38.** $(-\frac{78}{7}, -\frac{48}{7})$
39. $x + y = 48$; 17, 31   **40.** $x - y = 18$; 6, 24   **41.** $R = 0.08x$; over \$10,000   **42.** $x + y = 6$; 3 liters of each
$\quad\ \ y = 2x - 3$ $\qquad\qquad\qquad\ x = 4y$ $\qquad\qquad\ R = 500 + 0.03x$ $\qquad\qquad\quad 0.3x + 0.5y = 0.4(6)$
43. $x + y = 600$; 565 mph plane, 35 mph wind   **44.** $d = 2(x + 2)$; 18 mph, 40 mi
$\quad\ \ x - y = 530$ $\qquad\qquad\qquad\qquad\qquad\qquad\quad d = 2.5(x - 2)$

45.   **46.** 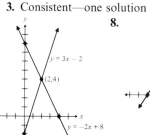   **47.** 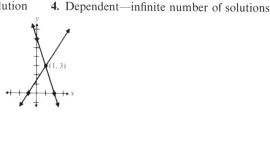   **48.**

Chapter 8 Practice Test

1. b   **2.** Inconsistent—no solution   **3.** Consistent—one solution   **4.** Dependent—infinite number of solutions
5. no solution   **6.** one solution   **7.**   **8.**

$y = 3x - 2$
$(2,4)$
$y = -2x + 8$

$(1, 3)$

9. $(\frac{7}{2}, -\frac{5}{2})$ **10.** $(3, 1)$ **11.** $(5, -5)$ **12.** $(\frac{44}{19}, \frac{48}{19})$ **13.** $c = 40 + 0.13x$; 100 mi
$\qquad\qquad\qquad\qquad\qquad\qquad\qquad\qquad\qquad\qquad c = 45 + 0.08x$

14. $x + y = 20$; $13\frac{1}{3}$ lb of peanuts, $6\frac{2}{3}$ lb of cashews   **15.**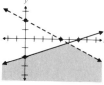
$\quad\ \ 4x + 2.5y = 3(20)$

CHAPTER 9

1. 1   **3.** 0   **5.** -9 **7.** 11   **9.** -4 **11.** 12   **13.** 13   **15.** -1 **17.** 9   **19.** -11 **21.** $\frac{1}{2}$ **23.** $\frac{3}{4}$
25. $-\frac{2}{5}$ **27.** 2.828   **29.** 3.873   **31.** 8.944   **33.** 9.000   **35.** 9.849   **37.** 1.732
39. It is nonnegative. The square root of a negative number is not a real number.
41. True   **43.** True   **45.** False   **47.** True

Exercise Set 9.2

1. 4 **3.** $2\sqrt{2}$ **5.** $4\sqrt{6}$ **7.** $4\sqrt{2}$ **9.** $4\sqrt{10}$ **11.** $4\sqrt{3}$ **13.** $6\sqrt{3}$ **15.** $2\sqrt{39}$ **17.** 16 **19.** 30 **21.** y^3
23. xy^2 **25.** $x^4y^6\sqrt{x}$ **27.** $ab^2\sqrt{c}$ **29.** $x\sqrt{3x}$ **31.** $5xy\sqrt{2y}$ **33.** $10y^2z^6\sqrt{2y}$ **35.** $9qb\sqrt{3bc}$
37. $8xz^2\sqrt{2xyz}$ **39.** $5x^2\sqrt{10yz}$ **41.** $2\sqrt{6}$ **43.** $3\sqrt{6}$ **45.** $15\sqrt{2}$ **47.** $x\sqrt{15}$ **49.** $2x^2\sqrt{10x}$ **51.** $6xy^2\sqrt{2x}$
53. $3xy^2\sqrt{6xy}$ **55.** $3xy^3\sqrt{10yz}$ **57.** $6x^3y^5$ **59.** $4x$ **61.** $13x^4y^6$

Exercise Set 9.3

1. 2 **3.** 3 **5.** 3 **7.** $\frac{1}{5}$ **9.** $\frac{3}{7}$ **11.** $\frac{1}{7}$ **13.** $2x\sqrt{5}$ **15.** $\sqrt{3xy}$ **17.** $\sqrt{5}/3y$ **19.** $1/xy^2$ **21.** $2x^2$
23. $4/5xy$ **25.** $(y\sqrt{5})/z$ **27.** $(3\sqrt{2})/2$ **29.** $\sqrt{2}$ **31.** $\sqrt{10}/2$ **33.** $\sqrt{10}/5$ **35.** $\sqrt{5}/5$ **37.** $(x\sqrt{2})/2$
39. $(x\sqrt{2})/4$ **41.** $(x^2\sqrt{5})/5$ **43.** $(x^3\sqrt{15y})/15y$ **45.** $(x\sqrt{y})/2y$ **47.** $\sqrt{6}/5x^2y^2z$ **49.** $(3\sqrt{5x})/xy^2$

Exercise Set 9.4

1. $2\sqrt{3}$ **3.** $-2\sqrt{7}$ **5.** $5-4\sqrt{3}$ **7.** $5\sqrt{x}$ **9.** $3\sqrt{x}$ **11.** $3+2\sqrt{y}$ **13.** $x+\sqrt{x}+4\sqrt{y}$ **15.** $3-2\sqrt{x}$
17. $2\sqrt{2}-2\sqrt{3}$ **19.** $4\sqrt{2}$ **21.** $7\sqrt{5}$ **23.** $16\sqrt{2}$ **25.** $-30\sqrt{3}+20\sqrt{5}$ **27.** $25\sqrt{10}-36\sqrt{5}$ **29.** $64-4\sqrt{6}$
31. 7 **33.** 31 **35.** $x-9$ **37.** $6-x^2$ **39.** $x-y^2$ **41.** $y-x^2$ **43.** $4(2-\sqrt{3})$ **45.** $-3(\sqrt{2}-5)/23$
47. $-2(\sqrt{2}-\sqrt{3})$ **49.** $-8(\sqrt{5}+2\sqrt{2})/3$ **51.** $2(6-\sqrt{x})/(36-x)$ **53.** $6(4+\sqrt{y})/(16-y)$ **55.** $4(\sqrt{x}+y)/(x-y^2)$
57. $x(\sqrt{x}-\sqrt{y})/(x-y)$ **59.** $(\sqrt{5x}-x)/(5-x)$

Exercise Set 9.5

1. 64 **3.** No solution **5.** 4 **7.** No solution **9.** 4 **11.** 4 **13.** No solution **15.** 7 **17.** 7 **19.** 2
21. 9 **23.** $\frac{14}{3}$ **25.** 1 **27.** 1, 9 **29.** No solution **31.** -2 **33.** 1, 9
35. Because there may be extraneous roots

Just for Fun

1. 9 **2.** $\frac{9}{16}$ **3.** 9

Exercise Set 9.6

1. $\sqrt{119}=10.91$ **3.** $\sqrt{164}=12.81$ **5.** $\sqrt{175}=13.23$ **7.** $\sqrt{41}=6.40$ **9.** $\sqrt{149}=12.21$ **11.** $\sqrt{128}=11.31$
13. $\sqrt{17,240.89}=131.30$ yd **15.** $\sqrt{18.25}=4.27$ m **17.** $\sqrt{10}=3.16$ **19.** $\sqrt{288}=16.97$ **21.** $\sqrt{144}=12$ in.
23. $\sqrt{25.48}=5.05$ ft **25.** $\sqrt{225}=15$ in. **27.** $\sqrt{5120}=71.55$ ft/sec **29.** $\sqrt{80,000}=282.84$ ft/sec
31. $6.28\sqrt{1.25}=7.02$ sec **33.** $6.28\sqrt{0.3125}=3.51$ sec **35.** $0.2(\sqrt{1418})^3=10,679.34$ days
37. $\sqrt{19.5(6,370,000)}=11145.18$ m/sec

Just for Fun

1. 9 in by 12 in **2.** 32.66 ft/sec **3.** 3.25 ft **4.** 40.93 in

Exercise Set 9.7

1. 2 **3.** -2 **5.** 2 **7.** 3 **9.** -1 **11.** 4 **13.** $3\sqrt[3]{2}$ **15.** $2\sqrt[3]{2}$ **17.** $3\sqrt[3]{3}$ **19.** $2\sqrt[4]{2}$ **21.** $2\sqrt[3]{5}$
23. x **25.** y^3 **27.** x^4 **29.** y **31.** x^5 **33.** 16 **35.** 8 **37.** 1 **39.** 27 **41.** 8 **43.** 625 **45.** $\frac{1}{9}$
47. $\frac{1}{32}$ **49.** $x^{7/3}$ **51.** $x^{4/3}$ **53.** $y^{15/4}$ **55.** $y^{21/4}$ **57.** $x^{2/3}$ **59.** x **61.** x^4 **63.** x^2

Just for Fun

1. xy **2.** $3x^2y$ **3.** $\sqrt[4]{2}$ **4.** $3x\sqrt[3]{3y}$ **5.** $\sqrt[3]{4}/2$ **6.** $\sqrt[3]{x^2}/x$

Chapter 9 Review Exercises

1. 5 **2.** 6 **3.** 1 **4.** 10 **5.** -9 **6.** -12 **7.** $4\sqrt{2}$ **8.** $2\sqrt{11}$ **9.** $3x^2y^2\sqrt{5x}$ **10.** $5x^2y^3\sqrt{5}$
11. $x^2z\sqrt{15xyz}$ **12.** $4b^2c^2\sqrt{3ac}$ **13.** $4\sqrt{6}$ **14.** $5x$ **15.** $6x\sqrt{y}$ **16.** $5xy\sqrt{3}$ **17.** $10xy^3\sqrt{y}$ **18.** $2xy^2\sqrt{6xy}$
19. 4 **20.** $\frac{1}{5}$ **21.** $\frac{1}{2}$ **22.** $(3\sqrt{5})/5$ **23.** $\sqrt{15x}/6$ **24.** $\sqrt{6x}/6$ **25.** $(x\sqrt{2})/2$ **26.** $(x^2\sqrt{2x})/4$ **27.** $(y\sqrt{15})/x^2$
28. $(x\sqrt{2y})/y^2$ **29.** $(3\sqrt{5x})/2x^2y$ **30.** $(z\sqrt{14x})/7x$ **31.** $-3(1-\sqrt{2})$ **32.** $5(3+\sqrt{6})/3$ **33.** $(2\sqrt{3}-\sqrt{3x})/(4-x)$
34. $2(\sqrt{x}+5)/(x-25)$ **35.** $(\sqrt{5x}-\sqrt{15})/(x-3)$ **36.** $4\sqrt{3}$ **37.** $-\sqrt{2}$ **38.** $-2\sqrt{x}$ **39.** 0 **40.** $\sqrt{2}$

41. $12\sqrt{10}$ **42.** $-10\sqrt{2}$ **43.** $26\sqrt{2}$ **44.** $16\sqrt{3} + 20\sqrt{5}$ **45.** 81 **46.** No solution **47.** 39 **48.** 8 **49.** 9
50. 4 **51.** 0 **52.** 7 **53.** 3 **54.** 10 **55.** $\sqrt{88}$ **56.** $\sqrt{12}$ **57.** $\sqrt{61}$ **58.** $\sqrt{135} = 11.62$ ft
59. $\sqrt{261} = 16.16$ in. **60.** 10 **61.** $\sqrt{153} = 12.37$ **62.** 11 ft **63.** 2 **64.** -3 **65.** 2 **66.** $2\sqrt[3]{2}$ **67.** $2\sqrt[3]{3}$
68. $2\sqrt[4]{2}$ **69.** $2\sqrt[3]{6}$ **70.** $3\sqrt[3]{2}$ **71.** $2\sqrt[4]{6}$ **72.** x^5 **73.** x^4 **74.** y^4 **75.** y^5 **76.** 4 **77.** 4 **78.** $\frac{1}{9}$
79. 16 **80.** $\frac{1}{8}$ **81.** 125 **82.** $x^{5/3}$ **83.** $x^{10/3}$ **84.** $y^{9/4}$ **85.** $x^{5/2}$ **86.** $y^{11/2}$ **87.** $x^{7/4}$ **88.** x
89. $\sqrt[3]{x^2}$ **90.** x^3 **91.** x^2 **92.** x^2 **93.** x^2 **94.** x^3 **95.** x^6

Chapter 9 Practice Test

1. $x + 3$ **2.** $2\sqrt{15}$ **3.** $4\sqrt{6}$ **4.** $2x\sqrt{3}$ **5.** $4x^2y^2\sqrt{2y}$ **6.** $4xy\sqrt{3x}$ **7.** $5x^2y^2\sqrt{3y}$ **8.** $\frac{1}{5}$ **9.** $y/4x$
10. $\sqrt{2}/2$ **11.** $(2\sqrt{5x})/5$ **12.** $(2\sqrt{15x})/3xy$ **13.** $-3(2 - \sqrt{5})$ **14.** $6(\sqrt{x} + 3)/(x - 9)$ **15.** $11\sqrt{3}$ **16.** $-3\sqrt{x}$
17. 76 **18.** 4, 8 **19.** $\sqrt{106} = 10.30$ **20.** $\sqrt{74} = 8.60$ **21.** $1/81$ **22.** x^5

CHAPTER 10

Exercise Set 10.1

1. 4, -4 **3.** 10, -10 **5.** 6, -6 **7.** $\sqrt{10}, -\sqrt{10}$ **9.** $2\sqrt{2}, -2\sqrt{2}$ **11.** 2, -2 **13.** $\sqrt{17}, -\sqrt{17}$
15. 3, -3 **17.** $\sqrt{15}/2, -\sqrt{15}/2$ **19.** $\sqrt{195}/5, -\sqrt{195}/5$ **21.** 1, -3 **23.** 7, -1 **25.** 2, -10
27. $1 + 2\sqrt{3}, 1 - 2\sqrt{3}$ **29.** $-6 + 2\sqrt{5}, -6 - 2\sqrt{5}$ **31.** 3, -7 **33.** 19, -1 **35.** $(-3 + 3\sqrt{2})/2, (-3 - 3\sqrt{2})/2$
37. $(-1 + 2\sqrt{5})/4, (-1 - 2\sqrt{5})/4$ **39.** $(6 + 3\sqrt{2})/2, (6 - 3\sqrt{2})/2$ **41.** $w = 2\sqrt{10} \approx 6.32$ ft, $l = 4\sqrt{10} \approx 12.65$ ft

Exercise Set 10.2

1. 1, -3 **3.** 5, -1 **5.** $-2, -1$ **7.** 5, 3 **9.** -3 **11.** $-2, -3$ **13.** $-3, -6$ **15.** 7, 8
17. 6, -2 **19.** $-1 + \sqrt{7}, -1 - \sqrt{7}$ **21.** $-3 + \sqrt{3}, -3 - \sqrt{3}$ **23.** $(5 + \sqrt{57})/2, (5 - \sqrt{57})/2$ **25.** 1, -3
27. $(-9 + \sqrt{73})/2, (-9 - \sqrt{73})/2$ **29.** $-8, -3$ **31.** $(-5 + \sqrt{31})/2, (-5 - \sqrt{31})/2$ **33.** $-1 + \sqrt{3}, -1 - \sqrt{3}$
35. 0, -4 **37.** 0, 2 **39.** 1, -4 **41.** 0, -6 **43.** 4, 7

Just for Fun
1. $(-3 + \sqrt{59})/10, (-3 - \sqrt{59})/10$ **2.** $(5 + \sqrt{70})/15, (5 - \sqrt{70})/15$
3. $(-1 + \sqrt{193})/12, (-1 - \sqrt{193})/12$

Exercise Set 10.3

1. Two solutions **3.** No solution **5.** Two solutions **7.** Single solution **9.** Two solutions **11.** No solution
13. Two solutions **15.** Two solutions **17.** 1, 2 **19.** 4, 5 **21.** $-8, 3$ **23.** 4, 9 **25.** 5, -5 **27.** 0, 3
29. 3, 4 **31.** $(7 + \sqrt{17})/4, (7 - \sqrt{17})/4$ **33.** $\frac{1}{3}, -\frac{1}{2}$ **35.** $(2 + \sqrt{6})/2, (2 - \sqrt{6})/2$ **37.** No solution **39.** $\frac{5}{4}, -1$
41. $\frac{9}{2}, -1$ **43.** 3, $\frac{5}{2}$ **45. (a)** none **(b)** one **(c)** two **47.** 4 ft, 7 ft **49.** 3 ft

Just for Fun
1. 300 ft long by 50 ft wide, or 100 ft long by 150 ft wide

Exercise Set 10.4

1. $x = -1, (-1, -8)$, up **3.** $x = \frac{5}{2}, (\frac{5}{2}, \frac{1}{4})$, down **5.** $x = \frac{5}{6}, (\frac{5}{6}, \frac{121}{12})$, down **7.** $x = -1, (-1, -8)$, down
9. $x = \frac{1}{3}, (\frac{1}{3}, \frac{5}{3})$, up **11.** $x = -\frac{3}{2}, (-\frac{3}{2}, -14)$, up **13.** **15.**

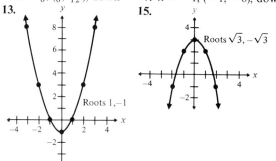

17.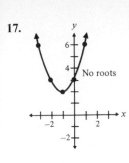
No roots

19.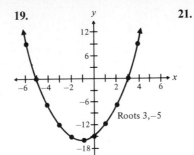
Roots 3, −5

21.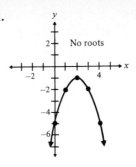
No roots

23.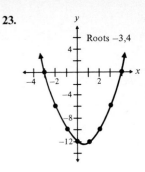
Roots −3, 4

25.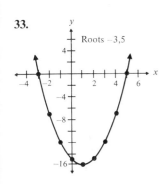
Root 3

27.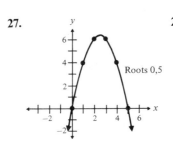
Roots 0, 5

29.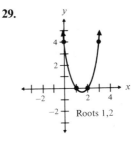
Roots 1, 2

31.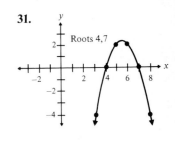
Roots 4, 7

33.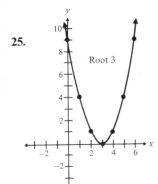
Roots −3, 5

35.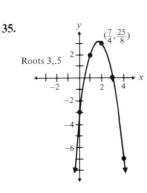
Roots 3, .5
$\left(\frac{7}{4}, \frac{25}{8}\right)$

37.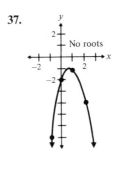
No roots

39.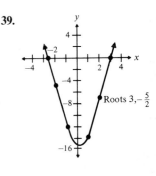
Roots 3, $-\frac{5}{2}$

41. $(-b/2a, (4ac - b^2)/4a)$ **43.** The values of x where the graph crosses the x axis.
45. (a) two   **(b)** none   **(c)** one (double root)

Chapter 10 Review Exercises

1. 5, −5   **2.** $2\sqrt{2}, -2\sqrt{2}$ **3.** $\sqrt{6}, -\sqrt{6}$ **4.** $\sqrt{6}, -\sqrt{6}$ **5.** $2\sqrt{5}, -2\sqrt{5}$ **6.** $\sqrt{7}, -\sqrt{7}$ **7.** $2\sqrt{2}, -2\sqrt{2}$
8. $3 + 2\sqrt{3}, 3 - 2\sqrt{3}$ **9.** $(-4 + \sqrt{30})/2, (-4 - \sqrt{30})/2$ **10.** $(5 + 5\sqrt{2})/3, (5 - 5\sqrt{2})/3$ **11.** 2, 8   **12.** 3, 5
13. 1, 13   **14.** 2, −3   **15.** 9, −6   **16.** 1, −6   **17.** $-1 + \sqrt{6}, -1 - \sqrt{6}$ **18.** $(3 + \sqrt{41})/2, (3 - \sqrt{41})/2$
19. −4, 8   **20.** 5, −3   **21.** $\frac{3}{2}, -2$ **22.** $\frac{5}{3}, \frac{3}{2}$ **23.** Two solutions   **24.** No solution   **25.** No solution
26. No solution   **27.** One solution   **28.** No solution   **29.** Two solutions   **30.** Two solutions   **31.** 2, 7
32. −10, 3   **33.** 2, 5   **34.** 2, $-\frac{3}{5}$ **35.** −2, 9   **36.** 6, −5   **37.** $\frac{3}{2}, -\frac{5}{3}$ **38.** $(-2 + \sqrt{10})/2, (-2 - \sqrt{10})/2$
39. $(3 + \sqrt{57})/4, (3 - \sqrt{57})/4$ **40.** $3 + \sqrt{2}, 3 - \sqrt{2}$ **41.** No solution   **42.** $(3 + \sqrt{33})/3, (3 - \sqrt{33})/3$ **43.** 0, $-\frac{3}{2}$
44. 0, $\frac{5}{2}$ **45.** 3, 8   **46.** 7, 9   **47.** 5, −8   **48.** −9, 3   **49.** −6, 10   **50.** 7, −6   **51.** 1, −12   **52.** 5, −5
53. 0, −6   **54.** $\frac{1}{2}, -3$ **55.** $\frac{5}{2}, 2$ **56.** $\frac{2}{3}, -\frac{3}{2}$ **57.** $(-3 + \sqrt{33})/2, (-3 - \sqrt{33})/2$ **58.** 2, $\frac{5}{3}$ **59.** 1, $-\frac{8}{3}$
60. $(3 + 3\sqrt{3})/2, (3 - 3\sqrt{3})/2$ **61.** 0, $\frac{5}{2}$ **62.** 0, $-\frac{5}{3}$ **63.** $x = 1, (1, -4)$, up   **64.** $x = 5, (5, -1)$, up
65. $x = -\frac{7}{2}, (-\frac{7}{2}, -\frac{1}{4})$, up   **66.** $x = -1, (-1, 16)$, down   **67.** $x = \frac{3}{2}, (\frac{3}{2}, -\frac{9}{4})$, up   **68.** $x = -\frac{7}{4}, (-\frac{7}{4}, -\frac{25}{8})$, up
69. $x = 0, (0, -8)$, down   **70.** $x = 1, (1, 9)$, down   **71.** $x = -\frac{1}{2}, (-\frac{1}{2}, \frac{81}{4})$, down   **72.** $x = -\frac{5}{6}, (-\frac{5}{6}, -\frac{121}{12})$, up

73.

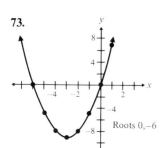

Roots 0,−6

74.

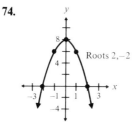

Roots 2,−2

75.

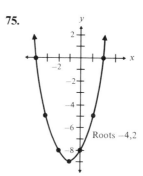

Roots −4,2

76.

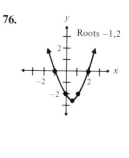

Roots −1,2

77.

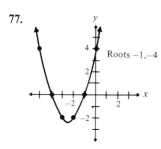

Roots −1,−4

78.

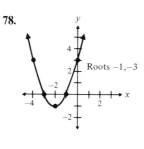

Roots −1,−3

79.

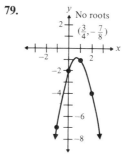

No roots
$\left(\frac{3}{4}, -\frac{7}{8}\right)$

80.

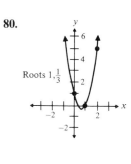

Roots $1, \frac{1}{3}$

81.

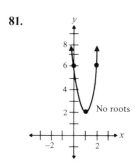

No roots

82.

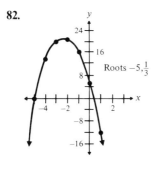

Roots $-5, \frac{1}{3}$

83. Roots $-3 - \sqrt{5}$,
$-3 + \sqrt{5}$

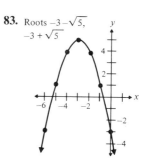

84.

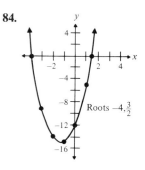

Roots $-4, \frac{3}{2}$

85. 8, 11   **86.** 7 ft, 9 ft

Chapter 10 Practice Test

1. $2\sqrt{5}$, $-2\sqrt{5}$ **2.** $(3 + \sqrt{35})/2$, $(3 - \sqrt{35})/2$ **3.** 10, -6 **4.** -4, 3 **5.** 6, -1 **6.** $(-4 + \sqrt{6})/2$, $(-4 - \sqrt{6})/2$
7. $0, \frac{5}{3}$ **8.** $-5, \frac{1}{2}$ **9.** No solution   **10.** $x = \frac{3}{2}$, $(\frac{3}{2}, \frac{41}{4})$, down   **11.**
12.   **13.** $w = 3$ ft, $l = 10$ ft

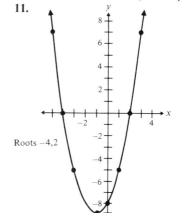

(3,0)

Root 3

Roots −4,2

(−1,−9)

Index

Chapter 7 Graphing Linear Equations

Linear equation in two variables: $ax + by = c$

A **graph** is an illustration of the set of points whose coordinates satisfy the equation.

Every **linear equation** of the form $ax + by = c$ will be a straight line when graphed.

To find the y intercept (where the graph crosses the y axis) set $x = 0$ and solve for y.

To find the x intercept (where the graph crosses the x axis) set $y = 0$ and solve for x.

$$\text{slope } (m) = \frac{\text{change in } y}{\text{change in } x} = \frac{y_2 - y_1}{x_2 - x_1}$$

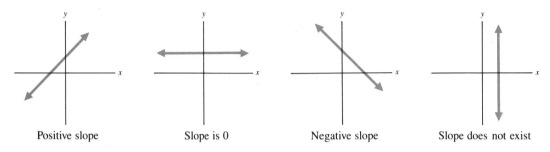

| Positive slope | Slope is 0 | Negative slope | Slope does not exist |

Slope intercept form of a line: $y = mx + b$, m is the slope and b is the y intercept.

Point slope form of a line: $y - y_1 = m(x - x_1)$, m is slope and (x_1, y_1) is a point on the line.

Chapter 8 Systems of Linear Equations

The **solution** to a system of linear equations is the ordered pair or pairs that satisfy all equations in the system. A system of linear equations may have no solution, exactly one solution, or an infinite number of solutions.

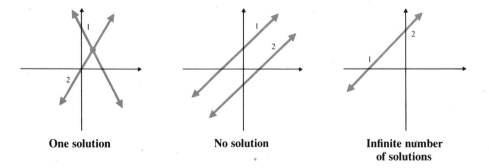

| One solution | No solution | Infinite number of solutions |

A system of linear equations may be solved graphically, or algebraically by the substitution method or by the addition (or elimination) method.